Studies in Systems, Decision and Control

Volume 477

The series "Studies in Systems, Decision and Control" (SSDC) covers both new developments and advances, as well as the state of the art, in the various areas of broadly perceived systems, decision making and control–quickly, up to date and with a high quality. The intent is to cover the theory, applications, and perspectives on the state of the art and future developments relevant to systems, decision making, control, complex processes and related areas, as embedded in the fields of engineering, computer science, physics, economics, social and life sciences, as well as the paradigms and methodologies behind them. The series contains monographs, textbooks, lecture notes and edited volumes in systems, decision making and control spanning the areas of Cyber-Physical Systems, Autonomous Systems, Sensor Networks, Control Systems, Energy Systems, Automotive Systems, Biological Systems, Vehicular Networking and Connected Vehicles, Aerospace Systems, Automation, Manufacturing, Smart Grids, Nonlinear Systems, Power Systems, Robotics, Social Systems, Economic Systems and other. Of particular value to both the contributors and the readership are the short publication timeframe and the world-wide distribution and exposure which enable both a wide and rapid dissemination of research output.

Indexed by SCOPUS, DBLP, WTI Frankfurt eG, zbMATH, SCImago.

All books published in the series are submitted for consideration in Web of Science.

Alla G. Kravets · Alexander A. Bolshakov ·
Maxim V. Shcherbakov
Editors

Cyber-Physical Systems Engineering and Control

Springer

Editors
Alla G. Kravets
CAD&RD
Volgograd State Technical University
Volgograd, Russia

Alexander A. Bolshakov
Peter the Great St. Petersburg Polytechnic University
St. Petersburg, Russia

Maxim V. Shcherbakov
CAD&RD
Volgograd State Technical University
Volgograd, Russia

ISSN 2198-4182 ISSN 2198-4190 (electronic)
Studies in Systems, Decision and Control
ISBN 978-3-031-33158-9 ISBN 978-3-031-33159-6 (eBook)
https://doi.org/10.1007/978-3-031-33159-6

This Springer imprint is published by the registered company Springer Nature Switzerland AG
The registered company address is: Gewerbestrasse 11, 6330 Cham, Switzerland

Preface

Nowadays, Cyber-Physical Systems, combining physical and digital components, become widely used in different domains. They help increase the efficiency of manufacturing both material and intangible products and optimize asset operation. Scientific society suggests new approaches to engineering and optimization of cyber-physical systems. However, there are still open questions that need to be covered by research and development.

The book focuses on open issues of cyber-physical system engineering and control. Also, it discusses the implementation of cyber-physical systems and their components, including AI-based and robotics.

The book contains 23 chapters joined into three sections. The first section "Cyber-Physical Systems: AI & Robotics Engineering" suggests new findings regarding the engineering of cyber-physical systems with artificial intelligence and robotics components covering both fundamental and application levels. The next section, named "New Materials Engineering for Cyber-Physical Systems," provides interesting and novel results on how to design new materials and optimize asset operations for enhancing cyber-physical systems. Chapters in the last section "Cyber-Physical Systems: Computing and Control" highlight the essential aspect of high-performance computing issues for the design, monitoring, control, and maintenance of cyber-physical systems.

The book is intended for practitioners, enterprise representatives, scientists, students, and Ph.D. and master's students conducting research in the area of cyber-physical system engineering and control and its implementation in various domains.

We are grateful to the authors and reviewers for their ideas and contribution which make this book solid and valuable.

Volgograd, Russia — Alla G. Kravets
St. Petersburg, Russia — Alexander A. Bolshakov
Volgograd, Russia — Maxim V. Shcherbakov
March 2023

Contents

New Materials Engineering for Cyber-Physical Systems

Cyber-Physical Systems Computing and Control

Cyber-Physical Systems: AI and Robotics Engineering

Attention-Based Random Forests and the Imprecise Pari-Mutual Model

Lev V. Utkin, Andrei V. Konstantinov, and Natalia A. Politaeva

Abstract A modification of the attention-based random forest (RF) called ABRF-PM is proposed. In contrast to the original attention-based RF, the ABRF-PM model uses the imprecise pari-mutual model in place of the Huber's contamination model for representing the attention weights and trainable parameters. Moreover, a general approach to applying various imprecise models to the attention-based RF, which is based on considering extreme points of sets of probability distributions produced by the imprecise models as sets of attention weights, is proposed such that the pari-mutual model can be regarded as a special case of the sets. The general approach ensures the linear relationship between attention weights and the corresponding attention trainable parameters, which leads to the standard quadratic optimization problem for computing the trainable parameters. Numerical experiments with real datasets illustrate the proposed ABRF-PM model.

Keywords Attention mechanism · Random forest · Nadaraya-Watson regression · Contamination model · Pari-mutual model · Regression

1 Introduction

One of the approaches improving many models of machine learning is the attention mechanism which has been successfully used in many tasks starting from the natural language processing models to vision transformers [1–5]. The main idea behind the attention mechanism is to concentrate on some important examples or sets of important features in order to enhance the classification and regression accuracy and capability. In other words, the attention mechanism assigns large weights to important examples or features and small weights to other examples or features.

In spite of efficiency and flexibility of the attention mechanism, it is mainly implemented as a component of neural networks. Hence, the corresponding models have

L. V. Utkin (✉) · A. V. Konstantinov · N. A. Politaeva
Peter the Great St. Petersburg Polytechnic University, Saint-Petersburg, Russia
e-mail: lev.utkin@gmail.com

A. G. Kravets et al. (eds.), *Cyber-Physical Systems Engineering and Control*,
Studies in Systems, Decision and Control 477,
https://doi.org/10.1007/978-3-031-33159-6_1

problems inherent in neural networks, namely, overfitting, a large amount of training data, expensive computations due to use of gradient-based algorithms, etc. Utkin and Konstantinov [6] proposed a new attention-based model called the attention-based random forest (ABRF), which allows us to use the attention mechanism in the framework of random forests (RFs). The main peculiarity of ABRF is that it is implemented as a component of the RF and does not use neural networks and gradient-based algorithms. Moreover, training attention parameters of the ABRF model are computed by solving the standard quadratic optimization problem or the linear optimization problem. The main idea behind ABRF is to implement the Nadaraya-Watson kernel regression model [7, 8] in the form of the RF and to assign attention weights to leaf nodes of all trees from the corresponding RF, which depend on the training or testing example, and have large values for the important examples to some extent. The second idea behind ABRF is to use the Huber's ϵ-contamination model [9] where the trainable parameters of attention weights are optimally selected from an arbitrarily adversary distribution.

The ABRF model has demonstrated its efficiency and higher accuracy of predictions for many regression datasets. At the same time, the Huber's ϵ-contamination model used in the model is one of the so-called imprecise models [10] which can be applied to implementation of the attention-based RFs. Therefore, it is interesting also to study other imprecise models which include the imprecise pari-mutual model, the constant odds-ratio model, etc. In this chapter, we propose a general approach to the application of various imprecise models, which is based on considering extreme points of sets of probability distributions produced by the imprecise models. As a special case of the approach, the imprecise pari-mutual model and its application to develop a new modification of ABRF is studied. It is used in place of the Huber's ϵ-contamination model. The attention based random forest using the imprecise pari-mutual model for trainable parameters is called the ABRF-PM.

Numerical experiments with real data are performed for investigating the proposed ABRF-PM model. We compare the proposed model with the original RF and with ABRF using the Huber's ϵ-contamination model [6].

The chapter is organized as follows. A brief introduction to the attention mechanism is given in Sect. 2. A description of the original attention-based random forest model is available in Sect. 3. A general approach based on applying extreme points is presented in Sect. 4. The ABRF-PM model using the pari-mutual model is provided in Sect. 5. Numerical experiments with real data illustrating ABRF-PM are provided in Sect. 6. Concluding remarks can be found in Sect. 7.

2 The Attention Mechanism

Let $S = \{(\mathbf{x}_1, y_1), (\mathbf{x}_2, y_2), \dots, (\mathbf{x}_n, y_n)\}$ be a dataset consisting of n examples such that $\mathbf{x}_i = (x_{i1}, \dots, x_{im}) \in R^m$ is a feature vector involving m features and $y_i \in R$ is the regression or classification labels. The attention mechanism is an approach to improve the machine learning model by introducing the trainable weights assigned to each training example. One of the ways for studying the weight assignment is to consider the well-known Nadaraya-Watson kernel regression model [7, 8]. It allows us to estimate the regression target value y associated with a new input vector $\mathbf{x}$ by computing the weighted average of target values as follows:

$$\tilde{y} = \sum_{i=1}^{n} \alpha(\mathbf{x}, \mathbf{x}_i) y_i. \tag{1}$$

Here $\alpha(\mathbf{x}, \mathbf{x}_i)$ is the attention weight depending on the distance between the input feature vector $\mathbf{x}$ and the i-th training example such that the greater the distance, the smaller the attention weight. Feature vector $\mathbf{x}$, vectors $\mathbf{x}_i$ and labels y_i in the framework of the attention mechanism are called as the *query*, *keys* and *values*, respectively [11].

The most common approach to define the attention weights is to use a scoring function in the form of a kernel $K(\mathbf{x}, \mathbf{x}_i)$. In particular, the weight can be written as:

$$\alpha(\mathbf{x}, \mathbf{x}_i) = \frac{K(\mathbf{x}, \mathbf{x}_i)}{\sum_{j=1}^{n} K(\mathbf{x}, \mathbf{x}_j)}. \tag{2}$$

In order to ensure the flexibility and efficiency of attention mechanism, attention weights are defined with trainable parameters, for example, if we use the Gaussian kernels with trainable parameters, then we get the softmax function as follows:

$$\begin{aligned}\alpha(\mathbf{x}, \mathbf{x}_i) &= \text{softmax}(\text{score}(\mathbf{x}, \mathbf{x}_i, \mathbf{W})) \\ &= \frac{\exp(\text{score}(\mathbf{x}, \mathbf{x}_i, \mathbf{W}))}{\sum_{j=1}^{n} \exp(\text{score}(\mathbf{x}, \mathbf{x}_j, \mathbf{W}))},\end{aligned}$$

where $\text{score}(\mathbf{x}, \mathbf{x}_i, \mathbf{W})$ is a score function, for example, [12]:

$$\text{score}(\mathbf{x}, \mathbf{x}_i, \mathbf{W}) = \mathbf{x}\mathbf{W}\mathbf{x}_i^{\mathrm{T}}. \tag{3}$$

Here $\mathbf{W}$ is the matrix of trainable parameters.

It should be noted that there exist many definitions of the attention weights, in particular, the additive attention [11] and multiplicative or dot-product attention [12, 13] can be regarded as the most popular models.

3 Attention-Based RF (ABRF)

In order to improve RFs, which can be regarded as one of the accurate machine learning models dealing with tabular data, we propose to incorporate the attention mechanism into the RF. In fact, we can consider the set of decision trees in the RF as a set of observations with labels obtained from the tree predictions, which is similar to the Nadaraya-Watson kernel regression.

Let the RF consist of T decision trees and $\mathcal{J}^{(k)}(\mathbf{x})$ be a set of indices of examples, which fall into the leaf node jointly with the example $\mathbf{x}$ after training the k-th tree. Let us also introduce a vector $\mathbf{A}_k(\mathbf{x})$ as the mean of training feature vectors, which fall into the i-th leaf of the k-th tree, i.e.,

$$\mathbf{A}_k(\mathbf{x}) = \frac{1}{\#\mathcal{J}^{(k)}(\mathbf{x})} \sum_{j \in \mathcal{J}^{(k)}(\mathbf{x})} \mathbf{x}_j. \tag{4}$$

Here $\#\mathcal{J}^{(k)}(\mathbf{x})$ means the number of elements in the set $\mathcal{J}^{(k)}(\mathbf{x})$. Vector $\mathbf{A}_k(\mathbf{x})$ can be regarded as the k-th observation in the Nadaraya-Watson kernel regression. However, in contrast to the original Nadaraya-Watson regression, the k-th attention weight is defined by the distance between $\mathbf{x}$ and $\mathbf{A}_k(\mathbf{x})$ in place of the distance between $\mathbf{x}$ and $\mathbf{x}_k$. The Euclidean distance is used here for computing the distance.

Let us introduce parametric attention weights assuming that every weight has one trainable parameter. This assumption is not strong and many trainable parameters can be used, but the reason to use one parameter is to avoid overfitting when the number of trees is large.

Denote the attention weight assigned to the k-th tree as $\alpha(\mathbf{x}, \mathbf{A}_k(\mathbf{x}), w_k)$. It depends on the distances between $\mathbf{x}$ and $\mathbf{A}_k(\mathbf{x})$ and the trainable parameter w_k. The sum of attention weights assigned to all trees is assumed to be equal to 1. If the k-th tree has a prediction $y_k^*(\mathbf{x})$ for example $\mathbf{x}$, then the Nadaraya-Watson regression model for computing the RF prediction $y^*(\mathbf{x})$ can written as follows:

$$y(\mathbf{x}) = \sum_{k=1}^{T} \alpha(\mathbf{x}, \mathbf{A}_k(\mathbf{x}), w_k) \cdot y_k^*(\mathbf{x}). \tag{5}$$

To find optimal trainable parameters and the attention weights, the loss function is optimized over the set $\mathcal{W}$ of all trainable parameter $\mathbf{w} = (w_1, \ldots, w_T)$, i.e., we solve the following optimization problem:

$$\mathbf{w}_{opt} = \arg\min_{\mathbf{w} \in \mathcal{W}} \sum_{s=1}^{n} L\left(y^*(\mathbf{x}_s), y_s, \mathbf{w}\right). \tag{6}$$

By using L_2-norms for the loss function, we can rewrite the objective function as follows:

$$\sum_{s=1}^{n} L\left(y^*(\mathbf{x}_s), y_s, \mathbf{w}\right) = \sum_{s=1}^{n} \left(y_s - \sum_{k=1}^{T} y_k^*(\mathbf{x}_s) \alpha(\mathbf{x}_s, \mathbf{A}_k(\mathbf{x}_s), w_k) \right)^2. \tag{7}$$

In order to avoid using gradient-based algorithms for computing optimal trainable parameters, a simple representation of attention weights α based on the ϵ-contamination model is proposed in [6]. The ϵ-contamination model represents a form of the contaminated probability distribution as $(1-\epsilon)\cdot P+\epsilon\cdot R$, where P is the discrete probability distribution contaminated by another distribution R with contamination parameter $\epsilon\in[0,1]$, which controls the degree of the contamination. If we assume that vector $\mathbf{w}$ of trainable parameters forms the distribution R and the score function (the softmax function) of the attention model produces the distribution P, the attention weight assigned to the k-th tree can be represented as the contaminated probability distribution, i.e., there holds

$$\alpha(\mathbf{x}_s,\mathbf{A}_k(\mathbf{x}_s),w_k)=(1-\epsilon)\cdot\text{softmax}\left(\|\mathbf{x}_s-\mathbf{A}_k(\mathbf{x}_s)\|^2\right)+\epsilon\cdot w_k. \tag{8}$$

Here the vector $\mathbf{w}$ of trainable parameters belongs to the unit simplex because it corresponds to the probability distribution R, i.e., set $\mathcal{W}$ is the unit simplex.

It can be seen from the above definition of attention weights α that they are linearly depend on trainable parameters w_k. This implies that the optimization problem for computing trainable parameters is reduced to a standard quadratic optimization problem with linear constraints defined by the unit simplex.

4 A General Approach with Extreme Points

It turns out that attention weights can be obtained by using other imprecise statistical models different from the Huber's ϵ-contamination model. We consider a general approach to deal with arbitrary models or convex sets of probability distributions. This approach is based on using the extreme points of convex sets of probability distributions as functions of softmax functions.

Suppose that we have a model for attention weights $\alpha(\mathbf{x},\mathbf{A}_k(\mathbf{x}),w_k)$, $k=1,\dots,T$, where $\mathbf{x}$, $\mathbf{A}_k(\mathbf{x})$, w_k have been defined in the previous sections. Let us denote for short the attention weight distribution as $\mathbf{a}_s=\left(\alpha_1^{(s)},\dots,\alpha_T^{(s)}\right)$ and $\mathbf{p}_s=\left(p_1^{(s)},\dots,p_T^{(s)}\right)$, where $\alpha_k^{(s)}=\alpha(\mathbf{x}_s,\mathbf{A}_k(\mathbf{x}_s),w_k)$ and $p_k^{(s)}=\text{softmax}\left(\|\mathbf{x}_s-\mathbf{A}_k(\mathbf{x}_s)\|^2\right)$. A set of attention weights $\mathcal{M}_s$ for an example $\mathbf{x}_s$ can can be represented as

$$\mathcal{M}_s=\left\{\boldsymbol{a}_s:\alpha_k^{(s)}=\zeta h_k(\boldsymbol{p}_s)+\gamma g_k(\boldsymbol{w}),\ \forall k=1,\dots,T\right\}, \tag{9}$$

where $\mathbf{p}_s=\left(p_1^{(s)},\dots,p_T^{(s)}\right)$, h and g are functions of vectors $\mathbf{p}$ and $\mathbf{w}$, respectively, ζ and γ are non-trainable parameters of the model.

This is a general representation of many models of the probability distribution sets. At that, the probability distribution $\mathbf{p}_s$ is fixed for every example $\mathbf{x}_s$, i.e., it it not trainable, but the distribution $\mathbf{w}$ may be arbitrary in the unit simplex, i.e., it is trainable. In particular, the ϵ-contamination model can be represented in the same form under condition that $h_k(\mathbf{p}_s) = p_k^{(s)}$, $g_k(\mathbf{w}) = w_k$, $\zeta = 1 - \epsilon$, $\gamma = \epsilon$. It is important to note that the probability distribution $\mathbf{p}_s$ depends on example $\mathbf{x}_s$, but the parameter vector $\mathbf{w}$ does not depend on examples, and it is computed for the whole dataset.

The set $\mathcal{M}_s$ can be regarded as a convex polytope in the unit simplex $S(1, T)$ of probabilities. Its bias from the center of the unit simplex is defined by $h_k(\mathbf{p}_s)$, its form and its size are defined by $g_k(\mathbf{w})$, ζ and γ. The convexity of the set $\mathcal{M}_s$ implies that it is defined by a set $\mathcal{E}(\mathcal{M}_s)$ of M extreme points denoted as $\mathbf{e}_{s.1}, \dots, \mathbf{e}_{s,M}$ such that every extreme point belongs to the unit simplex. Then the attention weight $\mathbf{a}_s$ associated with example $\mathbf{x}_s$ can be represented as the convex combination of extreme points, i.e., there holds

$$\mathbf{a}_s = \sum_{i=1}^{M} v_i \mathbf{e}_{s,i}, \tag{10}$$

where $v_1, \dots, v_M$ are coefficients satisfying the following condition: $v_1 + \dots + v_M = 1$.

It should be noted that extreme points depend on example $\mathbf{x}_s$ and, therefore, on the probability distribution $\mathbf{p}_s$ whereas coefficients $v_1, \dots, v_M$ do not depend on on examples. The vector $\mathbf{v} = (v_1, \dots, v_M)$ is a set of trainable parameters which defines the attention weights.

If we could express every extreme point $\mathbf{e}_{s,i}$ through functions $h_k(\mathbf{p}_s)$, $k = 1, \dots, T$, then we can replace the trainable parameters $\mathbf{w}$ with new trainable parameters $\mathbf{v} = (v_1, \dots, v_M)$. It follows from the convex combination of extreme points that attention weight $\mathbf{a}_s$ linearly depends on $\mathbf{v}$. Hence, we can construct the quadratic optimization problem for computing the coefficients $\mathbf{v}$ in the unit simplex $S(1, M)$ of dimension M as follows:

$$\begin{aligned}\mathbf{v}_{opt} &= \arg\min_{\mathbf{v}\in S(1,M)} \sum_{s=1}^{n} L\left(y_k^*(\mathbf{x}_s), y(\mathbf{x}_s), \mathbf{v}\right) \\ &= \arg\min_{\mathbf{v}\in S(1,M)} \sum_{s=1}^{n} \left(y(\mathbf{x}_s) - \sum_{k=1}^{T} y_k^*(\mathbf{x}_s) \sum_{j=1}^{M} v_j \mathbf{e}_{s,j} \right)^2 \end{aligned} \tag{11}$$

We have obtained a general approach for constructing the attention-based RF model. Every model is defined by the definition of set $\mathcal{M}_s$ and by expressions for extreme points as functions of softmax functions $\mathbf{p}_s$. However, there is an important problem of using the approach. The number of extreme points for many models of sets of probability distributions may depend on parameters ζ and γ and on values of $\mathbf{p}_s$. Two examples of $\mathcal{M}_1$ and $\mathcal{M}_2$ defined by means of the pari-mutual model [10] with different numbers of extreme points and $T = 3$ are illustrated in Fig. [f:simplex_ pm_0]. For different examples, vector $\mathbf{p}_s$ may produce different numbers of extreme points. In this case, the proposed general approach cannot be used. It can be seen from Fig. [f:simplex_pm_0] that the difference of the extreme point numbers is due to locations of points $\mathbf{p}_1$ and $\mathbf{p}_2$, which are centers of the small simplices (see the left unit simplex in Fig. [f:simplex_pm_0]), and due to different parameters ζ and γ (see the right unit simplex in Fig. [f:simplex_pm_0]). If parameters ζ and γ can be fixed, then vectors $\mathbf{p}_s$ may be arbitrary. Moreover, if we know the used dataset, then we do not know test examples. Therefore, we have to find a way to overcome the above difficulty. The main idea behind this way is to restrict the set of parameters ζ and γ and the set of possible vectors $\mathbf{p}_s$ such that the number of extreme points remains the same for all examples (training and testing). This can be implemented by considering the well-known specific models of the probability distribution sets.

The first well-known model is the constant odds ratio model [10]. For this model, the set of probability distributions defined as the neighborhood of a given distribution $\mathbf{p}_s$ is

$$\mathcal{M}_{\mathrm{OR},s} = \left\{ \mathbf{a}_s : \alpha_k^{(s)} / \alpha_i^{(s)} \geq (1 - \nu) p_k^{(s)} / p_i^{(s)} \right\}, \tag{12}$$

where $\nu \in [0, 1)$ is a model parameter.

The constant odds ratio model is an interesting model and can be regarded as one of the ways for implementing the attention-based RF model. However, according to [14], the number of extreme points is $2^T - 2$. This implies that we get a huge amount of extreme points such that the number of parameters will be also enormous. Therefore, the constant odds ratio model cannot be used in real applications when the number of trees in RF is rather large even if the depth of the corresponding trees is small. In contrast to the constant odds ratio model, we consider another interesting model called the pari-mutual model.

5 Pari-Mutual Model

The pari-mutuel model is a betting scheme originated in horse racing [10]. However, it has been used in various fields, including economics, risk analysis, life insurance [15, 16]. According to the pari-mutual model, a set of probability distributions $\mathcal{M}_{\mathrm{PM},s}$ is defined as:

$$\mathcal{M}_{\mathrm{PM},s} = \left\{ \mathbf{a}_s : \alpha_k^{(s)} \leq (1 + \theta) p_k^{(s)} \right\}. \tag{13}$$

The model can be represented in the form of ([RF_Att_116]) as follows:

$$\mathcal{M}_{\mathrm{PM},s} = \left\{ \mathbf{a}_s : \alpha_k^{(s)} = (1 + \theta) p_k^{(s)} - \theta w_k \right\}, \tag{14}$$

where $h_k(\mathbf{p}_s) = p_k^{(s)}$, $g_k(\mathbf{w}) = w_k$, $\zeta = 1 + \theta$, $\gamma = -\theta$.

It is proved in [17] that the set $\mathcal{M}_{\text{PM},s}$ is a small flip simplex defined by $M = T$ extreme points under condition that $\theta \leq (T-1)^{-1}$. The j-th extreme point has $T - 1$ elements equal to $(1+\theta)p_k^{(s)}$ and the j-th element equal to $(1+\theta)p_j^{(s)} - \theta$, i.e., the extreme point is of the form:

$$\mathbf{e}_{s,j} = \left((1+\theta)p_k^{(s)}, \ldots, \left((1+\theta)p_j^{(s)} - \theta\right)_j, \ldots, (1+\theta)p_k^{(s)}\right). \tag{15}$$

Moreover, the set $\mathcal{M}_{\text{PM},s}$ has T extreme points if the small simplices are entirely included in the unit simplex. The left unit simplex in Fig. 1 illustrates two sets $\mathcal{M}_{\text{PM},s}$. One can see that the first simplex is entirely included in the unit simplex and it has three extreme points whereas the simplex $\mathcal{M}_2$ intersects the bound of the unit simplex, therefore, it is no longer a simplex and has four extreme points. This example implies that we have to determine the boundary sets $\mathcal{M}_{\text{PM},s}$ included into the unit simplex. These sets show how to restrict probability distributions $\mathbf{p}_s$ to have T extreme points always.

Figure 2 illustrates boundary small simplices $\mathcal{M}_1$, $\mathcal{M}_2$, $\mathcal{M}_3$ depicted by the dashed lines. These sets are produced by the pari-mutual model. Their centers are probability distributions $\mathbf{p}_s$ produced by the softmax functions. If these distributions are inside the simplex $\mathcal{P}(\lambda)$, then all small simplices of the same size will be entirely located in the unit simplex. This implies that we have to find parameters of the simplex $\mathcal{P}(\lambda)$. Note that this simplex is nothing else but the λ-contamination model with parameter λ. We use another notation of the parameter of the contamination model in order to avoid confusion with the same model for attention weights.

Let us look at the first small simplex $\mathcal{M}_1$. Two vertices among three ones have one zero element, i.e., these vertices in the general case of dimension T are of the form (see (13)):

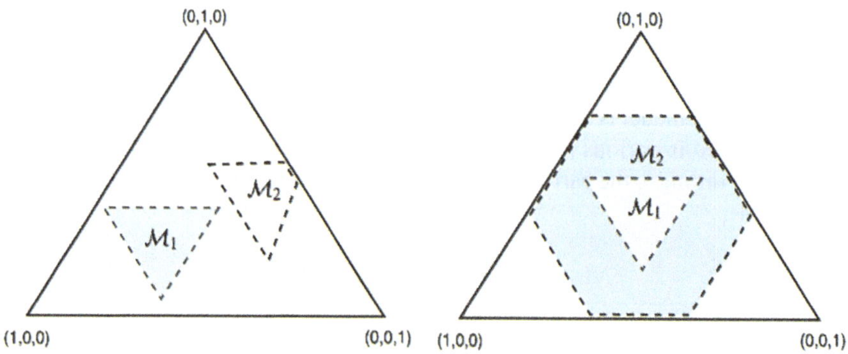

Fig. 1 Two cases of location of the attention weight sets $\mathcal{M}_1$ and $\mathcal{M}_2$ defined by the pari-mutual model with different numbers of extreme points and q = 3

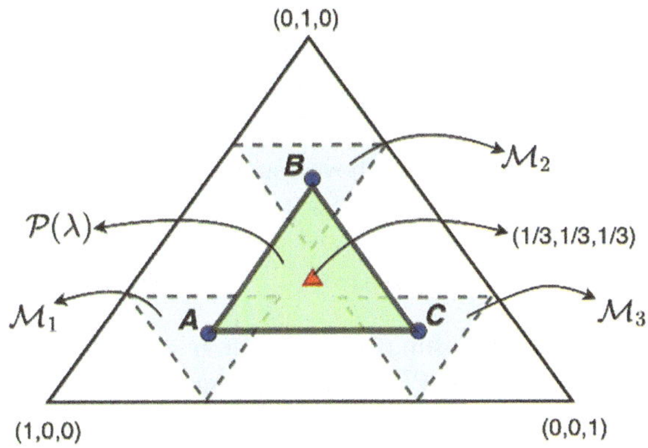

Fig. 2 Illustration of bounds for parameters of the pari-mutual model in order to have $M = T$ extreme points for all instances

$$\mathbf{e}_{s,j} = \left((1+\theta)p_k^{(s)}, \ldots, 0_j, \ldots, (1+\theta)p_k^{(s)}\right). \tag{16}$$

The j-th element of the vertex is 0, i.e., $(1+\theta)p_j^{(s)} - \theta = 0$. This implies that $p_j^{(s)} = \theta/(1+\theta)$ for all j. However, there is a single vertex, say t-th one, which does not have zero elements. In this case, the t-th probability $p_t^{(s)}$ can be calculated as

$$p_t^{(s)} = 1 - \sum_{j=1, j\neq t}^{T} p_j^{(s)} = 1 - (T-1)\frac{\theta}{1+\theta}. \tag{17}$$

In sum, we can write vertices of the simplex $\mathcal{P}(\lambda)$ as

$$\left(\frac{\theta}{1+\theta}, \ldots, \left(1 - \frac{(T-1)\theta}{1+\theta}\right)_t, \ldots, \frac{\theta}{1+\theta}\right), \tag{18}$$

where $t = 1, \ldots, T$.

Hence, $\mathcal{P}(\lambda)$ can be represented as follows (the λ-contamination model):

$$\mathcal{P}(\lambda) = \left\{\mathbf{r} : r_k = (1-\lambda)\frac{1}{T} + \lambda \pi_k\right\}, \tag{19}$$

where $(1/T, \ldots, 1/T)$ is the center of $\mathcal{P}(\lambda)$, $(\pi_1, \ldots, \pi_T)$ is an arbitrary probability distribution.

Our aim is to find the parameter λ which defines the size of $\mathcal{P}(\lambda)$ and restricts the set of $\mathbf{p}_s$. It is simply to carry out if we take one of the extreme points, for example, the point A and consider it as a point belonging two models $\mathcal{M}_{\mathrm{PM},s}$ and $\mathcal{P}(\lambda)$ with $\pi_1 = 1$. The index of element π_1 is used here without loss of generality. Hence, we get two equal probability distributions $\mathcal{M}_{\mathrm{PM},1}$ and $\mathcal{P}(\lambda)$:

$$\left(1-(T-1)\frac{\theta}{1+\theta}, \frac{\theta}{1+\theta}, \ldots, \frac{\theta}{1+\theta}\right) = \left(\frac{1-\lambda}{T}+\lambda, \frac{1-\lambda}{T}, \ldots, \frac{1-\lambda}{T}\right). \tag{20}$$

It follows from the above equality that

$$\lambda = \frac{\theta(T-1)-1}{(1+\theta)(T-1)}. \tag{21}$$

Finally, the softmax function for an arbitrary example is transformed in order to be in the simplex $\mathcal{P}(\lambda)$ and to guarantee the same number of extreme points for the pari-mutual model as follows:

$$\begin{aligned} p_k^{(s)} &= (1-\lambda)\frac{1}{T} \\ &+ \lambda \cdot \text{softmax}\left(\left\|\mathbf{x}_s - \mathbf{A}_k(\mathbf{x}_s)^2\right\|\right), \\ k &= 1, \ldots, T, \end{aligned} \tag{22}$$

where λ is computed from (19).

After substituting the values $p_k^{(s)}$ into (13), we get the set of T extreme points $\mathbf{e}_{s,j}$, $j = 1, \ldots, T$, which are used for solving the quadratic optimization problem (12) and obtaining optimal parameters $v_1, \ldots, v_M$.

6 Numerical Experiments

The proposed approach was investigated and compared with original RF and the ABRF model by using datasets from open sources. A brief introduction about these data sets are given in Table 1 where m and n are numbers of features and examples, respectively. A detailed information can be found in the corresponding R Packages for Diabetes, at site https://www.stat.berkeley.edu/breiman/bagging.pdf for Friedman 1, 2 3, in package "Scikit-Learn" for Regression and Sparse datasets, from the UCI Machine Learning Repository [18] for Wine, Boston, Concrete, Yacht, Airfoil.

The coefficient of determination denoted R^2 and the mean absolute error (MAE) are used for the regression evaluation. The greater the value of the coefficient of determination and the smaller the MAE, the better results we get. In all tables, we compare R^2 and the MAE for three cases: the original RF, the ABRF model with the attention-based RF using contamination model [6], the proposed ABRF-PM model. Every RF consists of 100 decision trees. To evaluate the average accuracy measures, we perform a cross-validation with 100 repetitions, where in each run, we randomly select $n_{\text{tr}} = 4n/5$ training data and $n_{\text{test}} = n/5$ testing data.

Measures R^2 and MAE for three cases (RF, ABRF, ABRF-PM) are provided in Table 2. The best results in Table 2 are shown in bold. The results are obtained by training the RF and the parameter vector $\mathbf{w}$ on the regression datasets under condition

Table 1 A brief introduction about the regression data sets

Data set	Abbreviation	m	n
Diabetes	Diabetes	10	442
Friedman 1	Friedman 1	10	100
Friedman 2	Friedman 2	4	100
Friedman 3	Friedman 3	4	100
Scikit-Learn Regression	Regression	100	100
Scikit-Learn Sparse Uncorrelated	Sparse	10	100
UCI Wine red	Wine	11	1599
UCI Boston Housing	Boston	13	506
UCI Concrete	Concrete	8	1030
UCI Yacht Hydrodynamics	Yacht	6	308
UCI Airfoil	Airfoil	5	1503

that trees are built to ensure at least 10 examples in each leaf. It can be seen from Table 2 that ABRF-PM is comparable with ABRF and outperforms the original RF. The worse results provided by ABRF-PM for several datasets in comparison with ABRF do not mean that the approach using the pari-mutual model should not be used. It follows from Table 2 that the ABRF-PM model is better for several datasets. Therefore, it should be used among other models.

Table 2 Measures R^2 and MAE for comparison of models (the RF, ABRF, ABRF-PM) trained on regression datasets

		R^2			MAE	
Data set	RF	ABRF	ABRF-PM	RF	ABRF	ABRF-PM
Diabetes	0.416	0.424	**0.427**	44.92	44.66	**44.20**
Friedman 1	0.459	0.470	**0.474**	2.540	2.540	**2.535**
Friedman 2	0.841	**0.877**	0.794	111.7	**102.0**	124.4
Friedman 3	0.625	0.686	**0.693**	0.154	0.144	**0.141**
Regression	0.380	**0.450**	0.414	109.1	**100.8**	102.2
Sparse	0.470	**0.529**	0.516	1.908	**1.790**	1.893
Wine	0.823	**0.843**	0.840	2.203	**2.070**	2.114
Boston	0.814	**0.823**	**0.823**	2.539	**2.494**	2.502
Concrete	0.845	**0.857**	0.841	4.855	**4.694**	4.912
Yacht	0.433	0.423	**0.439**	0.451	0.459	**0.449**
Airfoil	0.981	**0.989**	**0.989**	1.004	**0.787**	**0.787**

7 Conclusion

A new attention-based RF model using the imprecise pari-mutual statistical model producing a specific set of probability distributions has been proposed. Moreover, the pari-mutual model is incorporated as a special case of the proposed general approach for training the ABRF model based on the representation of attention weights and trainable parameters by means of extreme points. ABRF-PM can be regarded as an additional attention-based model which extends a set of models using the attention mechanism for improving the regression and classification predictions. It should be also point out that the proposed general approach based on extreme points can be applied to a more complex cases of attention weights. The approach can be viewed as a starting point for developing a class of attention-based models for tabular data.

We have studied a case when attention weights linearly depend on trainable parameters. This case leads to solving the standard quadratic optimization problem. However, it is interesting to consider cases when trainable parameters are included in the softmax operations. Similar cases have been studied in ABRF [6] when the Huber's ϵ-contamination model was used for defining the set of attention weights. However, they have not investigated with the pari-mutual statistical model. This is a direction for further research. Another interesting direction is to study the proposed model by using the gradient boosting machine [19, 20] in place of the RF. The use of this model can also significantly improve the classification and regression quality of models.

Acknowledgements This work is supported by the Russian Science Foundation under grant 21-11-00116.

References

1. Chaudhari, S., Mithal, V., Polatkan, G., Ramanath, R.: An attentive survey of attention models. arXiv:1904.02874 (2019)
2. Correia, A., Colombini, E.: Attention, please! A survey of neural attention models in deep learning. arXiv:2103.16775 (2021)
3. Correia, A., Colombini, E.: Neural attention models in deep learning: survey and taxonomy. arXiv:2112.05909 (2021)
4. Lin, T., Wang, Y., Liu, X., Qiu, X.: A survey of transformers. arXiv:2106.04554 (2021)
5. Niu, Z., Zhong, G., Yu, H.: A review on the attention mechanism of deep learning. Neurocomputing **452**, 48–62 (2021)
6. Utkin, L., Konstantinov, A.: Attention-based random forest and contamination model. arXiv: 2201.02880 (2022)
7. Breiman, L.: Random forests. Mach. Learning **45**, 5–32 (2001)
8. Nadaraya, E.: On estimating regression. Theory Probability Appl. **9**, 141–142 (1964)
9. Watson, G.: Smooth regression analysis. Sankhya: The Indian Journal of Statistics, Series A, 359–372 (1964)
10. Huber, P.: Robust Statistics. Wiley, New York (1981)
11. Walley, P.: Statistical Reasoning with Imprecise Probabilities. Chapman and Hall, London (1991)

12. Bahdanau, D., Cho, K., Bengio, Y.: Neural machine translation by jointly learning to align and translate. arXiv:1409.0473 (2014)
13. Luong, T., Pham, H., Manning, C.: Effective approaches to attention-based neural machine translation. In: Proceedings of the 2015 Conference on Empirical Methods in Natural Language Processing, The Association for Computational Linguistics, 1412–1421 (2015)
14. Vaswani, A., Shazeer, N., Parmar, N., Uszkoreit, J., Jones, L., Gomez, A., Kaiser, L., Polosukhin, I.: Attention is all you need. In: Advances in neural information processing systems, 5998–6008 (2017)
15. Montes, I., Miranda, E., Destercke, S.: Unifying neighbourhood and distortion models: part i -new results on old models. Int. J. Gen. Syst. **49**, 602–635 (2020)
16. Montes, I., Miranda, E., Destercke, S.: A study of the pari-mutuel model from the point of view of imprecise probabilities. In: Proceedings of the Tenth International Symposium on Imprecise Probability: Theories and Applications. Volume 62 of Proceedings of Machine Learning Research., PMLR, 229–240 (2017)
17. Pelessoni, R., Vicig, P., Zaffalon, M.: Inference and risk measurement with the pari-mutuel model. Int. J. Approximate Reasoning **51**, 1145–1158 (2010)
18. Utkin, L., Wiencierz, A.: Improving over-fitting in ensemble regression by imprecise probabilities. Inf. Sci. **317**, 315–328 (2015)
19. Dua, D., Graff, C.: UCI machine learning repository (2017)
20. Friedman, J.: Greedy function approximation: a gradient boosting machine. Ann. Stat. **29**, 1189–1232 (2001)
21. Friedman, J.: Stochastic gradient boosting. Comput. Stat. Data Anal. **38**, 367–378 (2002)

The Formation of Metrics of Innovation Potential and Prospects

D. M. Korobkin, S. A. Fomenkov, A. R. Zlobin, G. A. Vereshchak, and A. B. Golovanchikov

Abstract In today's rapidly developing technological world, new ideas, inventions and developments appear daily. At the same time, individual technologies may have common features, through the use of similar methods, modification and expansion of existing technologies, or solving a common problem for which developments are being created. Patenting is often used to preserve these ideas and protect intellectual property. During the development of the program for the analysis of the patent array to obtain criteria assessments of innovation potential and prospects, enshrined in patent high–tech technical systems and technologies, the subject area—patent array was investigated, methods for analyzing texts in natural language and various options for determining the criteria of innovation potential were considered. As a result, algorithms were developed to obtain the following criteria for assessing the innovative potential of a patent: the mass character of the subject of this technology for the current year and the estimated frequency of occurrence for the next, the economic characteristics of the patent holder's company and the potential citation of the patent. These criteria are determined based on the analysis of texts and data of patents using clustering, classification, regression analysis and normalization of the name of the patent holder. The developed algorithms were tested on patents issued by the US Patent and Trademark Office, as well as on Google Patents.

Keywords Cyber-physical systems · Patents · Fact extraction

1 Introduction

In today's rapidly developing technological world, new ideas, inventions and developments appear daily. Patents are often used to preserve these ideas and protect intellectual property. The patent document describes all the key features of the invention, thanks to which it is unique, and also describes the processes necessary for its

D. M. Korobkin (✉) · S. A. Fomenkov · A. R. Zlobin · G. A. Vereshchak · A. B. Golovanchikov
Volgograd State Technical University, d. 28, Av. Lenina, Volgograd 400005, Russia
e-mail: dkorobkin80@mail.ru

A. G. Kravets et al. (eds.), *Cyber-Physical Systems Engineering and Control*,
Studies in Systems, Decision and Control 477,
https://doi.org/10.1007/978-3-031-33159-6_2

reproduction. According to statistics, the number of patented technologies is only increasing every year, so it is becoming increasingly difficult to find among the many patent documents those that have a technology with high innovative potential and prospects.

Analysis of the existing patent base [1, 2] is often used to simplify the process of generating new ideas, finding analogues for the developed technology and analyzing technological trends. Speaking about the task of determining the prospects and innovative potential of the technology enshrined in the patent, we can say that there is no standard for solving and automating such a task.

Companies with tangible assets are interested in effective investments in new high-potential developments. Experts determine the potential of new technologies, however, given the number of patents issued, this work requires enormous labor costs. The program for determining innovative potential and prospects can reduce labor costs.

The paper [3] describes a method for determining the innovation potential in the form of a questionnaire on a patent. The authors proposed an "Innovation Radar", which, by and large, is a questionnaire, filling out which for a specific patent, the authors receive an assessment of its potential on a 100-point scale, judging by which all innovations can be divided into 3 large groups—innovations with high, medium and low potential. The method is a method of analyzing and synthesizing data on a patent, first of all, about the readiness of a project using a patented technology, as well as about its creators.

In [4], the authors use an interesting approach to determine the innovation potential—the analysis of the natural language in which user reviews of the patent are written, from these reviews it is also easy to get information about the technology that is specified in the patent and the color of user reactions.

In [5], the authors consider the approach of analyzing the technology presented in the patent by extracting matrix vectors consisting of SAO structures on the example of one of the graphene technologies, in this work the proximity of the technology to the technological trend is also checked. And in [6], the authors extract SAO structures (using a set of regular expressions using a parser) from the texts of patents on the subject of smart homes, defining and combining semantically similar structures and creating a term-document matrix based on them, where patents are documents and SAO structures are terms. The innovation potential in this chapter is presented according to one criterion—the relative size of the thematic cluster. Based on the results of this work, it was concluded that such clustering is 3% more accurate than clustering by keywords.

The use of the SAO model to extract various concepts for processing English-language patents is reflected in various systems [7–9]. To improve the quality of parsing, a syntactic analysis of the tree with a separate identification of the subject, action and object in the work was used [9]. According to the authors, the average values of accuracy and completeness of extraction are as follows: 0.8058 and 0.8446, respectively. At the same time, the implementation is carried out in the GATE system. Based on the rules using the Stanford parser software, the SAO structures are extracted in [8]. In [7, 10], patents are processed using linguistic markers (specific verbs and nouns) and lexico-syntactic templates, while [10] notes the need to study the structure of patent documents to improve the quality of data extraction.

The authors [11] have created a technique called "TOD" (Technology Opportunity Discovery—the process of identifying technologies with high potential). This is a good example of work using the approach of clustering patents by keywords, and the technology in this work is not tied to a separate patent and gets the form of a set of keywords. The disadvantage of this work is that it is impossible to evaluate a single patent or technology through their methodology, since they do not offer any numerical system for evaluating patents. The advantage lies in the fact that the authors are able to single out a technology with high potential from their multitude.

The purpose of the work is to increase the efficiency of the formation of information support for the processes of technology forecasting and determining the quality of patents through automated analysis of patent arrays.

2 Materials and Methods

2.1 Patent Array Analysis

The input of the algorithm is a directory with a patent selection containing data on English-language patents issued by the US Patent and Trademark Office. A total of 11,322 patent documents were collected, 4008 of which were directly from the USPTO website for 2020–2021, and 7314 from Google Patents for 2016–2019. Selected patents containing the patent classification «Machine learning» («G06N20/00»). The patents of this classification are a sample that simulates the general patent array, since machine learning methods are used in all branches of science, the resulting patent array according to this classification will have the property of diversity. Data from Google Patents is required for advanced data acquisition functionality and to obtain information on the number of patent citations. Figure 1 shows the algorithm for analyzing the patent array.

Keywords and phrases (from two or three frequently combined words) are extracted from the texts of patent documents after preprocessing the text, which includes:

- lemmatization of words based on WordNet definitions and contextual parts of speech;
- removal of general and thematic stop words;
- reduction to lowercase;
- segmentation of sentences by punctuation marks;
- removing punctuation and separation marks
- removing numbers and words less than two characters long;
- text tokenization;

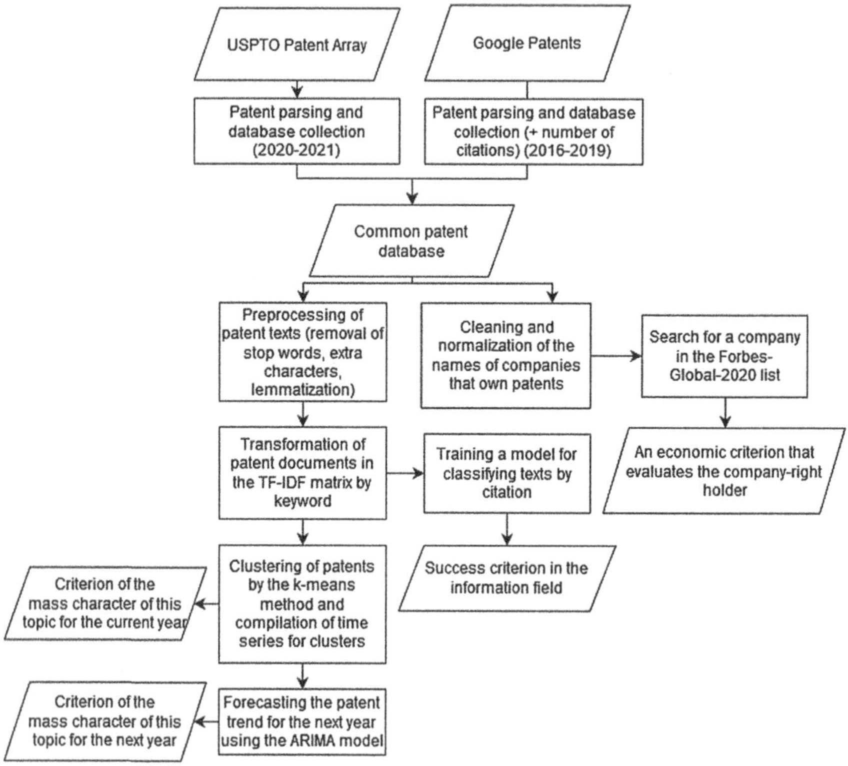

Fig. 1 Analysis of the patent array

- definition of phrases (bigrams and trigrams).

The result of the work of the algorithm for analyzing the patent array is the assessment of the innovative potential of the technology enshrined in the patent according to four different criteria.

2.2 *Criteria-Based Assessments of the Patent's Innovation Potential*

The criterion of the mass character of the topic for the current year. The criterion of the mass character of the topic for the current period of time determines the proximity of the technology fixed in the patent to the technological trend. Since the novelty of the technology is a mandatory criterion when registering a patent, the relative size of the thematic cluster to which the patent belongs shows the popularity of this topic at the current time. The higher the popularity of the topic, the more likely it is to be relevant and the more capital relates to this topic.

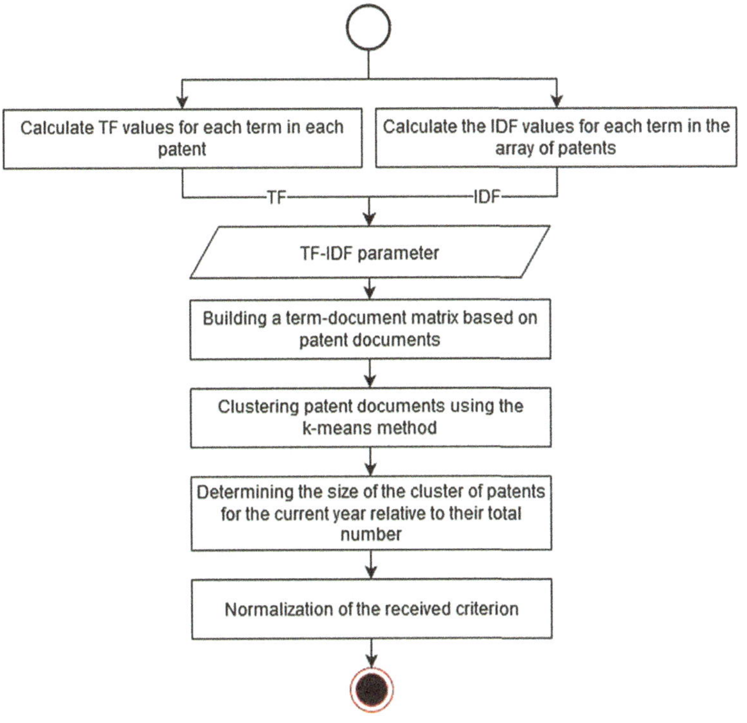

Fig. 2 Algorithm for determining the criterion of mass character of the topic for the current year

After the text preprocessing steps described above, it is necessary to implement the algorithm included in the method for determining this criterion, the algorithm is shown in Fig. 2.

The TF-IDF metric is defined as the product of the frequency of occurrence of a term within a document and the logarithm of the ratio of the number of documents in the collection to the number of documents with a given term.

On the constructed term-frequency vectors for documents, clustering by topics is performed using the k-means method, then the size of the largest thematic cluster is determined and the numerical value of the criterion being determined is calculated as the ratio of the cluster size to which the evaluated patent document belongs to the size of the largest cluster rounded to a ten-point scale.

To determine the number of clusters, the elbow method is used, which implies multiple cyclic execution of the algorithm with an increase in the number of selected clusters, as well as subsequent postponement of the clustering score on the graph—the quadratic sum of errors. The inflection point of the graph is the optimal number of clusters.

The criterion of the mass character of the topic for the next year. This criterion is determined based on clustering data obtained at the stage of determining the first criterion. Within each cluster, its size is calculated for each of the five years under consideration. These five points are used to determine the estimated sixth point—the size of this thematic cluster for the next year. This criterion is close to the first criterion, it is defined similarly—as the size of the cluster related to the patent to the size of the largest cluster, but for the next period of time. The ARIMA model («AutoRegressive Integrated Moving Average») is used to make a forecast for the next year for five points. This is a model class that captures a set of different standard time structures in time series data [12].

This model is a combination of autoregression (determination of the sequential dependence of time series elements) and the moving average process. The parameters of this model are:

- p: the number of lagged observations included in the model, also called the lag order.
- d: the number of times the original observations differ, also called the degree of difference.
- q: the size of the moving average window, also called the order of the moving average.

The criterion of success in the information field. Such a criterion is found in many IP management systems, however, it is often applied only to patents that have already existed for a relatively long time. It is determined by the number of patents.

The diagram shown in Fig. 3 shows the distribution by the number of patent citations.

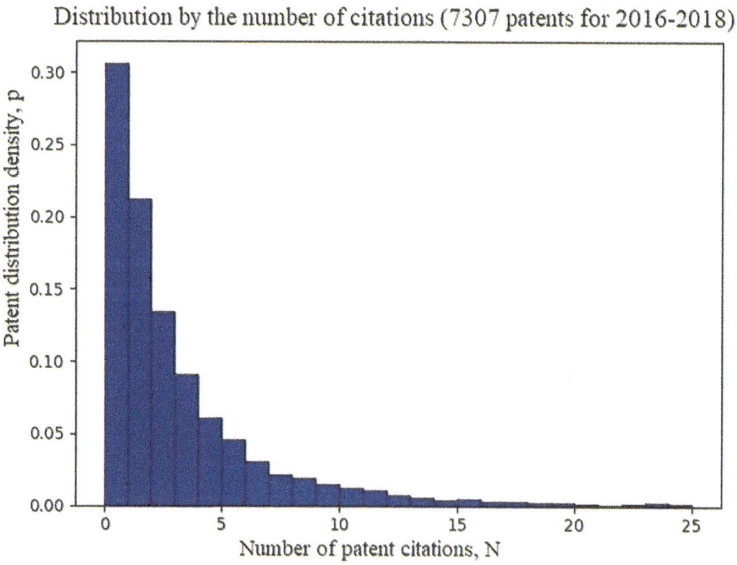

Fig. 3 Distribution by number of citations

As can be seen from the above diagram, more than half of all patents studied for citation have a citation of 0 or 1. It was decided to divide all patents into two citation classes—high citation and low. The first category includes all patents whose citation is greater than 2, and the first category includes patents whose citation is less than or equal to 2.

After the text preprocessing process, words with a frequency of occurrence less than 5 in the document are excluded from the total mass of the list of words. Then a training sample is built—the processed text of the patent and the corresponding classification by citation. This sample is divided into test and training.

To solve the problem of classifying patents by citation, the CBOW («Continuous Bag of Words») model is used using TF-IDF metrics. This is a way to represent text data when modeling text using machine learning algorithms. The «bag of words» model is easy to understand and implement and has great success in problems such as language modeling and document classification.

For the task of classifying patents by citation, this model was chosen for three reasons:

- Term-frequency vectors for patent documents have already been constructed earlier as part of the clustering task by topic;
- This model is great for tasks with a large amount of text data;
- This model has sufficient accuracy with low training costs.

The neural network itself, used for classification—TensorFlow, contains input, hidden and output layers. Term-frequency vectors are fed to the input, and the citation class is fed to the output.

As a result, the success criterion in the information field is a binary criterion that can take the values 0 or 1, for few and many cited patents, respectively.

Economic criterion. This criterion is an assessment of the company- the copyright holder of the patent. The criterion itself is complex—it may include such indicators as the capitalization of the company, the measure of the company's readiness for innovation, the technical and technological level of the organization, the share of the organization in the number of patented technologies on the subject to which this patent relates, intellectual and financial property of the company.

There are various ratings of technology companies, scientific universities and state scientific organizations. In this chapter, two different existing ratings of companies are used:

1. «The Global Innovation 1000 study» from Strategy & PwC for 2018 (only this version is publicly available) [13];
2. «The Global 2000 2020» by Forbes [14].

If the company is in the first list, two parameters are taken from there—«R&D Intensity» and «R&D Expense» for the last year. The first parameter indicates the ratio of income and expenses for research, and the second parameter indicates the total amount of research costs. Further, classification is carried out according to these parameters on five–point scales, and the sum of these parameters is the final result—an economic criterion.

If the company was not found in the first list, the second is considered, from where the parameters «Profits» And «Market value» are considered by analogy, the first describes the company's profit, the second—its total capitalization.

To determine this criterion, two actions are required—normalization of the name of the company that holds the patent and a search for the normalized name in the above lists.

The task of normalization arises due to the fact that the name of the same company can be specified differently in the patent and there are no ideal standards when writing the name.

For normalization, a ready–made open solution is used, using a database of company names from the private non-profit organization NBER («National Bureau of Economic Research»—«National Bureau of Economic Research») This database contains the normalized names of more than 3000 companies and their corresponding possible names with errors or their alternative spellings. As statistics show, about 75% of all existing variations of the spelling of the names of various companies are determined directly through this database, for the rest of the cases, the Levenshtein distance method is used—the character-by-character difference between two string values, to determine the nearest possible company name from the NBER list. The calculation of the nearest name/name written with an error is performed throughout the database. If the distance divided by the length of the string is more than half, we assume that this company is not in the database. Just for such cases, manual expansion of the NBER database and tables with capitalization data is required.

The study of data tables by capitalization and by research costs made it possible to create a data classification scale. Table 1 shows the ranges of criteria on a five-point scale, these ranges were obtained based on the distribution of numerical data presented in the table for all companies.

If a company is found in the table «TheGlobalInnovation 1000 study», two parameters are used—«R&D Intensity» and «R&D Expense» for the last year, and the final criterion is the sum of the parameters transferred according to the Table 1. For the table «The Global 2000 2020» everything is similar, but other parameters are considered—«Profits» and «Marketvalue».

3 Results

To determine the criteria of innovative potential and prospects of the developed high-tech technical systems and technologies, a program consisting of 4 blocks was designed and implemented:

- A block of patent parsing and text preprocessing;
- The block for determining the criteria of mass;
- The block of definition of informational and economic criteria;

Table 1 Conversion of criteria to a five-point scale

Parameter	Range (x—parameter value)	Grade
R&D Intensity	$0 < x <= 0,20$	1
	$0,20 < x <= 0,4$	2
	$0,4 < x <= 0,6$	3
	$0 < x <= 0,8$	4
	$x > 0,8$	5
R&D Expense	$0 < x <= 5$	1
	$5 < x <= 10,5$	2
	$10,5 < x <= 16$	3
	$16 < x <= 20$	4
	$x > 20$	5
Profits	$x < 0$	1
	$0 < x <= 238$	2
	$238 < x <= 654$	3
	$654 < x <= 965$	4
	$x > 965$	5
Marketvalue	$0 < x <= 11531$	1
	$11531 < x <= 22133$	2
	$22133 < x <= 38142$	3
	$38142 < x <= 73452$	4
	$x > 73452$	5

- Display unit.

Figure 4 shows a generalized structure of the program being developed with an indication of data flows between blocks.

The legend of the structure shown in Fig. 4:

1. The display block displays the user interface when the program starts, the user sends a command with the XML text of the patent or a link to Google Patents. The data is transmitted to the parsing and preprocessing unit.
2. Structured patent data is sent for preprocessing.
3. Preprocessed and structured patent data are recorded in the database.
4. The patent data is sent to the block for determining the mass criteria.
5. Data on clusters of patents by year are transmitted for regression analysis.
6. The criteria scores for this cluster and the patent are recorded in the database.
7. Patent data is transferred to the block for obtaining information and economic criteria (the text of the patent).
8. Patent data is transferred to the block for obtaining information and economic criteria (the name of the company of the copyright holder and the country).
9. The information criterion evaluation of the patent being processed is recorded in the database.

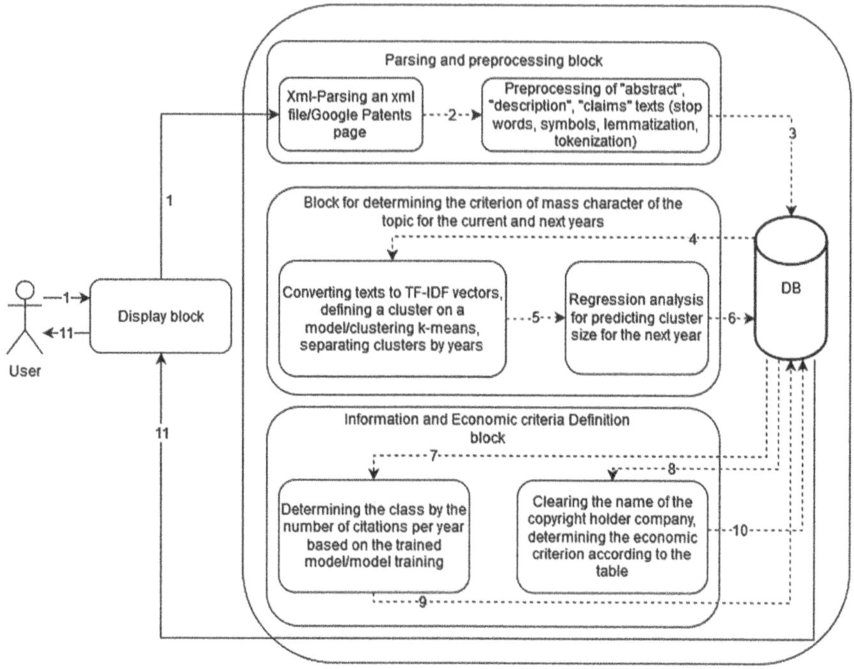

Fig. 4 General structure of the program

10. The economic criteria assessment of the patent being processed is recorded in the database.
11. Data on all estimates are transmitted for display to the user.

The program is implemented in the Python 3 programming language. SQLite [15] was chosen as the DBMS, since this system has a simple architecture and sufficient speed of operation. To facilitate work with the DBMS and protect against hypothetical SQL injections, the SQLAlchemy software library is used, the feature of which is the transition from database entities to Python class objects [16]. The graphical user interface was developed using a set of extensions of the PyQt graphical framework [17]. For most of the tasks of text preprocessing, a set of NLTK libraries is used [18].

The quality of clustering by subject is assessed based on a comparison of clustering used in the program, and similar, but through the international classification of each CPC patent (Cooperative Patent Classification). Term-frequency vectors are replaced by classification-frequency vectors and then a similar clustering is performed by the K-means method. The clusters corresponding to the maximum number of matches are considered. Then the number of patent documents that did not fall into the corresponding cluster is determined, this number is divided by the total number of patent documents under study. Patents were selected from the total sample for 2020–2021, a total of 4008 patents, testing results showed that the percentage of clustering differences is 4% (221 clustering differences).

Table 2 Testing of forecasting the size of a thematic cluster for the next year

Name	Value
Total patents in the sample	10,325
Number of thematic clusters	54
Average cluster size for 2020 (for 3011 patents)	55.8
The standard deviation of the cluster size forecast for 2020	2,71
The ratio of the standard deviation to the average cluster size	0.049

Table 3 Error matrix for the classification process

		Real citation	
		1	
Predicted citation	1	407	262
	0	214	517

To check the prediction of the size of a thematic cluster for the next year (regression models), clustering is performed using the method established in the program and clusters are divided into time series, then data for 2020 is removed from the sample and regression analysis is performed to determine the size of clusters for 2020. The results are presented in Table 2.

Verification of the definition of the information criterion consists in dividing a sample of patents with a known citation into test and training samples, after training the classification model, this model is checked on a test sample. The error matrix for the classification process is presented in Table 3. The accuracy was 66%.

During the verification of the normalization of the names of companies-patent holders for 50 patents, normalized patent names were manually determined. 47 out of 50 names were successfully identified during the work of the program, three companies were not found in the database of company names, for such companies the economic criterion is lowered, the accuracy of determination is 94% (to increase the accuracy of determination, it is necessary to expand the databases of companies).

4 Discussion

The chapter developed and described a method for analyzing patent arrays [19] to obtain criteria-based assessments [20, 21] of the innovative potential and prospects of the developed high-tech technical systems and technologies [22]. There are four main criteria reflecting various aspects of the patented development—economic, informational and criteria of mass character for the current and next years. Considering the totality of these criteria, we can say that a patent document that has received high marks according to these criteria is a promising patented technology.

The method of obtaining these criteria was automated and implemented in the form of a program with a visual interface that allows determining the potential estimates of the downloaded patent document.

5 Conclusion

The theoretical value of this work lies in the developed method of obtaining unique criteria that make up the innovative potential and prospects of the technology enshrined in the patent document. The unique criteria include the criterion of the mass character of this topic for the current year (construction of a term-document matrix with the TF/IDF metric, clustering of K-means), the criterion of the mass character of the topic for the next year (analysis of the time series using the ARIMA model), the criterion of success in the information field (classification by citation by TensorFlow), the economic criterion (normalization of the name of the patent holder company).

Acknowledgements The study was supported by the grant of the Russian Science Foundation No. 22-21-20125, https://rscf.ru/en/project/22-21-20125/, and the Administration of the Volgograd Region.

References

1. Korobkin, D., Fomenkov, S., Fomenkova, M., Vayngolts, I., Kravets, A.: The software for computation the criteria-based assessments of the morphological features of technical systems. In: Cyber-Physical Systems, pp. 161–172. Springer (2021)
2. Korobkin, D., Fomenkov, S., Vereschak, G., Kolesnikov, S., Tolokin, D., Kravets, A.: The formation of morphological matrix based on an ontology "patent representation of technical systems" for the search of innovative technical solutions. In: Cyber-Physical Systems, pp. 149–160. Springer (2021)
3. De Prato, G., Nepelski, D., Piroli, G.: Innovation radar: identifying innovations and innovators with high potential in ICT FP7. CIP & H2020 projects. JRC Scientific and Policy Reports-EUR. 27314, pp. 11–15 (2015)
4. Roh, T., Jeong, Y., Jang, H., Yoon, B.: Technology opportunity discovery by structuring user needs based on natural language processing and machine learning. PloS one. Public Library of Science. San Francisco, CA USA. 14(10), p. e0223404 (2019)
5. Yang, C., Huang, C., Su, J.: An improved SAO network-based method for technology trend analysis: a case study of graphene. J. Inf. **12**(1), 271–286. Elsevier (2018)
6. Kim, S., Park, I., Yoon, B.: SAO2Vec: Development of an algorithm for embedding the subject–action–object (SAO) structure using Doc2Vec. Plos one. Public Library of Science. San Francisco, CA USA **15**(2), p. e0227930 (2020)
7. Guo, J., Wang, X., Li, Q., Zhu D.: Subject–action–object-based morphology analysis for determining the direction of technological change. Technological Forecasting and Social Change **105**, pp. 27–40. Elsevier (2016)

8. Wang, X., Qiu, P., Zhu, D., Mitkova, L., Lei, M., Porter, A.: Identification of technology development trends based on subject--action--object analysis: the case of dye-sensitized solar cells. Technological forecasting and social change. 98, pp. 24–46. Elsevier (2015)
9. Yang, C., Zhu, D., Wang, X.: SAO semantic information identification for text mining. Int. J. Comput. Intell. Syst. Atlantis Press BV **10**(1), p. 593 (2017)
10. Souili, A., Cavallucci, D., Rousselot, F., Zanni, C. Starting from patents to find inputs to the problem graph model of IDM-TRIZ. Procedia Eng. **131**, 150–161. Elsevier (2015)
11. Feng, L., Niu, Y., Liu, Z., Wang, J., Zhang, K.: Discovering technology opportunity by keyword-based patent analysis: a hybrid approach of morphology analysis and USIT. sustainability. MDPI **12**(1), 136 (2019)
12. Brownlee, J.: How to Create an ARIMA Model for Time Series Forecasting in Python. https://machinelearningmastery.com/arima-for-time-series-forecasting-with-python/. Accessed 1 Nov 2022
13. The Global Innovation 1000 study. https://www.strategyand.pwc.com/gx/en/insights/innovation1000.html. Accessed 1 Nov 2022
14. Global 2000. https://www.forbes.com/lists/global2000/#4e109b4a5ac0. Accessed 1 Nov 2022
15. SQLite serverless relational database management system Documentation. https://sqlite.org/docs.html. Accessed 1 Nov 2022
16. SQLAlchemy 1.4. https://docs.sqlalchemy.org/en/14. Accessed 1 Nov 2022
17. PyQt5 Project description. https://pypi.org/project/PyQt5. Accessed 1 Nov 2022
18. NLTK 3.6.2 documentation. https://www.nltk.org. Accessed 1 Nov 2022
19. Korobkin, D., Fomenkov, S., Kravets, A., Kolesnikov, S.: Prior art candidate search on base of statistical and semantic patent analysis. In: Multi Conference on Computer Science and Information Systems, p. 231 (2017)
20. Korobkin, D., Fomenkov, S., Zlobin, A., Shabanov, D., Golovanchikov, A.: The software for formation of technical function assessments based on the patent analysis. In: Cyber-Physical Systems, pp. 137–147. Springer (2021)
21. Korobkin, D., Fomenkov, S., Golovanchikov, A.: Method of identification of patent trends based on descriptions of technical functions. In: Journal of Physics: Conference Series. IOP Publishing, p. 032065 (2018)
22. Viet, N.T., Kravets, A.G.: The new method for analyzing technology trends of smart energy asset performance management. Energies **15**(18), 6613 (2022). https://doi.org/10.3390/en15186613

Extraction of Information Fields in Administrative Documents Using Constellations of Special Text Points

Oleg A. Slavin and Igor M. Janiszewski

Abstract This chapter considers administrative documents with a simple flexible structure. To analyze the content of documents we propose a model based on constellations of special text points. A special text point is understood as a word unique for some neighborhood of the text. The descriptors of points extracted from the document images by OCR are considered. Local text singularities are defined. A modified normalized Levenshtein distance for comparison of words considering recognition errors is described. To resolve conflicts when compared words are distant in meaning but differ by a small Levenshtein distance, we propose penalty functions for character mismatches in a subset of symbols in words. Constellations are defined as sets of special text points grouped into terms and chains. The representation of keywords and terms as a set of alternatives is specified. An algorithm for matching the set of words of a recognized document and the set of constellation models is described. The experimental verification of the implemented algorithm on private test datasets of administrative documents is provided. The developed method can be applied in modern CAD systems to analyze the content of recognized textual files.

Keywords Administrative documents · Document recognition · Levenshtein distance · Special text point · Constellation of points · Term

1 Introduction

Administrative documents are used to exchange information within an organization and between organizations. Administrative document includes a template part (static and graphic elements) and fields (content). Static elements primarily consist

O. A. Slavin (✉)
Federal Research Center "Computer Science and Control" of the Russian Academy of Sciences, 9 Prospect 60-Letiya Oktyabrya, Moscow 117312, Russia
e-mail: oslavin@isa.ru

O. A. Slavin · I. M. Janiszewski
LLC Smart Engines Service, 9 Prospect 60-Letiya Oktyabrya, Moscow 117312, Russia

A. G. Kravets et al. (eds.), *Cyber-Physical Systems Engineering and Control*, Studies in Systems, Decision and Control 477,
https://doi.org/10.1007/978-3-031-33159-6_3

of static text. The field is defined as text bounded by static elements or dividing lines. Document design allows for template structure augmentation via.

- fonts and font sizes formatting;
- line spacing adjustment;
- moving text to the next line;
- moving text to the next page;
- deleting words of static text;
- replacing words of static text.

Such augmentations are inherent to flexible documents. The recognition of the latter is an urgent task in systems that input large flows of documents into the computer. Recognition of administrative documents allows for the automatization of the organization's workflow processes [1].

Extracting information from recognized administrative documents has several nuances. For example, a dictionary of allowed keywords is short, and algorithm should account for a significant number of recognition errors when comparing words. The recognition problem is defined as follows. Based on the recognized text objects and graphic primitives, the task is to extract information from areas that correspond to the bounded fields. It is required to extract information from the maximum number of fields with the smallest number of errors for each field. This chapter considers a two-pass method of document recognition using description of fields. During the first pass, after text recognition and extraction of graphic primitives, the boundaries of the fields are predicted. The boundaries of the fields are defined employing the word boundaries of the static text. Static text words are grouped for reliable identification. The invariance to document distortions is provided by the constellation model, which takes into account a large number of possible recognition errors. The purpose of the field prediction is to specify the parameters of the employed Artificial Neural Networks (ANNs) and the dictionary, which are used in the second pass. During the second pass, due to the recognition parameterization the quality of the field recognition within identified boundaries is improved.

This chapter considers noisy document images. Currently, this topic is of great interest [2, 3].

2 Background

In this chapter, we continue to consider the model for flexible administrative documents based on special text points [4, 5]. A special text point corresponds to the definition of an image keypoint which is:

- distinct from neighboring image points,
- resistant to noise,

- resistant to geometric transformations [6].

The simplest examples of image keypoints are corners, ends of segments and other topological features of the morphological structure of images. In document recognition, the keypoints mechanism is used for the classification and localization of a document or its part by comparison with a reference. Localization of documents with templates of fixed geometry can be performed using this method, even while capturing the said document with a camera, i.e. in the presence of projective distortions. In this case, the algorithms of the RANSAC family [7], are usually employed to compare the constellations of keypoints, which allows for a complete recognition of the internal coordinate system of the document in the captured image.

Special text point W, defined in [5], includes several components:

$$\mathrm{W} = \{\mathrm{T(W)}, \mathrm{B(W)}, \mathrm{L(W)}\} \tag{1}$$

where

- T(W)—a core of the special text point, i.e. a sequence of characters s1, s2… sn, from some alphabet;
- B(W)—a boundary, which consists of the coordinates of the following quadrilaterals: a quadrilateral which serves as a boundary of the special text point's image; quadrilaterals which serve as boundary of each character's image. The coordinates are scaled by the height and width of the normalized page;
- L(W)—parameters of the modified Levenshtein distance, L(W) = {d(W), FA(W), M(W)}.

A pair {T(W), B(W)} defines the descriptor of special text point. The detector of a special text point is the OCR recognition procedure.

The comparison of two special text points is performed with modified Levenshtein distance ρLEV [8]. To compare special text point W and recognized word WREC the parameter d(W) is employed. The latter is a component of parameters array L(W). If ρLEV(T(W), WREC) < d(W), then the recognized word WREC and special text point W are identical. The word identification mechanism is used in many tasks based on comparing words with the alphabet in the database [9, 10].

The modified Levenshtein distance is calculated according to the recurrence formula:

$$\rho^0_{\mathrm{LEV}}(\mathrm{S}_1, \mathrm{S}_2) = \mathrm{M}_{\mathrm{LEV}}(|S_1|, |S_2|) + \mathrm{Pen}(\mathrm{S}_1, \mathrm{S}_2) + \mathrm{Pen}(\mathrm{S}_2, \mathrm{S}_1),$$

$$\forall j \mathrm{M}_{\mathrm{LEV}}(0, j) = 0, \forall i \mathrm{M}_{\mathrm{LEV}}(i, 0) = 0 \tag{2}$$

$$\begin{aligned}\mathrm{M}_{\mathrm{LEV}}(i, j) = \min(&\mathrm{M}_{\mathrm{LEV}}(i, j-1) + 1, \mathrm{M}_{\mathrm{LEV}}(i-1, j) \\ &+1, \mathrm{M}_{\mathrm{LEV}}(i-1, j-1) + \mathrm{substCost}(s_{1i}, S_{2i})),\end{aligned}$$

then $\rho\ ^0{}_{\mathrm{LEV}}(\mathrm{S}, \mathrm{S})$ is normalized, e.g. as defined in [11]:

$$\rho_{LEV}(S_1, S_2) = \frac{2 \cdot \rho^0(S_1, S_2)}{|S_1| + |S_2| + \rho^0(S_1, S_2)}.$$

In (2), function substCost(s_{1i}, s_{2i}) computes the cost of substitution of character s1i with character s2i. In the general case, boundaries of words, B(W), are unknown and brute-force algorithms of words recognition via boundaries of separate characters are needed.

There are pairs of words for which most of the characters coincide, and the difference is observed in a small number of characters in certain positions. There may be cases of false identification of similar words with a common root but different suffixes or endings. To prohibit false identification of such words, we employ M(W) templates which are defined as follows:

$$\begin{aligned} &b_1b_2\ldots b_k\ldots m_1m_2\ldots m_p\ldots e_1e_2\ldots e_q \\ &\ldots m_1m_2\ldots m_p\ldots e_1e_2\ldots e_q, \\ &b_1b_2\ldots b_k\ldots e_1e_2\ldots e_q, \\ &b_1b_2\ldots b_k\ldots e_1e_2\ldots e_q, \\ &\ldots m_1m_2\ldots m_p\ldots, \end{aligned}$$

This template specifies mandatory characters at the beginning (b_1 b_2...b_k), in the middle (m_1 m_2... m_p), or at the end (e_1 e_2...e_q) of the word. Symbol * stands for a sequence of arbitrary characters. If during comparison characters of the recognized word do not match the template, then the Levenshtein distance is increased by some penalty Pen(M(S_1), S_2) for mismatch between word S_2 and word S_1.

When computing function substCost, the cost of substitution for various characters depends on the specific characters. For example, substitution of similar characters such as Il or 8B, i.e. substCost(I, l) or substCost(B, 8), costs less than substCost(I, B). Function substCost is set by FA(W) table from parameters of special text point W.

Let us consider a more detailed representation of a special text point. Let us define a core as a sequence of character positions C_1, C_2,... C_n. Each character position C_k represents an array of alternatives (c_{kj}, w_{kj}):

$$(c_{k1}, w_{k1}), (c_{k2}, w_{k2}), \ldots \left(c_{k,m(k)}, w_{k,m(k)}\right), \tag{3}$$

where c_{kj} is a character code, w_{kj} is a score of character c_{kj} match to the image of position C_k, and $m(k)$ is a number of alternatives. The scores are in the range between 0 and 1, the best score being 0. The values of $m(k)$ can be different for different k.

We employ alternatives in the special text point descriptor to reduce the cost of substitution for similar characters. For example, the matrix of alternatives in Table 1 describes the words HOPE and HCPE as identical.

In the recognized word descriptor, the character positions can be represented in more detail as a matrix of alternatives:

Table 1 Example of alternatives matrix of the special text point's core

H(0)	O(0)	P(0)	E(0)
For other characters the substitution cost is 1	C(0)	For other characters the substitution cost is 1	For other characters the substitution cost is 1
	For other characters the substitution cost is 1		

$$\begin{array}{l} s_1(p_{11}), s_1(p_{21}), \ldots s_1(p_{n1}) \\ s_2(p_{12}), s_2(p_{22}), \ldots s_2(p_{n2}) \\ \ldots \\ S_m(p_{1,m}), S_m(p_{2,m}), \ldots S_m(p_{nm}) \end{array} \tag{4}$$

Here, p_{ij} is a score of recognized image of character position C_i match to character s_j from recognition method's alphabet. m is recognition method's alphabet length. The score p_{ij} is between 0 and 1, the best score being 0. For each character position the boundary $F(C_i)$ is known.

The representation of special text point's core (3) can be used when computing modified Levenshtein distance. For example, if core T(W) is a set of characters without scores $\{s_1, s_2 \ldots s_n\}$, then when computing penalty function for mismatched between character position C_k and character alternative s_i is calculated as follows

$$\text{substCost}(C_k, s_i) = 1 - \min(w_{kj}).$$

Using the recognized word matrix (4), function substCost is calculated as

$$\text{substCost}(C_k, s_i) = 1 - \min(w_{kj} \oplus p_{ij}).$$

Let us clarify the concept of a recognized word. Intuitively, the definition of a word as a sequence of characters bounded by spaces or delimiters (gaps) is understandable. However, when recognizing noisy and distorted images (Fig. 1a) errors in detection of gaps between words are possible for all recognition methods, including ANN-based [12]. Problems emerge when searching for gaps in short lines. When filling out administrative documents, there may be situations when there is no space between the static word and the field text (Fig. 1b).

Complex words require an algorithm that uses the boundaries of individual letters. The part of a line with characters without gaps between them is considered. There, the search for substrings that lack linking to special text points is provided by, for example, the modified bitup algorithm which is based on the Levenshtein distance [13]. The value of threshold $d(W)$ is decreased compared to the linking threshold used earlier in the strings linking step. This algorithm makes it possible to find keywords as substrings in a sequence of characters bounded by already identified words. In this case, for each found substring of the limited dictionary, the boundaries containing all its symbols, are known. This helps to avoid expensive brute-force search of possible substring by strict algorithms [14].

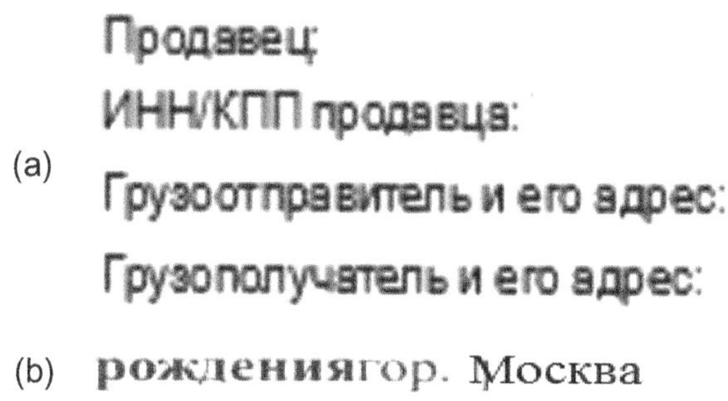

Fig. 1 Problems of difficult cases when searching for gaps between words

3 Constellation Model

A set of characters that differ from neighboring characters in some vicinity is called *local text feature*. For example, a recognized word that is identical to some local text feature allow for identification of neighboring recognized words.

The different connections between special text points (connections above-below, right-left, or geometric distance between boundaries) allows for combination of text points into constellations. Hereafter, we will consider text lines and paragraphs consisting of several lines as constellations. The constellation model L consists of a sequence of *chains*

$$\mathrm{L} = \{\mathrm{C}_1(\mathrm{L}), \mathrm{C}_2(\mathrm{L}), \ldots, \mathrm{C}_{+1}(\mathrm{L}), \mathrm{C}_{+2}(\mathrm{L}), \ldots, \mathrm{C}_{-1}(\mathrm{L}), \mathrm{C}_{-2}(\mathrm{L}), \ldots\}, \quad (5)$$

where each chain C(L), C + (L), C–(L), consists of a sequence of *terms*

$$\mathrm{C}(\mathrm{L}) = \{\mathrm{Tm}_1(\mathrm{L}, \mathrm{C}), \mathrm{Tm}_2(\mathrm{L}, \mathrm{C}), \ldots \mathrm{Tm}_n(\mathrm{L}, \mathrm{C})\}.$$

A term $\mathrm{Tm}_i(\mathrm{L}, \mathrm{C})$ is defined as a set of descriptor alternatives of special text points $\mathrm{W}_1(\mathrm{Tm}_i(\mathrm{L}, \mathrm{C}))$, $\mathrm{W}_2(\mathrm{Tm}_i(\mathrm{L}, \mathrm{C}))$,... and distances

$$d_{1_1}(\mathrm{Tm}_i(\mathrm{C})), d_{2_1}(\mathrm{Tm}_i(\mathrm{C})), d_{1_2}(\mathrm{Tm}_i(\mathrm{C})), d_{2_2}(\mathrm{Tm}_i(\mathrm{C})) \ldots$$

between given term Tm_i and the previous term Tm_{i-1} if $i > 1$. For each term the maximum allowed value for the generalized Levenshtein distance $d(\mathrm{Tm}_i(\mathrm{C}))$ is given. In the simplest case, a term consists of a single descriptor for which no connection to the previous term is specified. The descriptor alternatives represent the case when the corresponding words in the static text of the flexible document can be replaced by other words. Each of the specified thresholds $d_k(\mathrm{Tm}_i(\mathrm{L}, \mathrm{C}))$ is a parameter in the following conditions

$$\rho^{k_\mathrm{T}}(\mathrm{Tm}_{i-1}(\mathrm{L}, \mathrm{C}), \mathrm{Tm}_i(\mathrm{L}, \mathrm{C})) < d^{1_k}(\mathrm{Tm}_i(\mathrm{L}, \mathrm{C}))$$
$$\rho^{k_\mathrm{T}}(\mathrm{Tm}_{i-1}(\mathrm{L}, \mathrm{C}), \mathrm{Tm}_i(\mathrm{L}, \mathrm{C})) > d^{2_k}(\mathrm{Tm}_i(\mathrm{L}, \mathrm{C})) \quad (6)$$

where $>^k{}_T$ is one of the possible metrics between two terms. Examples of metrics are: the number of words between two terms or the distance calculated using the boundaries of special text points included in the terms. In the simplest case, a chain consists of a single term. The definition of a chain provides a description for a unique sequence of terms that differs from other chains in the context of a constellation. Unique chains are local textual features. The constellation is the neighborhood of the feature.

A constellation model is constructed in order to create a sequence of unique chains that allow for matching the constellation terms with the recognized words of the document image. In a constellation definition, some chains may not be unique in the context of a line. For example, a constellation pattern may contain several identical words.

Constellation *linking* is the matching of the constellation model and the sets of recognized objects (words, lines, paragraphs). A simple way to link a constellation is to use a bag of words. For administrative documents, this method is not effective for constellations where the bag of words is not unique compared to other co-stellations.

4 Algorithm for Coordinating the Set of Constellation Models and Lines of the Recognized Document

Let us describe an algorithm for coordinating a set of constellation models and strings of a recognized document. The constellation models L_i contain a sequence of detailed descriptions of terms and chains with possible alternatives. For each constellation, a quadrilateral boundary $F(L_i)$ is known. The boundaries order the set of constellations. In addition, for each constellation we can set several conditions such as:

- boundary $F(L_i)$ is lower than boundary $F(L_j)$;
- boundary $F(L_i)$ is higher than boundary $F(L_j)$;
- no constellation L_i exist, if there is constellation L_j or chain $C(Lj)$;
- no constellation L_i exist, if there is no constellation L_j or chain $C(Lj)$. (7)

During the first stage, the correspondence (*linking*) of the recognized words is established $\{W^{REC}{}_1, W^{REC}{}_2 \ldots\}$ by one or several constellation models $\{L2_1, L2_2, \ldots\}$. First, for each of the terms $Tm_i(L2_i, C_p)$ one of several the closest recognized words $W^{REC}{}_q$ are found:

$$\rho_{LEV}\left(T\left(Tm_i\left(L_i', C_p\right)\right), W_q^{REC}\right) < d\left(Tm_i\left(L_i', C_p\right)\right). \tag{8}$$

The candidate's $W^{REC}{}_q$ to be term $Tm_i(L2_i, C_p)$ are checked for correspondence to the term sequence in chain $C_p(L2_i)$. I.e. from all candidate's we pick the ones where the score for term sequence of chain $C_p(L2_i)$ is minimal:

$$^{\text{TM}}(\mathrm{C}_p) = \max()_{\mathrm{LEV}}\left(\mathrm{T}(\mathrm{Tm}_i(\mathrm{L2}_i, \mathrm{C}_p)), \mathrm{W}_{\substack{\mathrm{REC}\\ q}}\right)\Big) \rightarrow \min . \quad (9)$$

Calculating a score in the form (9) requires linking all chain terms. For severely noisy or highly distorted document images this is impossible. This problem is solved by applying the property of terms, which allows skipping some terms when calculating (9). Another way is to apply a weaker scoring:

$$\delta_2(\mathrm{C}_p) = \left(\rho_{\mathrm{LEV}}\left(\mathrm{T}(\mathrm{Tm}_i(\mathrm{L}'_i, \mathrm{C}_p)), \mathrm{W}_{\substack{\mathrm{REC}\\ \mathrm{q}}}\right) < d(\mathrm{Tm}_i(\mathrm{C}))\right) \rightarrow \max . \quad (9')$$

A brute-force algorithm is required to find the optimum (9) or (9′). The brute-forcing is simplified due to the limited dictionaries of possible words in administrative documents' static text. In such documents, the number of candidates for the majority of the terms is limited, for example, up to 1–3. Condition (6) significantly reduces the number of candidates. For this reason, a complete brute-force is acceptable. The result of chain linking is the selection of an ordered set of recognized words. Each of the latter corresponds to some term of the chain.

After linking, the chain linking score $^{\text{TM}}(\mathrm{C}_p(\mathrm{L2}_i))$ is known. For each term Tm_i of the chain, the linking score is calculated as follows:

$$^{\text{TM}}(\mathrm{Tm}_i) = \min()_{\mathrm{LEV}}\left(\mathrm{T}(\mathrm{Tm}_i), \mathrm{W}_{\mathrm{REC}_q}\right), ^{\text{TM}}(\mathrm{Tm}_{i+1})\Big), \quad (10)$$

where $\mathrm{W}^{\mathrm{REC}}{}_{\mathrm{q}}$ is each of the words which correspond to the term Tm_i. Conditions (8) must be met. The score $^{\text{TM}}(\mathrm{Tm}_{i+1})$ is computed for each $\mathrm{W}^{\mathrm{REC}}{}_{\mathrm{q}}$. Employing a minimum allows for choosing the chain with the best score of the term with the worst correspondence. The algorithm for linking chain C is as follows. For the first term $\mathrm{Tm}_1(\mathrm{C})$ the recognized words are found. Each of these words corresponds to core $\mathrm{T}(\mathrm{Tm}_1(\mathrm{C}))$ of the first term. For each candidate $\mathrm{W}^{\mathrm{REC}}{}_{\mathrm{q}}$ the linking of the rest of the chain $\{\mathrm{Tm}_2(\mathrm{C}), \ldots \mathrm{Tm}_n(\mathrm{C})\}$ is considered. Next, a similar linking of the term $\mathrm{Tm}_2(\mathrm{C})$ is considered. For each of these candidates for linking to the term $\mathrm{Tm}_2(\mathrm{C})$ the condition (6) is verified. That is, using condition (6) the connection between boundaries of biterm candidates $\{\mathrm{Tm}_1(\mathrm{C}), \mathrm{Tm}_2(\mathrm{C})\}$ is checked. For each of the tested candidates $\mathrm{W}^{\mathrm{REC}}{}_{\mathrm{q}}$ the linking of the rest of the chain $\{\mathrm{Tm}_3(\mathrm{C}), \ldots \mathrm{Tm}_n(\mathrm{C})\}$ is considered. The estimation (10) is employed. It evaluates the similarity of the keyword and the recognized word. Implementation of (6) allows for a significant reduction in the number of candidates to be biterm.

After linking some chains, some constellations are identified. Each of the chains forming the constellation (5) affects the constellation identification:

- $\{\mathrm{C}_1(\mathrm{L2}_i), \mathrm{C}_2(\mathrm{L2}_i), \ldots\}$—the estimation of the constellation linking is based on the estimations of the bound chains of this group $^{\text{TM}}(\mathrm{L2}_i) = \min\,(^{\text{TM}}(\mathrm{C}_p(\mathrm{L2}_i)))$;
- $\{\mathrm{C}^+{}_1(\mathrm{L2}_i), \mathrm{C}^+{}_2(\mathrm{L2}_i), \ldots,\}$—the worst estimation for the linking of the constellation $^{\text{TM}}(\mathrm{L})$ is set, if at least one chain from this group is not bound;

- $\{C^{-}_{1}(L2_i), C^{-}_{2}(L2_i), \ldots\}$—the worst estimation for the linking of the constellation $^{TM}(L2_i)$ is set, if at least one chain from this group is bound.

In the case of an ambiguous classification of constellations, the examination of their mutual arrangement is performed via (7).

The result of linking the constellations and the chains contained in them is a reliable classification of local textual features. This makes it possible to extract areas for further detailed examination, for example, for the recognition of fields restricted to a few keywords.

Let us consider several examples of chains and constellations.

A set of constellations $\{C_1, C_2, \ldots C_n\}$ can be used for classification of the document in terms of type. Each of these constellation corresponds to some class of documents, where the distance function between a document D and constellation C: r(D, C) is given. Constellations are ranked based on the distance between the document and each of the models, and the result is represented by the scores of all the models

$$\mathrm{r_1 = r(D, Ci1), r_2 = r(D, Ci2), \ldots r_n = r(D, Cin), 1\delta i_p\delta n, r_1\delta r_2\delta \ldots \delta r_n.}$$

After ranking, only constellations that are within a certain threshold m are considered:

$$\mathrm{r_1{=}r(D, Ci1), \ldots r_q{=}r(D, Ciq), 1\delta i_p\delta n, r_1\delta \ldots \delta r_q\delta m, 1\delta q\delta n,}$$

or the constellation Ci1 closest to the document is selected. The constellations have the following form:

$$\mathrm{L^{UP}, L, L^{DOWN}}$$

where chain L^{UP} limits the area of the constellation from above, L^{DOWN}—from below, and chain L consists of terms $Tm_1(L)$, $Tm_2(L)$, ..., each, starting from $Tm_2(L)$ is significantly more distant from the previous term. For example, the distance between quadrilateral boundaries of terms $Tm_{i-1}(L)$ and $Tm_i(L)$ should not be less than the set distance. In this case, condition (6) is as follows

$$\rho^{k_T}(\mathrm{Tm}_{i-1}(\mathrm{L}), \mathrm{Tm}_i(\mathrm{L})) > d_{\frac{2}{k}}(\mathrm{Tm}_i(\mathrm{L}))$$

It guarantees invariance of the connection order between two terms concerning possible transformations in the flexible document. The linking of the described constellations does not require the linking of all terms. The estimation of such linking is computed via (9′).

Figure 2 illustrates the example of the paragraph. Some words of a paragraph can be carried over to the next line. The following chain can be used for the paragraph linking:

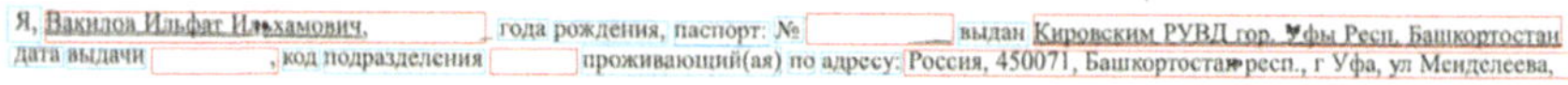
Я, Вакилов Ильфат Ильхамович, года рождения, паспорт: № выдан Кировским РУВД гор. Уфы Респ. Башкортостан, дата выдачи , код подразделения проживающий(ая) по адресу: Россия, 450071, Башкортостан респ., г Уфа, ул Менделеева,

Fig. 2 The examples of paragraph (blue boxes—static text, red boxes—filled content)

Fig. 3 The examples of a line without keywords (red box)

$$L^{UP}, L, L^{DOWN} \tag{11}$$

where chain L^{UP} limits the area of the paragraph from above, L^{DOWN}—from below, and chain L consists of terms $Tm_1(L)$, $Tm_2(L)$, …. For some biterms in the ideal case, the following is true:

$\rho^k{}_T(Tm_{i-1}(L), Tm_i(L)) < 1$.

Other biterms define field boundaries for extracting document attributes (рис. 2). Chains L^{UP} and L^{DOWN} can include paragraph words or be positioned outside the paragraph area.

Text lines are also described by (11). If a line differs from all the lines within a document or a document fragment, then one chain is enough to link it. This chain contains a sequence of terms based on mandatory and forbidden keywords (see chain definition (5)). If the line does not have unique words in the context of a document or a document fragment, chains L^{UP} and L^{DOWN} allows for limiting linking area to the recognized objects. Thus, the lines without keywords can be subjected to linking, see an example in Fig. 3.

5 Training

Training of the proposed method consists in forming chains that allow for a reliable classification of constellations within a single page or a single multipage document. Constellations can also be used to classify document types in a stream with multiple types.

An effective method of classification is the use of probabilistic topic models. Such models are designed to determine the topic of a collection of documents. In a topic model, each topic is represented by a discrete probability distribution of words. Each document in the model is represented by a discrete probability distribution of topics [15]. It is implicitly assumed that the document contains a sufficient number of words to construct a discrete word probability distribution. In [16], describes Additive Regularization on Topic Models (BigARTM), i.e., multi-criteria optimization of the weighted sum of the criteria. This optimization is necessary to determine and ensure the stability of the problem of constructing a topic model from a collection of documents. A useful approach to topic models for short text is a *biterm topic model* [17].

Machine learning techniques can be used to find unique keywords or simple chains, such as biterms. This requires a representative dataset. However, obtaining a representative dataset is not always feasible. Often a negligible number of samples (1–2) is provided for training. From such a number of samples, forming a representative sample seems to be a very difficult task. Acceptable training on an *ultra-small sample* can be achieved as follows:

- forming chains for error-free documents;
- augmenting the chains according to known reasons for static text augmentation in flexible documents.

The augmentation of the chains must be consistent with the possibilities for distorting the structure of administrative documents listed in the introduction of this chapter. For example, it is necessary to take into account the possibility of moving text in the case of long lines. Also, the augmentation of chains is required to take into account the errors of recognition. For example, when using the conditions (6) limiting the distance between terms in the recognized text, it makes no sense to specify $d^1{}_i(\mathrm{Tm}_i(\mathrm{C})) = 1$. The reason is the possibility of falsely recognized words between the two terms.

6 Fields Linking

Linking of fields is much more easier than constellations linking. A field F is defined via terms of two chains as follows:

- $L^{PREV}(F)$, F, L^{NEXT}—for a field placed between two chains;
- $L^{PREV}(F)$, F—for a field bounded by a chain to the left;
- F, L^{NEXT}—for a field bounded by a chain to the right.

Fields that are not bounded by $L^{PREV}(F)$ or L^{NEXT} are bounded by the constellation

$$L^{UP}, L, L^{DOWN}.$$

The exact prediction of the field boundary F is possible if the terms nearest to F are bounded. Specifically, the left boundary of the field F is accurately determined if the rightmost bounded term $\mathrm{Tm}^{PREV}(F)$ of chain $L^{PREV}(F)$ is found. Similarly, the right boundary of the field F is accurately determined if the leftmost bounded term $\mathrm{Tm}^{NEXT}(F)$ of chain $L^{NEXT}(F)$ is found. The upper and lower boundaries of field F are determined using the upper and lower boundaries of chains $L^{PREV}(F)$ and L^{NEXT}. Terms $\mathrm{Tm}^{PREV}(F)$ and $\mathrm{Tm}^{NEXT}(F)$ can specify both single-line and multi-line field boundaries.

The accuracy of field boundary prediction depends on document flexibility and document noisiness. To predict the field boundary it is reasonable to apply fuzzy logic statements over the formal description of the document structure [18, 19]. Fuzzy logic assertions are formed in the process of analyzing a training sample of documents. However, the small sample size may not allow for training the mechanism of field boundary prediction. In the latter case, the simplest statements are used for prediction. For example, for the cases of two words printed together without a gap, it is necessary to assume a zero distance between the field boundary and the boundary of the corresponding bounding term.

If the nearest bound term $Tm^{PREV}(F)$ of chain $L^{PREV}(F)$ is not found, it is possible to make a prediction using other bound terms of chain $L^{PREV}(F)$. Of course, the prediction accuracy is reduced in this case. For this purpose, the second step of the linking is used in the proposed method. During the second step, we try to find the boundaries of words recognized with a large number of errors, as well as words broken into parts or joined with other words.

7 Results of Experiments

The proposed method of constellation linking was tested on an private test dataset which includes images of «Consent» and «Power of Attorney» document classes. The documents were scanned at optical density from 100 to 300 dpi with different qualities of digitization. A private dataset was used because obtaining images of documents containing personal information is either difficult or impossible. This problem is highlighted in the work [20].

The accuracy of line and multiline paragraph linking was investigated. The SDK Smart Document Engine [7, 21] was used for recognition. The results are summarized in Table 2, which illustrates the efficiency of the proposed method. The decrease in linking quality on the «Power of Attorney» subset is associated with four images of documents exhibiting low digitization quality. If these four documents are removed from the «Power of Attorney» subset, then the linking error rate will be 0,78%.

The proposed constellation linking method was applied to classify document types. Constellations were created for two classes of documents $M_{consent}$ (32 special

Table 2 Точность привязки созвездий для тестовых наборов деловых документов

Document	Number of documents	Number of lines in document model	Number of paragraphs in document model	Number of linking errors	Linking error rate (%)
Consent, type 1	125	4	1	6	0,96
Consent, type 2	119	11	6	10	0,49
Power of Attorney	120	31	13	179	3,39

Table 3 Experimental results of documents classification

$M_{consent}$		$M_{attorney}$	
tp = 331	fp = 2	tp = 307	fp = 1
tn = 308	fn = 0	tn = 333	fn = 0
Precision = 0,99	Recall = 1	Precision = 0,99	Recall = 1

text points) and $M_{attorney}$ (30 special text points), each document contained a single-column text. The described method of fragment classification was applied to classify documents. In addition to the presence of positive detectors, the invariant vertical relations between the boundaries of the special text points were checked. The OCR Smart ID Reader was also used for recognition. The quality of the algorithm was evaluated according to the following criteria:

$$\text{Precision} = \text{tp}/(\text{tp} + \text{fp}),$$
$$\text{Recall} = \text{tp}/(\text{tp} + \text{fn}),$$

with the following values:

- the number of correctly classified the documents of within the class *tp*;
- the number of rejected classified the documents of within the class *fp*;
- the number of rejected classified the documents of another class *tn*;
- the number of falsely classified the documents of another class *fn*.

The obtained classification results are summarized in Table 3. The fp errors were associated with recognition errors in the overexposed regions.

On the same dataset the accuracy of field linking was evaluated using the proposed method. The accuracy was defined as the proportion of correctly bounded fields to the total number of fields in all dataset documents. Document "Consent" contained 11 fields, 4 of which were multi-line. Document "Power of Attorney" contained 15 fields, 4 of which were multi-line.

The linking accuracy was 95,7% for "Consent" document and 93,9% for "Power of Attorney" document. 1,2% of the fields within "Consent" and 0,3% of the fields within "Attorney" were not found. Linking errors were due to character recognition errors in overexposed areas.

8 Conclusion

The proposed constellation linking method applies to the content analysis of the recognized documents. The method is based on two assumptions:

- there may be a large number of recognition errors in the text;

- static text words and filled content words may be close to each other.

The effect is achieved by combining keyword sequences. Constellations and their constituent chains are constructed in such a way as to distinguish them from other sets of words in some areas of the document.

Constellations can be generated using Machine Learning on a representative sample of documents. In practice, however, it is impossible to get a representative sample. For small training samples, some of the work has to be done manually. Due to the small size of the training sample, it is impossible to use the frequency of encountered distortions in the text of the flexible form.

Development of the proposed method may imply the creation of probabilistic identification models of the ideal and the recognized word.

Acknowledgements The research was supported by the Russian Foundation for Basic Research (Project 20-07-00934 "Development, properties study and justification of anytime algorithms for computed tomography"). The authors are grateful to T. B. Chistyakova for her valuable comments and suggestions.

References

1. Rusiñol, M., Frinken, V., Karatzas, D., Bagdanov, A.D., Lladós, J.: Multimodal page classification inadministrative document image streams. IJDAR **17**(4), 331–341. https://doi.org/10.1007/s10032-014-0225-8 (2014)
2. Hjouji, A., EL-Mekkaoui, J., Jourhmane, M.: Image classification by mixed finite element method and orthogonal legendre moments. Pattern Recognit. Image Anal. **30**, 655–673. https://doi.org/10.1134/S1054661820040185 (2020)
3. Karkishchenko, A.N., Mnukhin, V.B.: On the metric on images invariant with respect to the monotonic brightness transformation. Pattern Recognit. Image Anal. **30**, 359–371 (2020). https://doi.org/10.1134/S1054661820030104
4. Slavin, O.A.: Using Special Text Points in the Recognition of Documents. Studies in Systems, Decision and Control. Springer Nature Switzerland AG., Vol. 259, pp. 43–53. https://doi.org/10.1007/978-3-030-32579-4_4 (2020)
5. Andreeva, E., Arlazarov, V.V., Slavin, O., Mishev, A.: Comparison of scanned administrative document images. Proceedings of SPIE, Vol. 11433: Twelfth International Conference on Machine Vision, ICMV 2019 (Amsterdam, Netherlands, pp. 16–18 November 2019). Art. ID: 114333A, pp. 1–8. https://doi.org/10.1117/12.2559369 (2020)
6. Rodehorst, V., Koschan, A.: Comparison and evaluation of feature point detectors
7. Skoryukina, N., Arlazarov, V., Nikolaev, D.: Fast method of ID documents location and type identification for mobile and server application. IEEE International Conference on Document Analysis and Recognition (ICDAR), pp. 850–857. https://doi.org/10.1109/ICDAR.2019.00141 (2019)
8. Deza, M.M., Deza, E.: Encyclopedia of distances. Springer, Berlin, xiv+590 pp (2009)
9. Viet, N.T., Kravets, A.G.: The new method for analyzing technology trends of smart energy asset performance management. Energies **15**(18), 6613 (2022). https://doi.org/10.3390/en15186613
10. Sabitov, A., Minnikhanov, R., Dagaeva, M., Katasev, A., Asliamov, T.: Text classification in emergency calls management systems. Cyber-Physical Systems, pp. 199–210. https://doi.org/10.1007/978-3-030-67892-0_17 (2021)

11. Yujian, L., Bo, L.: A normalized levenshtein distance metric. IEEE Trans. Pattern Anal. Mach. Intell. **29**(6), 1091–1095. https://doi.org/10.1109/TPAMI.2007.1078 (2007)
12. Chernyshova, Y.S., Sheshkus, A.V., Arlazarov, V.V.: Two-step CNN framework for text line recognition in camera-captured images. IEEE Access **8**, 32587–32600 (2020). https://doi.org/10.1109/ACCESS.2020.2974051
13. Ricardo, A., Baeza-Yates, G., Gonnet H.: A new approach to text searching. In: Communications of the ACM, Vol. 35, No. 10, pp. 74–82. https://doi.org/10.1145/135239.135243 (1992)
14. Aho, A.V., Corasick, M.J.: Efficient string matching: an aid to bibliographic search. Commun. ACM **18**(6), 333–340. https://doi.org/10.1145/360825.360855 (1975)
15. Rubin, T.N., Chambers, A., Smyth, P., Steyvers, M.: Statistical topic models for multi-label document classification. Mach. Learn. **88**, 157–208 (2012). https://doi.org/10.1007/s10994-011-5272-5
16. Vorontsov, K.V.: Additive regularization for topic models of text collections. Dokl. Math. **489**(3), 303–304 (2013). https://doi.org/10.1134/S1064562414020185
17. Yarn, X., Guo, J., Lan, Y., Cheng, X.A.: Biterm topic model for short texts. Proceedings of the 22Nd International Conference on World Wide Web, pp. 1445–1456. https://doi.org/10.1145/2488388.2488514 (2013)
18. Postnikov, V.V.: Flexible forms identification. Proceedings of the 5th German-Russian Workshop on Pattern Recognition and Image Understanding (GRWS98). Hamburg: Infix, (1999)
19. Postnikov, V.V.: Identification and recognition of documents with a predefined structure. Pattern Recogn. Image Anal. **13**(2), 332–334 (2003)
20. Awal, A.M., Ghanmi ,N., Sicre, R., Furon, T.: Complex document classification and localization application on identity document images. 14th IAPR International Conference on Document Analysis and Recognition (ICDAR), pp. 426–431. https://doi.org/10.1109/ICDAR.2017.77 (2017)
21. Smart Document Engine—automatic analysis and data extraction from business documents for desktop, server and mobile platforms. https://smartengines.com/ocr-engines/document-scanner. Accessed 22 April 2022

Method for Analyzing the Structure of Noisy Images of Administrative Documents

Oleg A. Slavin and Eugene L. Pliskin

Abstract The problem of recognition of images of administrative documents is considered. Administrative documents are used in an organization or group of organizations to accompany business processes. The structure of administrative documents contains static elements and fill areas (fields). The main stages of recognition are the linking stage (finding areas of filling) and the stage of recognition (extracting text or graphic information from these areas). The chapter considers the stage of linking field boundaries. The linking is carried out using a model based on static reference elements. The reference elements in the model are the line segments (separating lines) and the words of the text. The model consists of a set of descriptors. A simple text descriptor contains words and attributes to distinguish it from similar descriptors. A text descriptor has an order relationship with other text descriptors. It is proposed to use combined descriptors consisting of words and segments. Linking is carried out by superimposing descriptors and a set of recognized words and detected segments in the document page image. The SDK Smart Document Engine was used as OCR in the experiment. For our private dataset, consisting of universal transfer documents, it is shown that the extraction of object sets improves the accuracy of recognition of document fields during recognition by 16%, and the accuracy of extracting information from a document image by 15%.

Keywords Noisy image · Document recognition · Special text point · Descriptor

O. A. Slavin (✉)
Federal Research Center "Computer Science and Control" of the Russian Academy of Sciences, 9 Prospect 60-Letiya Oktyabrya, Moscow 117312, Russia
e-mail: oslavin@isa.ru

O. A. Slavin · E. L. Pliskin
LLC Smart Engines Service, 9 Prospect 60-Letiya Oktyabrya, Moscow 117312, Russia
e-mail: epliskin@isa.ru

A. G. Kravets et al. (eds.), *Cyber-Physical Systems Engineering and Control*,
Studies in Systems, Decision and Control 477,
https://doi.org/10.1007/978-3-031-33159-6_4

1 Introduction

The number of administrative documents printed on paper is constantly growing. Administrative documents accompany business processes in an organization or group of organizations. The structure of administrative documents contains static elements and fill areas. Filling areas (fields) contain parameters of the business process: personal data, identifiers, financial information, confirmation marks, signatures, stamps, and barcodes. The volume of flows of incoming and outgoing documents in large organizations can reach several million pages per day. Recognition of administrative documents allows you to automate the processes of office work of the organization [1].

We will define a document as a collection of fields and reference elements. The reference elements, first of all, are the words of the static text. Static text words are grouped into lines, headings, paragraphs, and paragraphs. Each field is bounded by several reference elements. The structure of a document can be simple (in the form of a single column) or complex. A complex structure suggests the presence of several fragments: columns or areas delimited by segments or large gaps.

Administrative documents have some of features. Chief among them is the small size of the vocabulary of valid keywords. It is assumed that in an administrative document, margins and reference elements differ from reference elements in some neighborhoods.

When extracting information from recognized noisy administrative documents, it is necessary to take into account a significant number of possible recognition errors. Based on the recognition of text objects and found graphic primitives, it is necessary to extract information from the areas corresponding to the field boundaries. It is required to extract information for the maximum number of fields with the least number of errors for each field. The chapter considers a two-pass method of document recognition using the description of the fields. On the first pass after text recognition and extraction of graphic primitives, field boundaries are predicted. Field boundaries are defined using the word boundaries of the static text and the segments found. Invariance to document distortions is provided by the constellation model. The constellation model is based on a group of words linked by order relationships. Linking of the set of recognized words of the document is carried out using the optimal overlay of constellations on subsets of recognized words. In the second pass, due to the parameterization of recognition, an improvement in the quality of recognition of fields within the found boundaries is achieved. A possible improvement comes from knowing the information about each of the bound fields. For example, it is possible to increase the accuracy of recognition if the alphabet is known, the type of text (printed or handwritten), and the rules for filling in the fields.

The chapter considers recognized noisy images of documents. The images of symbols, line segments (separating lines), and checkboxes are subjected to noise. This topic is currently actual [2–4].

The document image recognition scheme considered in more detail consists of the following steps:

- processing and normalization of the document page image;
- recognition of words and extraction of graphic primitives;
- rectification of a set of objects;
- search for local features;
- search for field boundaries using the boundaries of local singularities;
- extracting the contents of the fields within the boundaries found using field attributes;
- control of the extracted information using the rules for filling in the fields.

2 Background

It is possible to subdivide documents into rigid and flexible forms. A hard form is based on a single template for creating and printing a document. Rigid forms include documents printed on guilloche paper, such as passports, rights and state registration certificates. Rigid forms can also be created using a fixed template, such as a PDF. In rigid forms, the distance between reference documents is constant with some accuracy. The accuracy of comparison of reference elements is associated with digitization defects, as well as with normalization and processing of the document page image.

Other documents refer to flexible forms. In flexible forms, the relationship between reference documents can change through the following modifications:

- characteristics and sizes of fonts;
- line spacing;
- fields of text columns;
- words of static text;
- the mutual arrangement of the words of the text and the lines of the layout.

For rigid forms, local singularities are specified in the form of singular points of various types (SURF, YAPE, YOLO, SIFT, ASIFT). The structure of modern singular point descriptors can be quite complex [5, 6]. Field boundaries in rigid forms are anchored using a set of singular points that satisfy a set of relationships. This is adequate under the assumption that relationships between pairs of points are preserved in a rigid form after document digitization.

Relationships between pairs of corresponding points in a flexible document are not generally preserved. Similar to special points for rigid forms, it is possible to use text special points for flexible documents. In this chapter, a model for linking flexible administrative documents based on text special points is used [7, 8]. In this model, the kernel T(W) of a text singular point W, is considered, defined as a sequence of familiarity of characters belonging to some alphabet. Frame B(W) is also used, consisting of the coordinates of the quadrilateral that bounds the image W. The pair {T(W), B(W)} defines the descriptor of the text singular point W. The text singular point detector is the recognition procedure using OCR. A comparison of two text singular points is carried out using the modified Levenshtein distance [9]. The modified Levenshtein distance proposed in [9] is calculated using the following formulas:

$$\tilde{\rho}_{\mathrm{LEV}}(\mathrm{S}_1, \mathrm{S}_2) = \frac{2 \bullet \rho_{\mathrm{LEV}}(\mathrm{S}_1, \mathrm{S}_2)}{|\mathrm{S}_1| + |\mathrm{S}_2| + \rho_{\mathrm{LEV}}(\mathrm{S}_1, \mathrm{S}_2)}$$

where $\rho_{\mathrm{LEV}}(\mathrm{S}_1, \mathrm{S}_2) = \mathrm{M}_{\mathrm{LEV}}(|\mathrm{S}_1|, |\mathrm{S}_2|)$ is defined as follows:

$$\forall_{\mathrm{j}} \mathrm{M}_{\mathrm{LEV}}(0, j) = 0,\ \forall_i \mathrm{M}_{\mathrm{LEV}}(i, 0) = 0,$$

$$\begin{aligned}
&\mathrm{M}_{\mathrm{LEV}}(\mathrm{i}, \mathrm{j}) = \min(\mathrm{M}_{\mathrm{LEV}}(i, j-1) + 1, \\
&\mathrm{M}_{\mathrm{LEV}}(i-1, j) + 1, \\
&\mathrm{M}_{\mathrm{LEV}}(i-1, j-1) + \mathrm{substCost}(\mathrm{S}_{1i}, \mathrm{S}_{2i})) - \mathrm{Pen}(L(\mathrm{S}_1), \mathrm{S}_2).
\end{aligned}$$

The function $\mathrm{Pen}(\mathrm{M}(\mathrm{S}_1), \mathrm{S}_2)$ – is the penalty for the mismatch of the recognized word S_2, calculated using templates of the following form:

$$b_1 b_2 \ldots b_k \ldots m_1 m_2 \ldots m_p \ldots e_1 e_2 \ldots e_q,$$

These patterns specify required characters at the beginning, in the middle, or at the end of a word. If, when compared, the characters of the recognized word do not match the pattern, then the Levenshtein distance is increased by a predetermined penalty. The proposed modifications take into account the possibility of a large number of recognition errors in noisy document images. Also, the proposed modifications effectively distinguish between words that are close in terms of the classical Levenshtein metric.

A set of characters that differ from neighboring characters in some neighborhoods is called a local text feature. A recognized word that is identical to a word in some local text feature makes it possible to predict the boundaries of the nearest field. Some field boundary words are not local features. It is proposed to use compound descriptors (chains). Chain C consists of a sequence of generalized terms (terms)

$$\mathrm{C} = \{\mathrm{Tm}_1(\mathrm{C}),\ \mathrm{Tm}_2(\mathrm{C}),\ \ldots\ \mathrm{Tm}_n(\mathrm{C})\}.$$

The term $\mathrm{Tm}_i(\mathrm{C})$ is defined as a set of alternatives of text singular point descriptors $\mathrm{W}_1(\mathrm{Tm}_i(\mathrm{C}))$, $\mathrm{W}_2(\mathrm{Tm}_i(\mathrm{C}))$,... and several thresholds $d_1(\mathrm{Tm}_i(\mathrm{C}))$, $d_2(\mathrm{Tm}_i(\mathrm{C}))$ ... between given term Tm_i and the previous term Tm_{i-1} at $i > 1$. Relationships between terms are optional. Each of the indicated thresholds $d_k(\mathrm{Tm}_i(\mathrm{C}))$ is a parameter of the following condition

$$\rho_{\mathrm{T}}^k(\mathrm{Tm}_{i-1}(\mathrm{C}), \mathrm{Tm}_i(C)) < d_k(\mathrm{Tm}_i(\mathrm{C}))$$

where $\rho^k{}_T$ is one of the possible metrics between two terms. Examples of metrics are the number of words between two terms, or the distance calculated using the frames of special text points included in the terms. A term linking is defined as a selection from a set of words in a recognized document of words that are similar to a term. The linking is carried out using the modified Levenshtein distance. A set of relationships between terms greatly simplifies the linking procedure.

The simplest chain consists of a single generalized term. A chain definition allows you to describe a unique sequence of terms that is different from other chains in the context of the entire document. Also, a chain can be unique throughout the document or in a local neighborhood of the document, such as a line, paragraph, or fragment.

The chain descriptor and the recognized words are involved in the process of linking chain C and the recognized document. Recognized candidate words $W^{REC}{}_q$ for the place of the term $Tm_i(C)$ are checked against the sequence of terms in the chain C. Namely, from all candidates for the role of a term, such words are chosen such that the estimate of the sequence of terms in the chain C is minimized:

$$\delta(C) = \max\left(\rho_{LEV}\left(T(Tm_i(C)),\ W_q^{REC}\right)\right) \rightarrow \min .$$

A single-line or multi-line field F is described by a simple descriptor

$$\{C_1; F; C_2\}. \tag{1}$$

The left boundary of the field F is determined using the frames of the rightmost term of chain C_1 and the right boundary is determined by the frames of the leftmost term of chain C_2. Chains C_1 and C_2 are reference elements for field F. Field linking error is defined as the difference between the predictive frame and the boundaries of the real field in the document image. If the boundary terms of the chains C_1 and C_2 are not bound, then other terms of the same chains can be taken for linking. In this case, the field linking error can increase. The field descriptor may be missing one of the C_1 or C_2 nets. In such cases, the boundary of the corresponding field is given by the boundary of the fragment or document. The upper and lower boundaries of one of the lines of field F are set by the boundaries of the text lines included in the chains or located between the lower boundary of the C_1 chain and the upper boundary of the C_2 chain. For a line or paragraph containing multiple fields, the descriptor looks like this:

$$D_{TEXT}\left(F_1, F_2, \ldots, F_{p2}\right) = \begin{cases} C_1^{top}; C_2^{top}; \ldots C_{p1}^{top}; \\ C_1; F_1; C_2; F_2; \ldots C_{p2}; F_{p2}; C_{p2+1}; \\ C_1^{bottom}; C_2^{bottom}; \ldots C_{p3}^{bottom}. \end{cases} \tag{2}$$

The search region for reference C_1, C_2, … C_{p2+1} is limited from above by chains $C^{top}{}_1$, $C^{top}{}_2$,… $C^{top}{}_{p1}$. Similarly, the chains $C^{bottom}{}_1$, $C^{bottom}{}_2$, … $C^{bottom}{}_{p3}$ limit the search for reference chains from below. Descriptor linking score (2) of multiple fields is based on bound chain scores of this group

$$\delta\left(C_1, C_2, \ldots, C_{p2}\right) = \min(\delta(C_k)).$$

The linking algorithm is exhaustive. The linking algorithm is based on a search in a certain neighborhood of chains. In the fragment area, compared to the full document, there are significantly fewer words. For documents broken into fragments, the enumeration is carried out much faster.

3 Line segments Descriptor

In general, net linking is complicated by character recognition errors. However, the use of segments as reference elements simplify linking even for cases of highly noisy document images. In the example shown in Figs. 1 and 2, the linking can be made using segments of solid or dashed lines. In the case of poor recognition of the symbols of the reference chains, in some cases, the linking of the fields can be improved by using segments. In other cases, this is not possible, an example is the “date” field (see Fig. 1), the symbols of which are shifted relative to the dotted line. As already noted, this inconsistency with the design intent is inherent in flexible documents, if only because of the possibility for the user to change the design of the template.

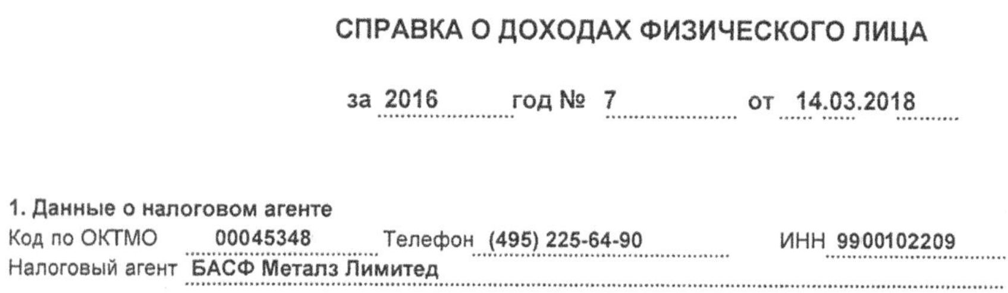
СПРАВКА О ДОХОДАХ ФИЗИЧЕСКОГО ЛИЦА

за 2016 год № 7 от 14.03.2018

1. Данные о налоговом агенте

Код по ОКТМО 00045348 Телефон (495) 225-64-90 ИНН 9900102209

Налоговый агент БАСФ Металз Лимитед

Fig. 1 An example of document fields, with reference elements in the form of segments (Cyrillic, dotted lines)

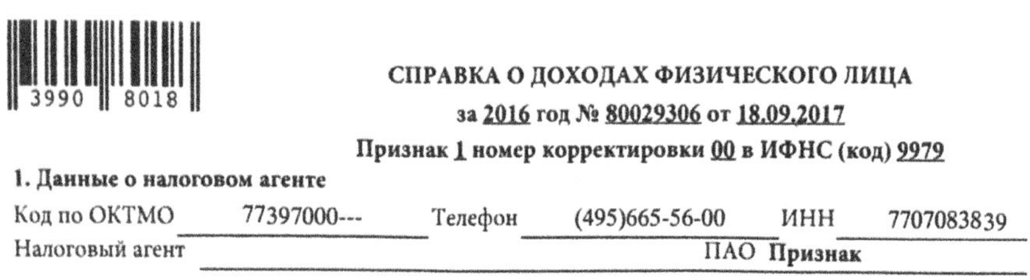
3990 8018

СПРАВКА О ДОХОДАХ ФИЗИЧЕСКОГО ЛИЦА

за 2016 год № 80029306 от 18.09.2017

Признак 1 номер корректировки 00 в ИФНС (код) 9979

1. Данные о налоговом агенте

Код по ОКТМО 77397000--- Телефон (495)665-56-00 ИНН 7707083839

Налоговый агент ПАО Признак

Fig. 2 An example of document fields, with reference elements in the form of segments (Cyrillic, solid lines)

Searching for document objects bounded by lines, in the cases discussed below, simplifies the linking of fields, table cells, or more complex forms. The attributes of the segment are:

- orientation (vertical, horizontal);
- type (solid, dotted, consisting of dots);
- segment frame;
- length and thickness of the segment;
- approximate placement in the image of the document or fragments.

Algorithms for extracting segments from a document image are known [10–12]. Algorithms are also known that form a table or a complex shape from an array of selected line segments.

One single segment cannot be considered a reference element. Consider sets of text words and segments. In Fig. 3 an example of a fragment of a universal transfer document is given. For this example, field linking using chains is difficult due to the loss of a large number of characters. This is due to the significant noise of the off-color image, which occurred due to the stamp and signatures. Recognition of lost characters of static text and fields is hardly possible. However, the linking of some fields in a highly noisy document image is of interest, including detecting the presence of signatures or detecting handwritten and printed fields.

For such cases, it is productive to use combined descriptors consisting of sets of chains and segments. The combined handle is the following object:

$$D(F_1, F_2, \ldots, F_{p2}) = \left\{ \begin{array}{c} C_1^{top}; C_2^{top}; \ldots C_{p1}^{top}; \\ C_1; \frac{F_1}{S_1}; C_2; \frac{F_2}{S_2}; \ldots C_{p2}; \frac{F_{p2}}{S_{p2}}; C_{p2+1}; \\ C_1^{bottom}; C_2^{bottom}; \ldots C_{p3}^{bottom}. \end{array} \right. \quad (3)$$

In the description (3), the segments $S_1, S_2, \ldots S_{p2}$ must be part of one line S. The distance to the line S of each of the frames of the terms of the chains $C_1, C_2, \ldots C_{p2+1}$ is limited. Elements (3) are selected and adjusted during training in such a way that

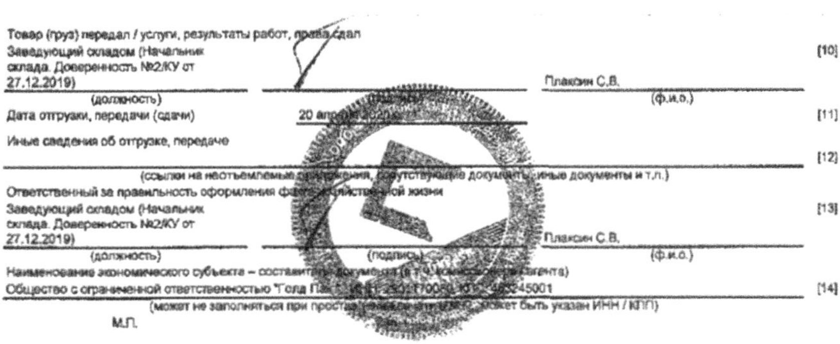

Товар (груз) передал / услуги, результаты работ, права сдал
Заведующий складом (Начальник склада. Доверенность №2/КУ от 27.12.2019) Плаксин С.В. [10]
(должность) (ф.и.о.)
Дата отгрузки, передачи (сдачи) 20 апр... [11]
Иные сведения об отгрузке, передаче [12]
(ссылки на неотъемлемые ... сопутствующие докум... иные документы и т.п.)
Ответственный за правильность оформления ф... ...ой жизни
Заведующий складом (Начальник склада. Доверенность №2/КУ от 27.12.2019) Плаксин С.В. [13]
(должность) (подпись) (ф.и.о.)
Наименование экономического субъекта – составит... ...гента)
Общество с ограниченной ответственностью "Гола Па... ...45001 [14]
(может не заполняться при проста... ...может быть указан ИНН / КПП)
М.П.

Fig. 3 Fragment of a universal transfer document with strong noise the descriptor is a local feature. Some of the chains and segments (3) may not be tied. The anchor score is based on the anchor scores of each chain and the anchor score of the span set. The set of segments is evaluated by the correspondence between the ratios of the sizes of the segments of the descriptor and the segments extracted from the document image. Multiple field descriptor linking scores are based on that group's bound chain scores and the span collection bound scores.

$$\delta\left(C_1, C_2, \ldots, C_{p2}\right) = \min(\delta(C_k)) - \delta\left(S_1, S_2, \ldots, S_{p2+1}\right).$$

Consider the case of a simplified representation of the descriptor (3):

$$D\left(F_1, F_2, \ldots, F_{p2}\right) = \begin{cases} C_1^{top}; C_2^{top}; \ldots C_{p1}^{top}; \\ \frac{F_1}{S_1}; \frac{F_2}{S_2}; \ldots \frac{F_{p2}}{S_{p2}}; \\ C_1^{bottom}; C_2^{bottom}; \ldots C_{p3}^{bottom}. \end{cases} \quad (4)$$

In descriptor (4), the reference elements are the upper and lower chains and the sequence of segments $S_1, S_2, \ldots, S_{p2+1}$. At the same time, in description (4) no chains are separating the fields F_1, F_2, ..., F_{p2+1}. Therefore, the horizontal boundaries of the fields F_1, F_2, ..., F_{p2+1} are given by the boundaries of the corresponding segments S_1, S_2, ..., S_{p2+1}.

Let us describe the linking algorithm with descriptor (4) under the assumption that the projections on the vertical axis of all segments are the same or slightly different. The linking process involves the extracted candidate segments $s_1, s_2, \ldots s_r$, located in the region bounded by the $C^{top}{}_1, C^{top}{}_2, \ldots C^{top}{}_{p1+1}$ и $C^{bottom}{}_1, C^{bottom}{}_2, \ldots C^{bottom}{}_{p3+1}$ chains. The coordinates of the candidate segments are normalized to the width of the fragment or the width of the document page. A cluster analysis of candidate segments $\{s_1, s_2, \ldots\}$. is carried out. The grouping is carried out using the proximity of the projections of the candidate segments on the vertical axis. Next, an assessment is made of the correspondence of each of the resulting clusters of segments with $Cl_t = \{s^t{}_1, s^t{}_2, \ldots\}$ and segments of the descriptor S_1, S_2, ..., S_{p2+1}. If the number of segments in the cluster coincided with the number of segments of the descriptor, then the cluster linking score is calculated as the sum of the residuals of the lengths $h(S_j)$ and $h(s_j)$ of pairs of segments:

$$\delta\left(Cl_t, D\left(F_1, F_2, \ldots, F_{p2+1}\right)\right) = \sum_j \left|h\left(S_j - h\left(s_j\right)\right)\right| \quad (5)$$

We also consider cases where several candidate segments $s^t{}_z$, $s^t{}_{z+1}$, ..., $s^t{}_{z+l(j)}$ corresponds to one segment S_j. In a similar case, several segments S_V, S_{V+1}, ..., $S_{V+r(z)}$ correspond to one candidate segment s_z. For these cases, unions of several segments or candidate segments are considered. The combination is based on an enumeration algorithm over descriptor segments and candidate segments. For the case of invariance of segment sizes (up to scaling), when printing documents, using

Fig. 4 Descriptor Configuration Examples (4)

the minimization of the penalty (5), sets of candidate segments Cl_t are selected, the boundaries of which are best aligned with the boundaries of S_1, S_2, ….

For document fragments like those shown in Fig. 3, linking using the described combined tags (3) and (4) is significantly improved compared to linking using a text tag (2). This happens for the following reasons:

- images of massive segments against the background of signatures and stamps are extracted more reliably than images of symbols and words;
- the use of several chains to limit the search area for a set of segments reliably limits the search area for segments.

The success of linking using the descriptor (3) and (4) is ensured in cases where, under conditions of a noisy document image, the segment extraction method reliably finds segments. The use of descriptors with dotted line segment types for noisy images may be less effective.

Generally speaking, it is not required that in the descriptor (3) in the sequence

$$C_1; \frac{F_1}{S_1}; C_2; \frac{F_2}{S_2}; \dots C_{p2}; \frac{F_{p2}}{S_{p2}}; C_{p2+1};$$

chains C_1, C_2, …, C_{p2+1} and segments S_1, S_2, …, S_{p2} constituted one test line. Consider the example in Fig. 4. The set of fields in Fig. 4 adequately describes the following descriptor.

$$\{C_1; \frac{F_1}{S_1}; \frac{F_2}{S_2}; C_2; \frac{F_3}{S_3}; \frac{F_4}{S_4}\}$$

in which chains C_1 and C_2 consist of two lines. Chains C_1 and C_2 are paragraphs whose boundaries change, and words can be carried over from the first line to the second. Let us pay attention to the absence of strict conditions on the correspondence of the placement of the chain of words C_1 and the segments S_1 and S_2 and, accordingly, the words of the chain C_2 and the segments S_3 and S_4. The correctness of the placement is checked by the proximity of the projection of the segments and frames of the corresponding chain in a wide range.

A simplified form of the combined descriptor is possible, similar to the simplest text descriptor (1):

$$\left\{C_1; \frac{F_1}{S_1}; \frac{F_2}{S_2}; \ldots \frac{F_p}{S_p}; C_2\right\} \text{ или } \left\{C_1; \frac{F_1}{S_1}; C_2\right\}. \tag{6}$$

In descriptor (6), as well as in descriptor (1), there may be no chains C_1 and C_2 that are unique for the document or fragment. In this case, it is possible to link the "Key-Value-Key" type to search for multiple instances of the F_1 fields or the F_1, F_2, … F_{p2} group in an arbitrary location. Configuration in Fig. 4 can be given by two descriptors

$$\{C_1; \frac{F_1}{S_1}; \frac{F_2}{S_2}\} \text{ и } \{C_2; \frac{F_3}{S_3}; \frac{F_4}{S_4}\}$$

for which the case is admissible when the upper boundary of the region $\{C_2; \frac{F_3}{S_3}; \frac{F_4}{S_4}\}$ is located below the lower boundary of the region $\{C_1; \frac{F_1}{S_1}; \frac{F_2}{S_2}\}$.

4 A Descriptor for Vertical Groups of Objects

In [7] descriptors of fields of the check-box type are considered. Some of them are described by a descriptor (2), some of the fields of which can be interpreted as a predictive checkbox frame. The specificity of the checkbox linking consists in using the frames of objects similar to checkboxes and detected simultaneously with segments detection. To detect checkboxes are used edge detection methods [13], CNN [14, 15] the Viola and Jones method [16, 17]. It is also necessary to take into account the possible distortion of the check-box frame due to handwriting. It is indicated that with significant noise of the document image due to a large number of errors in character recognition and detection of segments, some chains in descriptors (2), (3), (4) may not be attached. In such cases, descriptors of the following form can be applied to a group of vertical checkboxes:

$$D_{CB}(F_1, F_2, \ldots, F_{\rho 2}) = \begin{cases} C_1^{top}; C_2^{top}; \ldots C_{p1}^{top}; \\ C_1; \frac{F_1}{S_1}; C'_1; \\ C_2; \frac{F_1}{S_1}; C'_2; \\ \ldots \\ C_{p2+1}; \frac{F_1}{S_1}; C'_{p2+1}; \\ C_1^{bottom}; C_2^{bottom}; \ldots C_{p3}^{bottom} \end{cases} \tag{7}$$

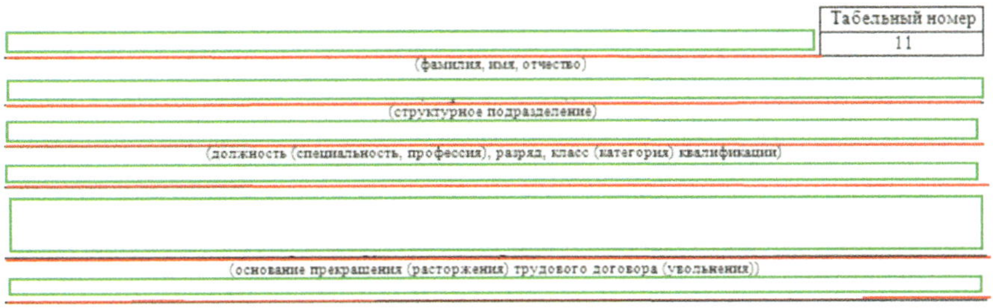

Fig. 5 Descriptor (8) Configuration Examples

In descriptor (7), the main elements are the top and bottom limiting chains $C^{top}{}_{1}$, $C^{top}{}_{2}$, … $C^{top}{}_{p1+1}$ and $C^{bottom}{}_{1}$, $C^{bottom}{}_{2}$, … $C^{bottom}{}_{p3+1}$ and a group of check-boxes arranged vertically. Chains C_1,C'_1, C_2,C'_2, … C_{p2+1},C'_{p2+1} are optional elements when linking a group of checkboxes. However, linking any of the chains C_1,C'_1, C_2,C'_2, … C_{p2+1},C'_{p2+1} simplifies linking in the case when not all checkboxes were detected.

A similar descriptor can be applied to a group of horizontal segments in a limited area. Configuration in Fig. 5 is given by the following descriptor.

$$D_{\mathrm{LH}}\left(\mathrm{F}_1, \mathrm{F}_2, \ldots, \mathrm{F}_{p2}\right) = \left\{ \begin{array}{c} C_1^{top}; C_2^{top}; \ldots C_{p1}^{top}; \\ \frac{F_1}{S_1}; \\ \frac{F_1}{S_1}; \\ \ldots \\ \frac{F_6}{S_6}; \\ C_1^{bottom}; C_2^{bottom}; \ldots C_{p3}^{bottom}. \end{array} \right. \tag{8}$$

The left, right, and bottom borders of each of the fields F_1, F_2, … F_6 are determined using the frames of the segments S_1, S_2, … S_6. The upper limits of the fields F_2, … F_6 can be determined using the following methods:

- using the frame of the upper segments S_1, S_2, … S_5;
- using a static text frame under the segments S_1, S_2, … S_5;
- using fixed field height values.

5 Using Images of Stamps and Barcodes as Reference Elements

Searching for stamps on images is possible using several methods:

- methods based on the search for geometric primitives [18], including those using the generalized Hough transform and the search and analysis of singular points in the image;
- methods based on spectral analysis of image pixels [19]. Spectral analysis is not applicable for non-color images.

Several reasons make it difficult to detect print in a highly noisy document image [20, 21]. These include physical fading over time, complex backgrounds, and loss of print integrity. Despite this, the stamp frame serves as a reliable reference element and can be used for linking in any of the above descriptors, complementing the C_1^{bottom}; C_2^{bottom}; ... C_{p3}^{bottom}. Reliability is determined by the way administrative documents are certified: stamps are placed at the bottom of the page.

The image of a barcode can act as a reference element similar to the image of a stamp. Of course, the barcode must be unique for some fragments of the page.

In some documents, the seal may be located in a rather arbitrary place. Figure 6 shows an example of such a document. In this example, the presence of a seal in a certain location indicates a change in some table cells.

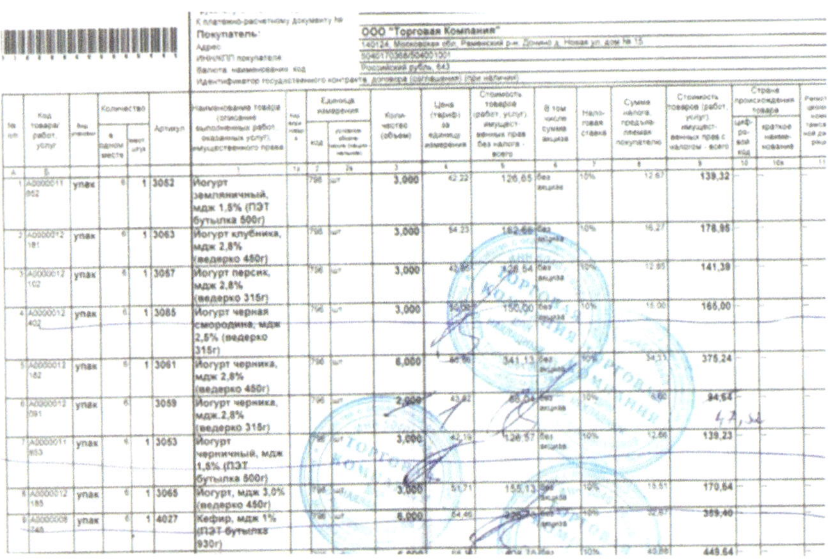

К платежно-расчетному документу №

Покупатель: ООО "Торговая Компания"

Адрес: 140124, Московская обл, Раменский р-н, Донино д, Новая ул, дом № 15

ИНН/КПП покупателя: 5040170368/504001001

Валюта: наименование, код: Российский рубль, 643

Идентификатор государственного контракта, договора (соглашения) (при наличии)

№ п/п	Код товара/работ, услуг	Вид упаковки	Количество в одном месте	Количество мест, штук	Артикул	Наименование товара (описание выполненных работ, оказанных услуг), имущественного права	Код вида товара	Единица измерения: код	Единица измерения: условное обозначение (национальное)	Количество (объем)	Цена (тариф) за единицу измерения	Стоимость товаров (работ, услуг), имущественных прав без налога - всего	В том числе сумма акциза	Налоговая ставка	Сумма налога, предъявляемая покупателю	Стоимость товаров (работ, услуг), имущественных прав с налогом - всего	Страна происхождения товара: цифровой код	Страна происхождения товара: краткое наименование	Регистрационный номер таможенной декларации
А	Б					1	1а	2	2а	3	4	5	6	7	8	9	10	10а	11
1	A00000011 862	упак	6	1	3062	Йогурт земляничный, мдж 1,5% (ПЭТ бутылка 500г)		796	шт	3,000	42,22	126,65	без акциза	10%	12,67	139,32	–	–	–
2	A00000012 181	упак	6	1	3063	Йогурт клубника, мдж 2,8% (ведерко 450г)		796	шт	3,000	54,23	162,68	без акциза	10%	16,27	178,95	–	–	–
3	A00000012 102	упак	6	1	3057	Йогурт персик, мдж 2,8% (ведерко 315г)		796	шт	3,000	42,85	128,54	без акциза	10%	12,85	141,39	–	–	–
4	A00000012 402	упак	6	1	3085	Йогурт черная смородина, мдж 2,5% (ведерко 315г)		796	шт	3,000	50,00	150,00	без акциза	10%	15,00	165,00	–	–	–
5	A00000012 182	упак	6	1	3061	Йогурт черника, мдж 2,8% (ведерко 450г)		796	шт	6,000	56,86	341,13	без акциза	10%	34,11	375,24	–	–	–
6	A00000012 091	упак	6		3059	Йогурт черника, мдж.2,8% (ведерко 315г)		796	шт	2,000	43,02	86,04	без акциза	10%	8,60	94,64 47,32	–	–	–
7	A00000011 853	упак	6	1	3053	Йогурт черничный, мдж 1,5% (ПЭТ бутылка 500г)		796	шт	3,000	42,19	126,57	без акциза	10%	12,66	139,23	–	–	–
8	A00000012 185	упак	6	1	3065	Йогурт, мдж 3,0% (ведерко 450г)		796	шт	3,000	51,71	155,13	без акциза	10%	15,51	170,64	–	–	–
9	A00000008 748	упак	6	1	4027	Кефир, мдж 1% (ПЭТ бутылка 930г)		796	шт	6,000	54,46	[illegible]	без акциза	10%	32,67	359,40	–	–	–
													без	10%	40,88	449,64	–	–	–

Fig. 6 Example of stamping table cells

6 Results of Experiments

The proposed method for analyzing the structure of documents was tested on our private test sets, consisting of images of documents of the type.

- Dataset-1: universal transfer document;
- Dataset-2: tax document.

The document was scanned at an optical density from 100 to 300 dpi with different digitization qualities. The accuracy of recognition and the accuracy of linking fields were studied. The case was considered with the use of segments as elements of combined descriptors and without the use of segments. SDK Smart Document Engine was used for recognition [15, 22]. The accuracy of field recognition was studied. We also studied the accuracy of linking fields, calculated as the proportion of found fields concerning the total number of fields.

Combined descriptors (3) and (4) were created for Dataset-1 documents. Simple descriptors were created for Dataset-2 documents (1). The results obtained are summarized in Tables 1 and 2 illustrating the effectiveness of the proposed method. Namely, the accuracy of linking and recognition of Dataset-1 documents increased by more than 10% due to the use of combined descriptors (3) and (4). This improvement is since without combined descriptors, the search for field boundaries in noisy fragments is difficult. The recognition accuracy of Dataset-2 documents increased by 2% due to the thinning of the right and left borders of the fields. At the same time, the accuracy of linking Dataset-2 documents remained the same. This is due to the design of the fields of Dataset-2 documents, for which the use of keywords is adequate.

Table 1 Linking accuracy for the fields of universal transfer documents in Dataset-1

Document	Number of documents	Number of fields	Fields recognition accuracy (%)	Fields linking accuracy (%)
With the employment of line segments	1732	55,648	68,82	83,34
Without the employment of line segments			52,23	68,44

Table 2 Linking accuracy for the fields of universal transfer documents in Dataset-2

Document	Number of documents	Number of fields	Fields recognition accuracy (%)	Fields linking accuracy (%)
With the employment of line segments	1575	55,869	91,76	94,34
Without the employment of line segments			89,66	92,34

7 Conclusion

The proposed in this chapter and similar descriptors are used to configure the linking of flexible documents in the SDK Smart Document Engine [15, 22]. A simple notation has been developed to create descriptors.

The proposed method for analyzing the structure of documents applies to recognizing highly noisy and distorted document images. In particular, the method makes it possible to detect marginal boundaries in black-and-white copies of color documents. Linking efficiency is achieved through the use of combined descriptors. Combined descriptors are based on simple descriptors of static text words, line cuts, checkboxes, stamp images, and barcodes.

The accuracy of the proposed method was evaluated on our private datasets of administrative documents. Linking accuracy when using combined descriptors ranges from 80 to 99% for various fields.

The worst results correspond to complex cases of noisy seals and signatures of word images of a black-and-white copy of a color administrative document. For such fields, the recognition accuracy is low. However, field border prediction is useful for detecting typing or handwriting in a field.

Acknowledgements The authors are grateful to T. B. Chistyakova for her valuable comments and suggestions.

References

1. Rusiñol, M., Frinken V., Karatzas, D., Bagdanov, A. D., Lladós, J.: Multimodal page classification in administrative document image streams. In: IJDAR. Vol. 17(4), pp. 331 Image Classification by Mixed Finite Element Method and Orthogonal Legendre Moments 341 (2014). https://doi.org/10.1007/s10032-014-0225-8
2. Jain, R., Wigington, C.: Multimodal Document Image Classification. pp.71–77 (2019). https://doi.org/10.1109/ICDAR.2019.00021
3. Qasim, S. Rukh., Mahmood, H., Shafait, F.: Rethinking table recognition using graph. Neural Networks, pp. 142–147 (2019). https://doi.org/10.1109/ICDAR.2019.00031
4. Vasiliev, S.S., Korobkin, D.M., Kravets, A.G., Fomenkov, S.A., Kolesnikov, S.G.: Extraction of cyber-physical systems inventions' structural elements of russian-language patents. Stud. Syst. Springer, Decis. Control, pp. 55–68. (2020). https://doi.org/10.1007/978-3-030-32579-4_5
5. Bellavia, F.: SIFT matching by context exposed. IEEE Trans. Pattern Anal. Mach. Intell. (2022).https://doi.org/10.1109/TPAMI.2022.3161853
6. Bay, H., Tuytelaars, T., Van Gool, Luc.: SURF: Speeded Up Robust Features. Computer Vision and Image Understanding - CVIU. Vol. 110(3), pp. 404–417. (2006)
7. Slavin, O.A., Pliskin, E.L.: Method for analyzing the structure of noisy images of administrative documents. Bulletin of the South Ural State University. Ser. Mathematical Modelling, Programming & Computer Software. Vol. 15(4), pp. 80–89 (2022). https://doi.org/10.14529/mmp220407
8. Slavin, O., Arlazarov, V., Tarkhanov, I.: Models and methods flexible documents matching based on the recognized words. In: Kravets, A.G., Bolshakov, A.A., Shcherbakov, M. (Eds.),

Cyber-Physical Systems: Advances in Design & Modelling. Springer Nature Switzerland AG. Vol. 350, pp. 173–184 (2021). https://doi.org/10.1007/978-3-030-67892-0_15
9. Deza, M.M., Deza, E., Encyclopedia of Distances. Springer, Berlin, 2009, xiv+590 pp.
10. Matas, J., Galambos, C., Kittler, J.: Robust detection of lines using the progressive probabilistic hough transform. Comput. Vis. Image Underst. **78**(1), 119–137 (2000). https://doi.org/10.1006/cviu.1999.0831
11. Grompone von Gioi, R., Jakubowicz, J., Morel, J.M. et al.: On straight line segment detection. J. Math. Imaging Vis. **32**, 313–347 (2008). https://doi.org/10.1007/s10851-008-0102-5
12. Grompone von Gioi, R., Jakubowicz, J., Morel, J.-M., Randall, G.: LSD: a fast line segment detector with a false detection control. IEEE Trans. Pattern Anal. Mach. Intell. **32**(4), 722–732 (2010). https://doi.org/10.1109/TPAMI.2008.300
13. Emaletdinova, L., Nazarov, M.: Construction of a fuzzy model for contour selection. Construction of a fuzzy model for contour selection. In: Kravets, A.G., Bolshakov, A.A., Shcherbakov, M. (eds.), Cyber-Physical Systems: Intelligent Models and Algorithms. Studies in Systems, Decision and Control, Vol. 417. pp. 243–246 (2022). https://doi.org/10.1007/978-3-030-95116-0_20
14. Zlobin, P., Chernyshova, Y., Sheshkus A., Arlazarov V.V.: Character sequence prediction method for training data creation in the task of text recognition. In: Proceedings SPIE 12084, Fourteenth International Conference on Machine Vision (ICMV 2021), 120840R (2022). https://doi.org/10.1117/12.2623773
15. Arlazarov, V.L., Arlazarov, V.V., Bulatov, K.B., et al.: Mobile ID document recognition–coarse-to-fine approach. Pattern Recogn. Image Anal. **32**, 89–108 (2022). https://doi.org/10.1134/S1054661822010023
16. Arlazarov, V., Voysyat, Ju. S., Matalov, D., Nikolaev, D., Usilin, S.A.: Evolution of the Viola-Jones Object Detection Method: A Survey. Bulletin of the South Ural State University. Mathematical Modelling, Programming and Computer Software. Vol. 14, pp. 52–23 (2021). https://doi.org/10.14529/mmp210401
17. Tereshin, A., Usilin, S., Arlazarov, V.V.: Performance improvement of multi-class detection using greedy algorithm for Viola-Jones cascade selection. In: Proceedings SPIE 10696, Tenth International Conference on Machine Vision (ICMV 2017), 106960D (2018). https://doi.org/10.1117/12.2310101
18. Roy, P.P., Pal, U., Llados J.: Seal detection and recognition: an approach for document indexing. In: Proceedings 10th International Conference on Document Analysis and Recognition. Vol. 2015, Article ID 367879. 15 p. (2015). https://doi.org/10.1109/ICDAR.2009.128
19. Katsuhiko, U.: Extraction of signature ad seal imprint from bankchecks by using color information. In: Proceedings of 3rd International Conference on Document Analysis and Recognition, pp. 665–668. (1995) https://doi.org/10.1109/ICDAR.1995.601983
20. Matalov, D., Usilin, S., Arlazarov, V.V.: Modification of the Viola-Jones approach for the detection of the government seal stamp of the Russian Federation. In: Proceedings SPIE 11041, Eleventh International Conference on Machine Vision (ICMV 2018), 110411Y (2019). https://doi.org/10.1117/12.2522793
21. Kravets, A., Salnikova, N., Mikhnev, I., Solovieva, N.: Relevant Image Search Method When Processing a Patent Array (2021). https://doi.org/10.1007/978-3-030-87034-8_6
22. Smart Document Engine—automatic analysis and data extraction from business documents for desktop, server and mobile platforms. URL: https://smartengines.com/ocr-engines/document-scanner. Accessed March 20, 2023

Application of Methods for Identification and Parrying the Threat of an Accident of a Helicopter

Alexander Bolshakov and Aleksey Kulik

Abstract The results of numerical simulation of the flight conditions of a helicopter-type aircraft with a change in variables affecting flight safety are presented. A comparative analysis of the assessment of the threat of an aviation accident was carried out, considering the varying degree of influence of the psychophysical state of the crew. Particular attention is paid to the modeling of the functioning of the decision support system for parrying the threat of an accident, which is performed based on data on the flight conditions of the aircraft and information from the knowledge base of the flight safety management system. The results obtained in the course of the work allow us to evaluate the effectiveness and efficiency of methods for assessing and parrying the threat of an aviation accident. These include the following: considering the impact of changes in the psychophysical state of the crew on flight conditions allows you to expand the completeness and reliability of identifying the threat of an aviation accident. Based on the simulation of this situation, recommendations for parrying the threat of an aviation accident were formed. The proposed methods can be implemented by software and hardware onboard systems for ensuring the safety of an aircraft flight. The results presented in the chapter are useful in designing using cyber-physical systems using mathematical modeling methods to evaluate the effectiveness of decision support systems for assessing and parrying helicopter-type aircraft accidents.

Keywords Flight safety · Expert system · Assessment of flight conditions · Decision support

A. Bolshakov
Peter the Great St. Petersburg Polytechnic University (SPbPU), 195251 St. Petersburg, Russian Federation

A. Kulik (✉)
National Research University "Moscow State Technical University", 111250 Moscow, Russian Federation
e-mail: kulikalekse@yandex.ru

A. G. Kravets et al. (eds.), *Cyber-Physical Systems Engineering and Control*, Studies in Systems, Decision and Control 477,
https://doi.org/10.1007/978-3-031-33159-6_5

1 Introduction

In the process of aircraft control, various kinds of collisions may appear that threaten the safety of the aircraft flight and the performance of the flight task. As a rule, the assessment of flight safety and parrying the threat of an aviation accident is carried out by the aircraft crew. However, the actions of the pilot are not always sufficient to ensure the safety of the flight. Therefore, modern systems are used onboard the aircraft to assess and parry the threat of an accident. For example, a flight safety system [1] is used, designed to improve the safety of aircraft flights from terrorism and human error, as well as to control the deviation of an aircraft flight beyond acceptable values. The aircraft safety system consists of an emergency subsystem, an automatic control and piloting subsystem, a system of monitoring devices, a computer for aviation equipment of security agencies, and a second aircraft control system. A feature of this system is the removal of the pilot from the control of the aircraft when the vessel is hijacked by terrorists or a suicide raft. The interaction of air traffic safety services and authorities with the aircraft is carried out through the secondary control system of the vessel and the computer of the aviation equipment of the security authorities.

However, the system considered above has the following significant shortcoming: there is no forecasting and assessment of the threat of an aviation accident, which consists in determining the conditions for the flight of an aircraft under the influence of external and internal influencing factors.

Also known as a system and method for ensuring flight safety and/or flight control of aircraft [2], designed to ensure the safety of air traffic of aircraft using a satellite navigation system and broadcast signals with a coverage area around the world. Signals are received by satellite receiving stations, which then transmit information to the flight safety center or other organizations. Thanks to the use of satellite data transmission from the aircraft to the ground, constant monitoring of the values of the parameters of its movement in the center of flight safety in various geographic latitudes of the Earth is ensured. The approach used makes it possible to perform constant control over the movement of aircraft in the airspace and ensure their flight safety.

The disadvantage of the system is the lack of direct control of the flight of the aircraft to counteract the threat of an accident, as well as the device for collecting and analyzing variables that affect the safety of the flight of the device.

Among the methods for assessing the threat of an aviation accident, one can single out the method of ensuring the "built-in" flight safety of an aircraft in difficult (multifactorial) conditions based on situational modeling [3, 4]. The chapter proposes an approach to assessing and predicting changes in the flight safety of an aircraft under the influence of heterogeneous factors affecting the flight conditions of the aircraft. The method consists in considering flight scenarios presented as events connected by a process that characterizes the behavior of the operator, the aircraft, and the operating environment. The author also introduced a "safety palette" that determines the ranges of controlled variables on a scale from normal to catastrophic values.

In addition, each controlled variable is described by fuzzy constraints. At the same time, an integrated flight safety assessment was introduced, representing the integral spectrum of safety, which determines the flight safety under the influence of several controlled variables, and if the specified value is exceeded by at least one variable, its value is assigned to the integral assessment.

The authors propose a macrostructural model that represents the results of the impact on the aircraft of various combinations of operational factors. The model has the form of a tree that combines related situations obtained as a variation of the base scenario under the influence of a given set of operational factors. Each situational tree is divided into 6 security levels. An integrated approach to assessing flight safety makes it possible to identify possible factors affecting the flight safety of an aircraft. The chapter also presents the conditions for restoring a safe flight mode to the point in time when the situation develops irreversibly. The scenario of the safest branch from the tree of the situational model is selected as the scenario. The script can be executed automatically or under the influence of an operator. Based on situational modeling, a knowledge base is formed for use as part of onboard intelligent systems, investigation of the causes of aviation accidents, etc.

However, the presented method has the following disadvantages: the integrated assessment of flight safety does not contain a classification of variables safe flight with division into groups of influencing factors, the assessment of the psychophysical state of the crew is not used when restoring flight safety, there is no identification of the threat of an aviation accident.

Thus, the improvement of methods for assessing and parrying the threat of an aviation accident is an urgent scientific and technical problem. The solution to this problem is achieved by an integrated approach to assessing the threat of an accident, taking into account the influence of the psychophysical state of the crew, the technical condition of the control object, and the weather conditions of the flight. At the same time, the assessment of the flight conditions of the aircraft is carried out by onboard means of ensuring flight safety in real-time. At the same time, as part of the onboard equipment, methods are being implemented to parry the threat of an aviation accident using artificial intelligence tools. As systems for assessing and parrying the threat of an aviation accident under the influence of a set of flight safety variables, a flight safety management system can be used [5]. The main elements of this system are devices for data preprocessing, decision support, and data transmission. At the same time, preliminary data processing consists in obtaining the system information and measurement data from the complex onboard equipment of the aircraft and in the formation of electrical signals for assessing flight conditions that characterize the excess of the specified values of factors affecting the control object. Then, based on the information about the change in the flight conditions of the aircraft, the decision support device generates a conclusion characterizing the degree of danger of the flight event and methods for its elimination. Checking the efficiency and operability of the system is carried out in the process of its testing on the semi-natural simulation bench under the conditions of correct operation and in case of failures of its interacting equipment concept of designing cyber-physical systems is an urgent scientific and technical problem.

2 Problem Statement

The purpose of this work is to evaluate the effectiveness of methods for identifying and parrying the threat of an accident during the flight of a helicopter-type aircraft.

To achieve this goal, the following tasks must be solved: (a) assess the flight conditions of the aircraft; (b) form a management decision to parry the threat of an aviation accident; (c) perform modeling of parrying the threat of an aviation accident.

3 Description of Aircraft Flight Conditions

Depending on the values of aircraft flight safety variables (psychophysical state of the crew, technical state of the control object, weather conditions), the conditions of its flight change, which can vary from accident-free to catastrophic [6, 7]. At the same time, the stage of flight and the type of aircraft have a significant impact on flight conditions and its safety. It is known that the most dangerous stages of a flight are the takeoff and landing of a vessel, which is due to the shorter available time for a pilot to parry an aircraft failure compared to the available time when flying along a route. Therefore, the landing approach requires the pilot to increase the concentration and accuracy of piloting the aircraft. As an example, a helicopter landing approach is shown in Fig. 1 [8].

Let us assume that during the landing approach of a helicopter in automatic mode, a failure of the automatic control system occurred, which was detected by the built-in control of the control system and transferred to the on-board equipment complex and the aircraft flight safety control system in the form of a one-time command. Simultaneously with the failure of the automatic control system, the psychophysical parameters of the crew decreased: absent-mindedness increased and attention decreased. Information from the registration means entering the aircraft flight safety management system.

By the method for assessing the flight conditions of an aircraft [9], the safety values of previously formalized input variables that affect flight safety are evaluated, which for the situation under consideration take on the safety values presented in Table 1.

Also, to assess the flight conditions of an aircraft, a matrix of precedents is used, based on which a knowledge base and a set of rules for determining the threat of an aviation accident are formed, which is presented in the form of Table 2.

The table shows that the situation under consideration is determined by rule 10, which characterizes the emergency flight conditions. At the same time, taking into account the failure of the aircraft's automatic control system, the threat of an accident can only be parried by the actions of the crew on the recommendations of the decision support system.

The effectiveness of the method for assessing the threat of an aviation accident can be verified by comparing the numerical simulation of aircraft flight conditions with

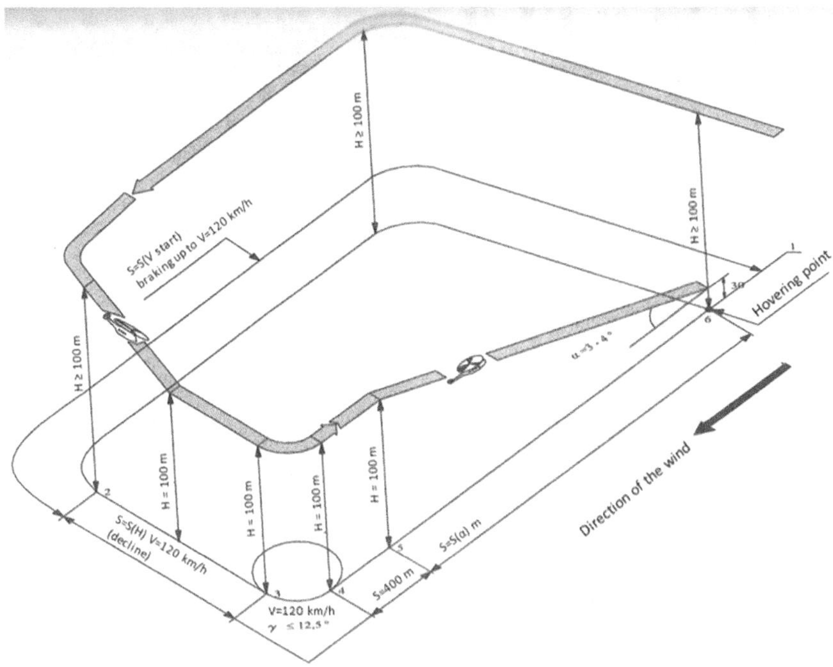

Fig. 1 Helicopter landing pattern

Table 1 Flight safety variable values

№	Group	Variable	Variable quantity	Linguistic variables
1	The psychophysical state of the pilot	Fatigue	$f_2 \in (-0.5 \div 0.5)$	Average
		Attention	$k_3 \in (-0.5 \div 0)$	Low
		Level of training (Competence)	$f_1 \in [1 \div 0.5)$	Hight
		Stress	$k_3 \in (0 \div 0.5)$	Average
2	Aircraft technical condition	Failure of functionally significant elements	$f_3 \in (0.5 \div 1]$	Significant
		Deformation of structural members	$k_1 \in [1 \div 0.5)$	Missing
		Aircraft controllability and stability	$f_1 \in (0.5 \div 1]$	Hight
		Error in the software of aircraft control systems	$k_2 \in (0.5 \div 0)$	Minor
3	External influencing factors	Head wind	$f_1 \in (0.5 \div 1]$	Week
		Visibility	$f_1 \in [-1 \div -0.5]$	Week
		Side wind	$k_1 \in (0.5 \div 1]$	Good

Table 2 Case matrix for accident threat assessment

№	The psychophysical state of the pilot				Aircraft technical condition				External influencing factors				Precedent (flight conditions)
	Fatigue	Attention	Competence	Stress	Failure of Functionally Elements	Decor.of Structural Members	Aircraft Con.& Stabil	Error in the Soft	Head Wind	Visibility	Side Wind	Visibility	
1	f_1	k_1	f_1	k_1	f_1	k_1	f_1	k_1	f_1	k_1	f_1	k_1	Accident-free
2	f_3	k_3	f_1	k_1	f_1	k_1	f_2	k_1	f_1	k_2	f_1	k_2	Difficult flight conditions (parrying the situation–automatic)
…	…	…	…	…	…	…	…	…	…	…	…	…	…
9	f_1	k_3	f_1	k_1	f_3	k_1	f_1	k_2	f_1	k_1	f_1	k_1	Difficult flight conditions (parrying the situation–automatic)
10	f_2	k_3	f_1	k_3	f_3	k_1	f_1	k_2	f_1	k_1	f_1	k_1	Emergency (parrying the situation–the actions of the crew)
…	…	…	…	…	…	…	…	…	…	…	…	…	…

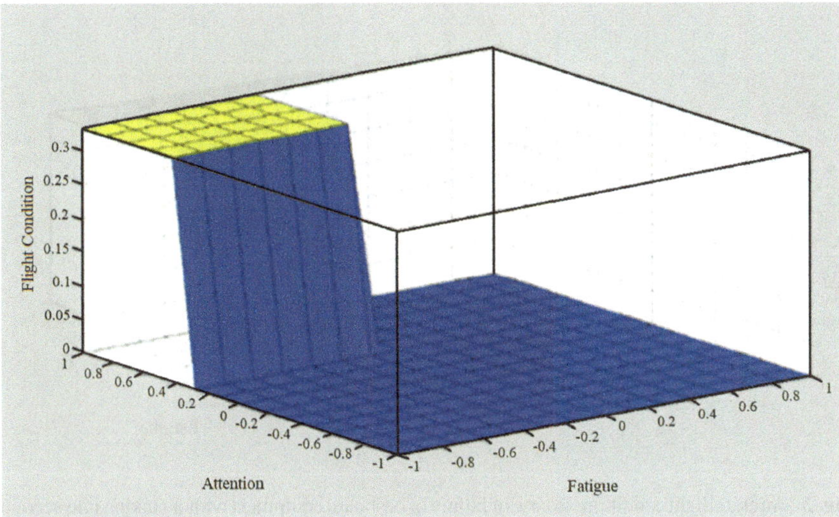

Fig. 2 Aircraft flight conditions in case of failure of on-board equipment, taking into account the deterioration of the parameters of the psychophysical condition of the crew

and without taking into account the influence of the psychophysical state of the crew on flight safety. Then we assume that during the helicopter landing approach, the automatic control system failed while maintaining high indicators of the psychophysical state of the crew and the weather conditions of the flight. This situation is characterized by rule 9 of Table. 2 and defines the difficult flight conditions in which the helicopter is landing.

Using the information from Table 2, the values of aircraft flight safety, numerical simulation was performed to determine the threat of an aviation accident. Applying approaches to data processing of cyber-physical systems using artificial intelligence tools [10–18], it is possible to assess the threat of an aviation accident. The results of mathematical modeling are shown in Figs. 2 and 3.

It can be seen from the figure that the assessment of aircraft flight conditions takes the value "−0.33", which corresponds to the emergency conditions of the flight.

It can be seen from the figure that the assessment of aircraft flight conditions takes the value "0.33", which corresponds to difficult flight conditions.

Thus, the introduction, together with indicators of the technical condition of the aircraft and external influencing factors affecting the flight safety of the aircraft, a group of variables characterizing the psychophysical state of the crew, makes it possible to increase the reliability of the assessment of the threat of an aircraft accident and, as a result, to develop more effective managerial decisions to counter it.

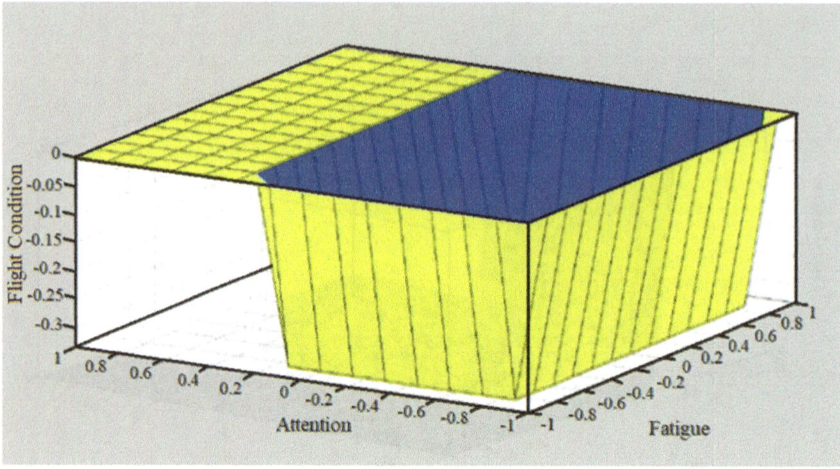

Fig. 3 Aircraft flight conditions in case of failure of on-board equipment without taking into account the influence of the psychophysical state of the crew

4 Parrying the Threat of an Aviation Accident

Parrying the threat of an aviation accident in the event of a failure of the automatic control system can only be carried out by the actions of the crew, taking into account the recommendations of the decision support system. The formation of recommendations to the crew on parrying the threat of an aviation accident is carried out based on data on flight conditions, flight safety variables, and the knowledge base of parrying the threat of an accident [19–22].

The formation of the decision support knowledge base is carried out on the basis of a set of rules, for an emergency which has the following form [23]:

$$\begin{aligned}
&\text{RULE } <1>\text{: IF} X_{1j} = \{f_1, k_3\} \& X_{2j} = \{f_3, k_4\} \& X_{3j} = \{f_2, k_1\} \text{THEN} Y = \{g_3\},\\
&\text{RULE } <2>\text{: IF} X_{1j} = \{f_1, k_1\} \& X_{2j} = \{f_2, k_4\} \& X_{3j} = \{f_3, k_1\} \text{THEN} Y = \{g_3\},\\
&\text{RULE } <3>\text{: IF} X_{1j} = \{f_3, k_4\} \& X_{2j} = \{f_2, k_3\} \& X_{3j} = \{f_1, k_1\} \text{THEN} Y = \{g_4\},\\
&\text{RULE } <4>\text{: IF} X_{1j} = \{f_2, k_2\} \& X_{2j} = \{f_1, k_1\} \& X_{3j} = \{f_3, k_2\} \text{THEN} Y = \{g_4\},
\end{aligned} \tag{1}$$

here g_3—signaling to the crew about failures onboard the control object, the threat of an accident, followed by a parry by the pilot on the recommendation of a speech translator; g_4—signaling to the crew about failures onboard the control object, the threat, followed by parrying the reconfiguration of the control system of the object and landing on the nearest suitable point.

For emergency flight conditions, accompanied by a failure of the aircraft automatic control system and a decrease in the psychophysical parameters of the crew, the following recommendations for parrying an accident should be followed:

- Disconnection of the automatic control system from the actuators of the helicopter;
- Issuance of recommendations to the crew on parrying the threat of an accident.

In the first case, the automatic control system is disconnected from the associated equipment, which is achieved by disconnecting the system from power sources.

In the second case, the decision support system issues the following information using voice alert means: "Failure of the ACS. Flight emergency conditions. Continue the flight according to the flight task. The flight altitude is more than 100 m. The flight is straight. Indicated speed 120 km/h. Altitude over 100 m. Course turn by 90 degrees. Flight straight. Indicated speed 120 km/h. Altitude over 100 m. Course turn by 90°. Decrease. Altitude 100 m. Indicated speed 120 km/h. Flight straight. Indicated speed 120 km/h. Height 100 m. Course turn 90°. with maintaining a roll of fewer than 12.5°. Altitude 100 m. Indicated speed 120 km/h. Flight straight. Range 400 m. Height 100 m. Indicated speed 120 km/h. Decrease to 30 m. Pitch angle 3 degrees. Hang up." The results of the derivation of recommendations for parrying the threat of an aviation accident are presented in Fig. 4.

Here H—height, m; V—speed, km/h.

It can be seen from the figure that the decision support system issues recommendations to the crew on how to control the aircraft in emergency flight conditions

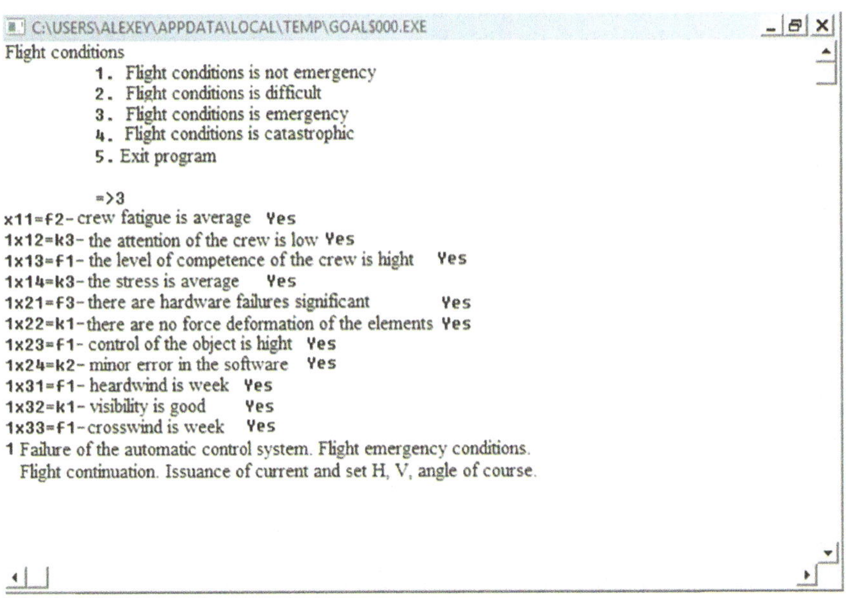

Fig. 4 Issuing recommendations for parrying the threat of an aviation accident

caused by the failure of the automatic control system and the decrease in the psychophysical state of the crew, which indicates the operability of the algorithm for the method of parrying the threat of an accident.

The issuance of recommendations to the crew on the control of the aircraft is carried out through the means of onboard signaling and indication, as well as speech informants.

From the information presented in Figs. 4, and 5, it follows that the decision support device can form recommendations to the crew on actions under the threat of an aviation accident through the means of aircraft display and signaling.

Based on the results of tests of the safety control system of the aircraft, it can be concluded that its software is capable and can be used in the further development of the system, as well as devices of this class.

At the same time, the implementation of methods and algorithms should be carried out based on artificial intelligence tools and using the concept of building the cyber-physical systems, which form the aircraft flight safety management system [24–27].

5 Conclusion

In the process of performing the work, a change in the flight conditions of a helicopter-type aircraft during the landing approach in case of failure of the automatic control system and a decrease in the psychophysical state of the crew was considered, which made it possible to identify emergency flight conditions. At the same time, without the influence of a change in the psychophysical state of the crew, the flight conditions are assessed as difficult. Therefore, the use of variables that characterize the psychophysical state of the crew to assess the flight conditions of the aircraft and the threat of an accident makes it possible to expand the completeness and reliability of identifying the threat of an aviation accident. Recommendations were also formed on parrying the threat of an aviation accident for the situation under consideration with subsequent modeling, which confirmed their performance. The results obtained in the process of performing the work can be used in the process of creating systems for ensuring the safety of a helicopter-type aircraft at various stages of its flight.

References

1. Garry, J.B.: Safety aircraft flight system Pat. US 7, 183, 946 B2 (2003)
2. Blomenhofer, H., Nolfeidt, H.: System and method for ensuring flight safety and/or flight control of aircraft. Pat. Russia Federation No 2471275 G08G 5/00 (2000)
3. Burden, I.Y.: Method and technology for ensuring the "built-in" flight safety of an aircraft in complex (multifactorial) conditions based on situational modeling. In: Proceedings of the VIII International Conference "System Identification and Control Problems (SICPRO-09)", pp. 1203–1248 (2009)

4. Burdun, I.Y.: Safety windows: knowledge maps for accident prediction and prevention in multifactor flight situation In: 27th International Congress of the Aeronautical Science (2010)
5. Kulik, A.A., Bolshakov, A.A., Sergushov, I.V., Scripal, E.N.: Development of a flight safety control system for a helicopter-type aircraft. Mechatron. Automat. Control. **2016**(10), 708–715 (2016)
6. Fedunov, B.E., Prokhorov, M.D.: Conclusion on precedent in knowledge bases of onboard intelligent systems. Artif. Intell. Decision-Making. **3**, 63–72 (2010)
7. Bahanov, L.E., Demkin, M.A., Fedunov, B.E.: Mathematical model of aircraft motion for knowledge bases of on-board operational advising expert systems In: Proceedings of the Russian Academy of Sciences. Theory and control systems, pp. 103–111 (2010)
8. Popov, A.N., Sergushov, I.V., Teterin, D.P.: Flight systems and navigation systems for helicopters. M.: Innovative mechanical engineering. p. 368 (2017)
9. Kulik, A.A., Bolshakov, A.A., Sergushov, I.V., Skripal, E.N.: Aviation accident threat intelligent assessment method. In: Herald of Computer and Information Technologies, pp. 3–9. (2018)
10. Khaitan, et al.: Design techniques and application of cyber-physical systems: a Survey. IEEE Syst J (2014)
11. Lee, E.A., Seshia, S.A.: Introduction to embedded systems—a cyber-physical systems approach. LeeSeshia.org (2011)
12. Cyber-Physical System Driving force for innovation in mobility, health, energy, and production. Acatech (2011)
13. Wolf, W.: Cyber-physical system. Computer **3**, 88–89 (2009)
14. Crop: a cyber-physical system architecture model in the field of precision agriculture. In: Conference Agriculture for Life, Life for Agriculture, vol. 6, pp. 73–79. (2009)
15. Lee, J., Bagheri, B., Kao, H.: A cyber-physical systems architecture for industry. 4.0-based manufacturing systems. Manufact. Lett. **3**, 18–23 (2015)
16. Chuprynousky, V.P., Namiot, D.E., Sinyakov, S.A.: A cyber-physical system as the basis for the digital economy. Int. J. Org. Open Inform. Technol. **4**(2), 18–25. ISSN: 2307–8162 (2016)
17. Averchenkov, V., Miroshnikov, V., Podvesovsky, A., Korostelyov, D.: Fuzzy and hierarchical models for decision support in software systems implementation. In: Knowledge-Based Software Engineering. pp. 410–422 (2014)
18. Kravets, A.G., Salnikova N., Dmitrenko K., Lempert M.: Industrial cyber-physical systems: risks assessment and attacks modeling. In: Kravets, A.G., Bolshakov, A.A., Shcherbakov, M.V. (eds.). Cyber-Physical Systems: Industry 4.0 Challenges Studies in Systems, Decision and Control, vol. 260, pp. 197–210. (2020)
19. Nejmark, M.S., Cesarskij, L.G., Filimonyuk, L.Y.: Decision support model for aircraft entering the airport's area of responsibility. Sci. Techn. J. "Flight" 31–37 (2013)
20. Dagaeva, M., Katasev, A.: Fuzzy rules reduction in knowledge bases of decision support systems by objects state evaluation cyber-physical systems. In: Kravets, A.G., Bolshakov, A.A., Shcherbakov, M.V. (eds.). Cyber-Physical Systems: Modelling and Intelligent Control, Studies in Systems, Decision and Control, vol. 338, pp. 113–123. Springer, (2021). https://doi.org/10.1007/978-3-030-66077-2_9
21. Reece, C.: The safety risk management of unmanned aircraft systems. http://www.researchgate.net/publication/255853556_The_Safety_Risk_Manegement_of_Unmanned_Aircraft_Systems
22. James, T.L., Trefor, P.: Williams integrated decision support for aviation safety inspectors. Finite Elem. Anal. Des. **23**, 381–403 (1996)
23. Kulik, A.A., Bolshakov, A.A.: Algorithm of a device designed to support decision making to counter the threat of an aviation accident. Herald of the Bauman Moscow State Technical University. Ser. Instrum. Eng. **3**(136), 46–59 (2021)
24. Bazhanov, A., Vashchenko, R., Rubanov, V., Bazhanova, O.: Development of an automated system for monitoring and diagnostic a guided robotic vehicle. In: Kravets, A.G., Bolshakov, A.A., Shcherbakov, M.V. (eds.) Cyber-Physical Systems: Industry 4.0 Challenges, pp. 93–107. Springer (2019)

25. Bolshakov, A.A., Kulik, A.A., Sergushov, I.V., Skripal, E.N.: Intelligent method for assessing the threat of an accident. Bullet. Comput. Inf. Technol **5**(167), 3–9 (2018)
26. Kulik, A.A., Bolshakov, A.A., Veshneva, I.V.: Increasing the safety of flights with the use of mathematical model based on status functions. In: International Conference on Information Technologies ICIT 2019: Recent Research in Control Engineering and Decision Making Springer. 29 January 2019, pp. 608–621 (2019)
27. Markov, N., Bogomolov, A., Shishov, A., Dvornikov, M.: Information technology concept of integration of computing resources and physical processes in cyber-physical systems for personalized information about the potential danger of an emergency situation in high-altitude flight: In: Kravets, A.G., Bolshakov, A.A., Shcherbakov, M.V. (eds.) Cyber-Physical Systems: Modelling and Intelligent Control, Studies in Systems, Decision and Control 338, pp. 205–214. Springer (2021). https://doi.org/10.1007/978-3-030-66077-2_16

Flight Mode Recognition Algorithms that Provide Validation of the Digital Twin of an Aircraft During Flight Tests

Aleksey V. Bogomolov, Aleksandr A. Osipov, and A. S. Soldatov

Abstract The article deals with issues related to digital modeling and design of prototypes of aviation equipment based on virtual automated systems. The expediency of using digital modeling in the process of conducting flight tests as an alternative to a full-scale experiment is substantiated. The features and problematic issues of the use of digital modeling during the flight experiment are considered. The essence of the concept of "digital twin" and its place in the system of flight tests of aviation equipment is revealed. The methods of validation of digital doubles in the conditions of flight tests of aviation equipment are described. The structure of the on-board validation system is presented, the features of the relationship of its components are described, and the principles of functioning of its internal algorithms are considered. The analysis of the principles of the algorithms of event expectation and recognition of flight modes is carried out, their mathematical description is presented. The difference between static and dynamic flight modes, as well as discrete and continuous parameters controlled during the flight experiment is described. The principle of constructing an a priori event space that allows the recognition of dynamic flight modes is considered, and the corresponding mathematical calculations are given. Conclusions are drawn about the importance of both the validation process of digital doubles during flight tests and algorithms that provide recognition of flight modes.

Keywords Aircraft · Flight tests of aviation equipment · Digital twin · Digital model validation · Recognition algorithms

A. V. Bogomolov · A. A. Osipov (✉)
Federal Research Center «Computer Science and Management» of the Russian Academy of Sciences, 44 Building 2, Vavilova Str., 119333 Moscow, Russian Federation
e-mail: AlAndr2263@gmail.com

A. A. Osipov · A. S. Soldatov
Saint Petersburg Federal Research Center of the Russian Academy of Sciences, 39, 14th line V.O., Saint Petersburg 199178, Russian Federation

A. G. Kravets et al. (eds.), *Cyber-Physical Systems Engineering and Control*,
Studies in Systems, Decision and Control 477,
https://doi.org/10.1007/978-3-031-33159-6_6

1 Introduction

The process of creating aviation equipment begins with calculating the required characteristics and parameters, designing a prototype of aviation equipment and further testing it [1–3]. At modern industrial enterprises, design is performed in virtual automated systems CAD/CAM (Computer Aided Design/Computer Aided Manufacturing) and CAE (Computer Aided Engineering), which allow modeling the required characteristics and parameters of the developed aircraft without significant costs for creation and full-scale modeling [4–6]. Such design systems are capable of modeling not only the object under development, but also external conditions that affect the operation of the prototype, which makes it possible to predict the behavior of the developed aircraft under the influence of external factors and the environment [7–9]. Thus, the full-scale performance of some types of tests of aviation equipment samples is replaced by computer simulation of their operating modes in a virtual environment [10–12].

However, even highly intelligent computer-aided design systems that use the computing power of modern supercomputers in their work are not able to completely replace a full-scale experiment on a real object and comprehensively evaluate its properties and characteristics [13–15]. The number of external and internal influencing factors and the level of their influence on the operation of systems and assemblies of an aircraft flying in near-Earth airspace cannot be exhaustively estimated and modeled in virtual space due to the significant number and complexity of the interaction of these factors [16, 17].

In this regard, computer modeling of the operating modes of an aircraft model is carried out with a certain level of accuracy and plausibility, which is determined by validating the digital model of the object being developed (digital twin) [18–20]. The validation process involves determining the adequacy of the mathematical model describing the object and the correspondence of its properties to the real characteristics of the full-scale sample. Validation of digital twins of the developed models of aviation equipment can be performed at the stage of testing the aircraft and its components using a specialized on-board validation system that implements the necessary algorithms [21].

2 Materials and Methods

Computer modeling of samples of developed aviation equipment is associated with calculating the characteristics and parameters of the equipment and the aircraft itself, as well as with finding the dependencies of the parameters obtained as a result of the calculation on time, external factors, and the dependencies of one calculated parameter on another. The specified dependencies in mathematical and graphical representation can be used to provide validation of digital models of developed aviation equipment by comparing them with real characteristics and dependencies of identical parameters obtained during test flights under the flight test program.

To register the characteristics and parameters of aircraft units and equipment during a test flight, an on-board measurement system (SBI) is installed on board that is capable of measuring tens of thousands of physical quantities in real time, converting the measured values into analog-to-digital data streams that are convenient for storage and further analysis, recording data on solid-state data drives and, if necessary, transmit them to ground-based data processing equipment via telemetric communication channels [22–24].

Each test flight is performed in order to obtain quantitative and qualitative assessments of the specific characteristics and properties of the aircraft being tested. Methods for obtaining such estimates are strictly regulated by standard methods for performing test flights from a set of regulatory and technical documentation and involve performing a flight under certain operating modes of aircraft units and equipment.

Based on standard flight test procedures and programs, flight tasks are developed that define the conditions for performing a test flight and the operating modes of aircraft units and equipment (flight modes) that must be achieved by the crew during piloting. In terms of content, flight tasks can be of several types: special, complex, and typical. Depending on the complexity of the test flight and the content of the flight task, the crew of the aircraft during the flight itself determines the sequence of flight modes, which, as a rule, are the performance of several typical maneuvers with certain parameters of the operation of the equipment and aggregates of the aircraft.

The transition from one specified flight mode to another is largely associated with the performance of a large number of intermediate maneuvers that are not specified in the flight task and are not taken into account by the onboard measurement system installed on board during the evaluation of the parameters and characteristics of the aircraft being tested. However, intermediate flight modes are still recorded and recorded by the onboard measurement system, although they do not participate in subsequent data analysis. Registration of flight modes is nothing more than determining the dependence of the parameters of aircraft units and equipment on time, external factors, and each other. Thus, a significant amount of information recorded by the on-board measurement system during the test flight can be used to provide validation of the digital model of the developed aircraft model.

For this purpose, an on-board validation system (BSV) has been developed [21, 25, 26], which assumes its installation on board the test aircraft when performing test flights. The structure of the onboard validation system is shown in Fig. 1.

The onboard validation system includes databases of validation (DV) and test (DT) flight modes. The database of flight validation modes contains the characteristics and parameters of the developed model of aviation equipment and their dependencies in mathematical and graphical form, obtained as a result of computer modeling. The Test Flight Modes database contains the flight modes that you want to achieve when performing a particular test flight.

During the flight, the on-board validation system uses information from the on-board measurement system about the current parameters of the aircraft's units and

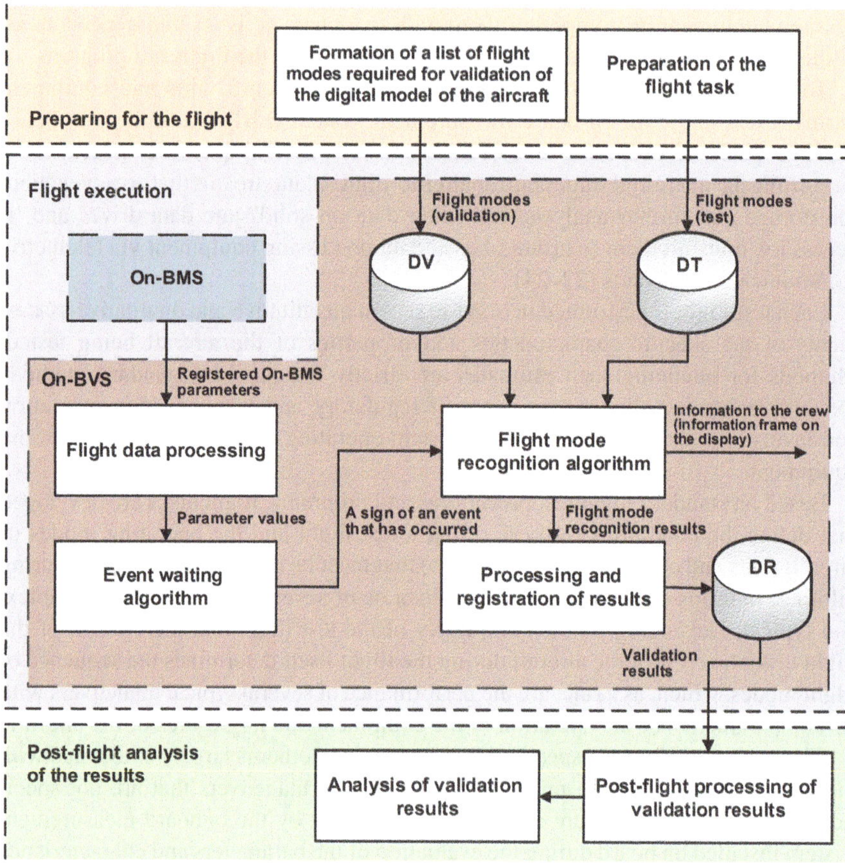

Fig. 1 Structure of the onboard validation system

equipment to compare it with the information contained in the database of flight validation modes. When registering identical dependencies, the on-board validation system makes a decision about the correspondence of the data obtained as a result of computer modeling and performing a real flight, and makes an entry about this in the results database (DR). The crucial rules and logic of information processing by the onboard validation system are as follows: internal algorithms for recognizing flight modes and waiting for events. At the same time, the on-board validation system can similarly compare the information received from the on-board measurement system with data from the database of test flight modes and perform a graphical display of the recorded parameters and dependencies on a multifunctional indicator to inform the crew about the achievement of a particular flight mode necessary for the successful completion of a test flight.

Thus, the on-board validation system provides validation of the digital model of the developed aircraft sample during the test flight and informs the aircraft crew about the achievement of the specified flight modes.

3 Results

In the process of flight mode recognition by the onboard validation system, there is a need to determine the boundaries of various types of events. Events are $c_1, c_2, c_3, \ldots$ understood as the facts of acceptance or non-acceptance by the parameters of the equipment and units of the aircraft, as well as external factors of the required values and values, and event boundaries are time indicators of the duration of the event itself or the process of waiting for the event to occur. It should be noted that the values of the recorded parameters, as well as time reports for determining the event boundaries, can be both continuous and discrete in nature [27–30].

The on-board validation system includes algorithms for waiting for events $F_{oi}(c_i), \quad i = (1, 2, \ldots)$, which ensure that the fact of the occurrence of an event is recorded, as well as determine its boundaries. Event waiting algorithms depend on an interconnected set of event occurrences, which is a vector consisting of a series of events $c = (c_1, c_2, c_3, \ldots)$.

Algorithms $F_{oi}(c_i), \quad i = (1, 2, \ldots)$ implemented in the onboard validation system are able to perceive 6 types of typical events: c_1—the time elapsed from the beginning of a certain operation or from a given fixed point in time; c_2—an event when the parameters of the aircraft's equipment and aggregates or external influencing factors begin to meet pre-set conditions; c_3—an event associated with the analysis of the state of discrete parameters, the values of which are determined by predefined functional dependencies; c_4—an event that characterizes the arrival of an initiative signal at the moment of an unexpected change in a discrete parameter; c_5, c_6—the beginning and end of a dynamic event, respectively. Thus, the event vector of algorithms $F_{oi}(c_i), \quad i = (1, 2, \ldots 6)$ takes the form $c = (\overline{c_1, c_6})$.

The algorithm that captures a typical event c_1 is based on the implementation of the procedure:

$$\tau \geq \tau_0 + \delta\tau$$

where τ_0—the specified initial time, $\delta\tau$—the time interval from the moment of occurrence, τ_0, τ—the real-time moment when checking the occurrence of an event c_1. Thus, the algorithm configuration vector $F_{o1}(c_1)$ will consist of two parameters and have the form $\alpha = (\tau_0, \delta\tau)$.

Monitoring the occurrence of the second type c_2 of event performed by the algorithm $F_{o2}(c_2)$ is described by the expression:

$$p \in \overline{p_n, p_v},$$

where p_n and p_v are the lower and upper limit values of the tolerance zone (the allowed range of the measured parameter value); p—the value of the controlled parameter at the time of checking the occurrence of the event. A parameter p can be understood as a parameter for setting up unified algorithms that form a parametric model of test technology; measured physical values of parameters; parameters of inter-module and inter-operational exchange; parameters that are quickly entered manually from video terminal devices and characterize non-measured parameters of the test object; calculated parameters of static and dynamic characteristics of aircraft equipment and aggregates, etc. $F_{o2}(c_2)$ in this case, the algorithm settings can be written as $\alpha = (p_n, \ p_v)$.

The third typical event c_3 is related to the analysis of the states of discrete parameters of equipment and aggregates of the test aircraft and is determined by algorithms of the form $F_{o3}(c_3)$ based on the procedure:

$$C(a_d) = \overline{C},$$

where $C(a_d)$ is the state of the discrete parameter corresponding to the position of the discrete equipment or unit with the identifier a_d during the next check; $\overline{C}$—the specified state of the monitored equipment or unit. The algorithm setup vector $F_{o3}(c_3)$ has the form$\alpha = (a_d, \overline{C})$.

Then, in accordance with previous expression, a typical event c_4 that characterizes the arrival of an initiative signal at the time of an unexpected change in the discrete parameter of the equipment or unit will be determined as follows:

$$C(a_i) = 1,$$

and be controlled by the algorithm $F_{o4}(c_4)$. Expression means the fact of registration of an initiative signal from the equipment or assembly of an aircraft with an identifier a_u. The algorithm tuning vector $F_{o4}(c_4)$ is a set of initiative signals of the form $a_u = (a_{u_1}, \ a_{u_2}, \ \ldots \ )$.

Typical events c_5, c_6 that determine the start and end of a dynamic event are determined by algorithms $F_{o5}(c_5)$, $F_{o6}(c_6)$ respectively. The construction of these algorithms is carried out taking into account fluctuations in the values of equipment and aggregates parameters caused by various interferences in the measurement channels and unstable physical processes occurring in the tested object of aviation equipment.

To ensure a high level of accuracy and plausibility of the values of fixed parameters in algorithms $F_{o5}(c_5)$, $F_{o6}(c_6)$ it is necessary to control the adoption of false decisions when gross measurement errors occur or when the value of the measured parameter exceeds the average value by a small amount of time.

Taking into account these comments, the analysis of the current values of the measured parameters and the algorithm's decision $F_{o5}(c_5)$ on the occurrence of an event c_5 should be made taking into account the nature of changes in a certain group of parameters of the tested object of aviation equipment. At the same time, the estimation of the nature of changes in each fixed parameter from this group should be based on a continuous comparative analysis of the current parameter values with the values of:

$$p_o = \frac{1}{k}\sum_{j=1}^{k} p_j, \sigma = \sqrt{\frac{1}{k-1}\sum_{j=1}^{k}(p_j - p_o)^2},$$

where p_o is the arithmetic mean, and σ—is the standard deviation of the parameter values measured in the current static mode, k is the number of measured parameters in the selected group.

Then, the event occurrence condition c_5 can be formulated as follows:

$$\vee j \in [\overline{j_1, j_2}] : ((\tau_{j_1} - \tau_{j_2} > \delta\tau) \wedge (p_j > p_o + K\sigma) \vee (p_j < p_o - K\sigma)).$$

The specified condition determines the one-way finding of all measured parameter values outside the area with boundaries $(p_o + K\sigma)$ and $(p_o - K\sigma)$ during the specified process time $\delta\tau$. In this case, the coefficient K is selected depending on the nature of the dynamic processes themselves and the processes of measuring the value p_j. However, while waiting for an event c_5 in the selected static mode, the parameter value may change, which may cause the condition specified in expression to be fulfilled before the event occurs c_5. In this regard, to eliminate possible errors, the values of the values of p_o they σ should be continuously updated.

The algorithm configuration vector $F_{o5}(c_5)$ can be expressed as follows $\alpha = (p, \ K)$, where p is the value of the measured parameter that determines the occurrence of the event c_5.

An event of this type c_6 determines the moment when the test object of aviation equipment passes from a dynamic state to a static one (the end of the test mode) and can be described by the following expression:

$$\vee j \in [\overline{j_1, j_2}] : ((\tau_{j_1} - \tau_{j_2} > \delta\tau) \wedge (|p_j - p_{j-1}|) \le K\sigma).$$

In this expression, the value $\delta\tau$ specifies the technological time during which the value of the measured parameter should fluctuate within no more than $K\sigma$ to determine the fact of completion of the dynamic state and transition to a static one, that is, the occurrence of an event c_6. The algorithm $F_{o6}(c_6)$ has settings that are identical to the algorithm settings $F_{o5}(c_5)$.

The considered algorithms $F_{o1}(c_1)$, $F_{o2}(c_2)$, $F_{o3}(c_3)$, $F_{o4}(c_4)$, $F_{o5}(c_5)$, $F_{o6}(c_6)$ provide determination of the fact of occurrence of one of $(\overline{c_1, c_6})$ the events and transfer of the specified information to the computational module of flight mode recognition algorithms.

Flight mode recognition algorithms provide comparison of parameters and their dependencies recorded by the on-board measurement system with values stored in the database of flight validation modes, in order to make a decision on whether the parameters of the current flight mode meet the requirements for validating the digital model of the aircraft being tested.

Analysis of the equipment and aggregates of the tested aircraft, as well as the technology of their testing, shows that to organize the process of recognizing the flight modes of an aircraft during flight tests, it is necessary to have a set of algorithms integrated into a computing module that allow solving the required recognition problems [31–33 and 34]. One of these tasks is the preliminary recognition of the subset to which the dynamic flight mode under study belongs, and the determination of the required measurement and parameter registration characteristics for it. To solve this problem, an algorithm is used $F_{p1}(\alpha)$, where α is the vector of parameters for configuring the algorithm operation.

The following algorithm $F_{p2}(\alpha)$ performs the final recognition of dynamic flight mode in order to find the necessary processing algorithms. In this case, the algorithm is tested after the measurement process is completed, using all the measurement information. The need to separate the recognition process of flight modes into preliminary and final ones is associated with strict time constraints on the recognition process and insufficient amount of a priori information at the beginning of the recognition process. At the same time, two-level recognition should be used only for those modes whose start is not associated with the arrival of initiative signals and whose recognition requires a comparative analysis of the measured and calculated parameters.

To recognize dynamic flight modes, the beginning of which is associated with changes in the discrete states of certain units and equipment of the test aircraft, an algorithm is used $F_{p3}(\alpha)$. The algorithm $F_{p4}(\alpha)$ provides recognition of static modes of operation of equipment or units of the tested aircraft.

Algorithms $F_{p1}(\alpha)$, $F_{p2}(\alpha)$ should provide recognition of those dynamic modes, the beginning of which is characterized only by a change in the values of analog parameters. At the same time, at the first level, the algorithm $F_{p1}(\alpha)$ performs preliminary recognition of the mode to the subset level m_r, where the identifiers $r_{\partial 1}, r_{\partial 2}, \ldots$ of modes that are identical in the list of measured parameters with the necessary measurement discreteness are taken as elements of the subset. The final recognition to the form of the dynamic mode name required for further processing and formalization is performed by the algorithm $F_{p2}(\alpha)$ at the second level of the end of the parameter measurement process.

In the process of recognition of flight modes by algorithms $F_{p1}(\alpha)$, $F_{p2}(\alpha)$, using the Bayes formula, probabilities are calculated $P(r_j|K)$ that the considered dynamic mode belongs to the subset m_r for the first level of recognition or is a mode $r_{d1}, r_{d2}, \ldots$ for the second level of recognition. The probability $P(r_j|K)$ depends on a set of attributes (events) $K = (k_1, k_2, \ldots)$ the appearance r_j of the operating mode, which includes information about the identifiers r_{cj}, directions h_j, and rates υ_j of change of the mode parameter for the initial and intermediate steady-state modes. Thus, the vector $K = (k_1, k_2, \ldots)$ is an a priori feature space (events) (Fig. 2).

In this case, the selection of the desired subset or mode is carried out according to the decisive rule:

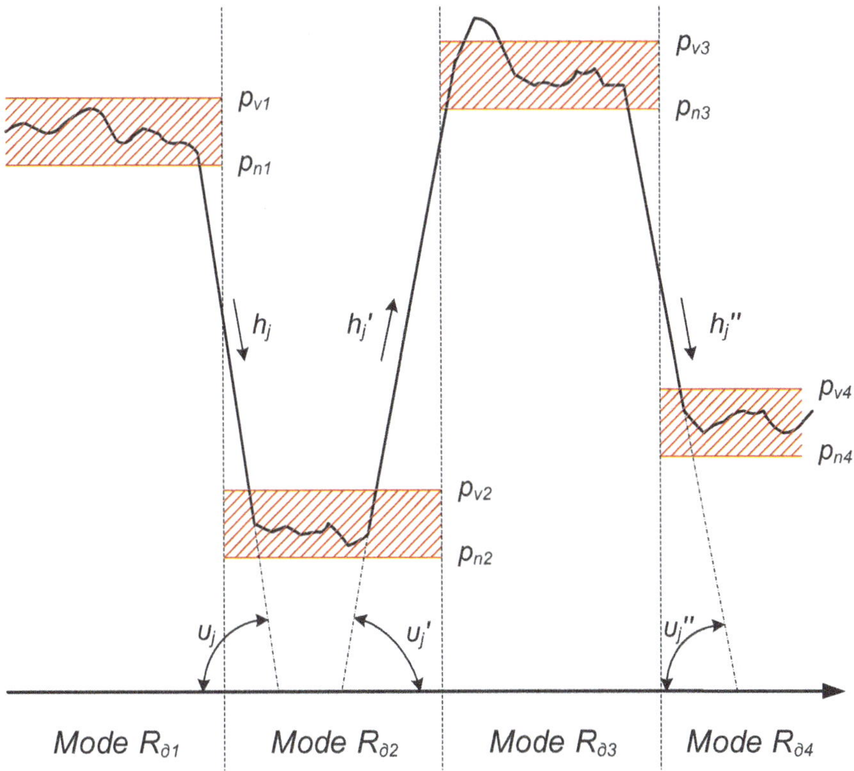

Fig. 2 Construction of a priori feature space for dynamic mode recognition

$$K \in r_j,\ \ P(r_j|K) > P(r_i|K),\ (i, j = 1,\ 2,\ \ldots)$$

In other words, the subset of modes or the dynamic mode itself is selected, for which the calculations resulted in the highest a posteriori probability value $P(r_j|K)$.

The algorithm adjustment vector $F_{p1}(\alpha)$ can be represented $\alpha = (\alpha_1, \alpha_2, \ldots \alpha_5)$ as: $\alpha_1 = p$—the identifier of the measured parameter, which is recognized by the nature of its change; $\alpha_2 = (m_{r1}, m_{r2}, \ldots)$—a subset of dynamic modes; $\alpha_3 = (p(m_{r1}), p(m_{r2}), \ldots)$—a set of a posteriori probabilities of occurrence of the corresponding subsets; $\alpha_4 = (p(k_1|m_{r1}), p(k_1|m_{r2}), \ldots, p(K_2|m_{r1}), p(k_1|m_{r2}), \ldots)$—a set of probabilities of occurrence of a feature k_j for a subset m_{rj} of dynamic modes; $\alpha_5 = (m_{r1}, b_1, m_{r2}, b_2, \ldots)$—a set of binary relations (m_{rj}, b_j) between subsets m_{rj} and characteristics b_j of the measurement process parameters in dynamic modes included in these subsets.

$F_{p2}(\alpha)$ the final recognition algorithm is configured using the configuration vector$\alpha = (\alpha_1, \alpha_2, \ldots \alpha_5)$, where the elements $\alpha_1, \ldots, \alpha_4$ contain information similar to the corresponding elements of the algorithm configuration vector $F_{p1}(\alpha)$ applied to the dynamic modes themselves. This parameter $\alpha_5 = (r_{d1}, r_1, r_{d2}, r_2, \ldots)$ contains a set of binary relationships (r_{dj}, r_j) between dynamic mode identifiers r_{dj} and characteristics r_j of the measurement results processing process.

The algorithms $F_{p2}(\alpha)$, $F_{p4}(\alpha)$ and $F_{p1}(\alpha)$, $F_{p2}(\alpha)$ are based on the use of probabilistic and logical methods for recognizing flight modes, and the description of the decision rules of these algorithms is reduced to an expression.

4 Conclusion

The use of digital twins of aircraft equipment and assemblies, which are widely used at present in their development and design, can significantly reduce the time and material costs of developing new products, but in its current form it cannot fully replace full-scale modeling and flight testing on real aircraft samples due to the insufficient level of plausibility and compliance of digital twins full-scale samples.

Validation of digital duplicates of aircraft equipment and units during flight tests helps to bring the accuracy of compliance of digital models with real analogues to the level required for certain types of tests (about 20–40% of the total content of the test program). Without using a full-scale experiment, as well as improve the quality of computer-aided design and development of prototypes using CAD/CAM systems by adjusting digital doubles based on validation results.

A significant role in this process is played by a promising on-board validation system and the algorithms used in its work for waiting for events and recognizing flight modes, the quality of which directly depends on the level of compliance of digital twins with full-scale samples of equipment and units of the developed aircraft and the aircraft as a whole.

Acknowledgements This research was supported by the grant of the President of the Russian Federation for leading scientific schools of the Russian Federation NSh 122.2022.1.6.

References

1. Krenitsky, E.I., Aleksandrovskaya, L.N., Melnikov, V.S., Maksimov, N.A.: Fundamentals of aircraft testing. In: Krinetsky, E.I. (ed.) Mashinostroenie. Moscow (1989)
2. Bolshakov, A.A., Kulik, A.A.: Algorithm of a device designed to support decision making to counter the threat of an aviation accident. In: Herald of the Bauman Moscow State Technical University. Ser. Instrum. Eng. **3**(136), 46–59 (2021)
3. Kizim, A.V., Kravets, A.G.: On systemological approach to intelligent decision-making support in industrial cyber-physical systems. Stud. Syst. Decis. Control **260**, 167–183 (2020)
4. Ezhilarasu, C.M., Skaf, Z., Jennions, I.K.: A generalised methodology for the diagnosis of aircraft systems. IEEE Access **9**, 11437–11454 (2021). https://doi.org/10.1109/ACCESS.2021.3050877

5. Lan, Y., Zhao, Z., Wang, J.Y., Shan, D., Yang, X.W., Cui, M.X.: Application of ICD management software in aircraft design and test. In: CSAA/IET International Conference on Aircraft Utility Systems (AUS 2022), pp. 563–568 (2022). https://doi.org/10.1049/icp.2022.1591
6. Potryasaev, S.A., Ronzhin, A.L., Sokolov, B.V., Zhao, VYu.D., Stepanov, P.V., Styskin, M.M.: Polymodel complex of a mobile service system designed to service aircraft. Inform. Commun. **6**, 113–118 (2020)
7. Larkin, E.V., Akimenko, T.A., Bogomolov, A.V.: Modeling the reliability of the onboard equipment of a mobile robot. News of the Saratov University. New series. Series: Mathematics. Mech. Inform. **21**(3), 390–399 (2021). https://doi.org/10.18500/1816-9791-2021-21-3-390-399
8. Shestakov, A.L., Zagrebina, S.A., Manakova, N.A., Sagadeeva, M.A., Sviridyuk, G.A.: Numerical optimal measurement algorithm under distortions caused by inertia, resonances, and sensor degradation. Autom. Remote Control **82**(1), 41–50 (2021). https://doi.org/10.31857/S0005231021010025
9. Pin, S., Belijar, G., Fetouhi, L., Somer, L., Van de Steen, C., Albert, L.: Ageing study in aircraft electromechanical chain: systems modeling for property evolution monitoring. In: IEEE Electrical Insulation Conference (EIC), pp. 79–84 (2020). https://doi.org/10.1109/EIC47619.2020.9158659
10. Larkin, E., Bogomolov, A., Antonov, M.: Modeling of increased rigidity of industrial manipulator. In: Interactive Collaborative Robotics: 5th International Conference, ICR 2020, St Petersburg, Russia, pp. 170–178 (2020). https://doi.org/10.1007/978-3-030-60337-3_17
11. Kornev, A.V., Kozelkov, A.S.: Application of domestic supercomputer technologies to create advanced models of aviation equipment. Modern Inform. Technol. IT Educ. **17**(2), 250–264 (2021)
12. Strelets, D.Y., Serebryansky, S.A., Shkurin, M.V.: Concept of creation of a digital twin in the uniform information environment of product life cycle. In: 13th International Conference "Management of large-scale system development" (MLSD), pp. 1–4 (2020). https://doi.org/10.1109/MLSD49919.2020.9247749.
13. Styskin, M.M., Stepanov, P.V., Zheltov, S.Y., Sokolov, B.V., Ronzhin, A.L.: Means of optical and radio frequency identification in the technological process of automated control of the turnover of mobile onboard equipment. Model. Optim. Inform. Technol. **10**(1), 26 (2022). https://doi.org/10.26102/2310-6018/2022.36.1.003
14. Shestakov, A.L., Zamyshlyaeva, A.A., Manakova, N.A., Sviridyuk, G.A., Keller, A.V.: Reconstruction of a dynamically distorted signal based on the theory of optimal dynamic measurements. Autom. Remote Control **82**(12), 2143–2154 (2021). https://doi.org/10.31857/S0005231021120084
15. Larkin, E., Bogomolov, A., Privalov, A.: Data buffering in information-measuring system. In: 2nd International Ural Conference on Measurements (UralCon), pp. 118–123 (2017). https://doi.org/10.1109/uralcon.2017.8120697
16. Chernov, A.V., Kozelkov, A.S., Shagaliev, R.M., Strelec, D.Y.: The current system of requirements in the field of virtual testing of samples of created products of aviation equipment. Qual. Life **4**(28), 61–71 (2020)
17. Fedorov, M.V., Rudakov, S.V., Podorozhnyak, A.A., Logachev, S.V.: Method for optimizing the measurement vector of an inertial-satellite control system based on radial basic artificial neural networks. Inform. Meas. Control Syst. **7**(5), 67–74 (2009)
18. Zhao, J., Chang, K., Li, S., Xie, Y.: Construction and application of a digital twin-based aircraft landing gear management environment. In: CSAA/IET International Conference on Aircraft Utility Systems (AUS 2022), pp. 440–443 (2022). https://doi.org/10.1049/icp.2022.1569
19. Wang, Y.: Cooperative localization and association of commercial-off-the-shelf sensors in three-dimensional aircraft cabin. IEEE Trans. Autom. Sci. Eng. **19**(4), 3508–3519 (2022). https://doi.org/10.1109/TASE.2021.3124207
20. Golosovskiy, M., Bogomolov, A.V., Balandov, M.: A model of continuous integration and deployment of engineering software. In: Silhavy, R., Silhavy, P., Prokopova, Z. (eds.) Data Science and Intelligent Systems, vol. 231, pp. 83–97. Springer, Cham, (2021). https://doi.org/10.1007/978-3-030-90321-3_64

21. Soldatov, A.S.: The concept of creating a promising information-measuring system for flight tests of aviation equipment using Industry 4.0 technologies. Aerospace Forces. Theory Pract. **20**, 167–177 (2021)
22. Adgamov, R.I., Berkheev, M.M., Zalyaev, I.A.: Automated testing in aircraft construction. Mashinostroenie, Moscow (1989)
23. Gu, J.: Dynamic measurement and data calibration for aerial mobile IoT. IEEE Internet Things J. **7**(6), 5210–5219 (2020). https://doi.org/10.1109/JIOT.2020.2977910
24. Makarenko, V.G., Bogomolov, A.V., Rudakov, S.V., Podorozhnyak, A.A.: Technology for building an inertial-satellite navigation system for vehicles control with neural network optimization of the composition of the measurement vector. Mechatron. Automation Control **1**, 39–44 (2007)
25. Ronzhin, A.L., Soldatov, A.S.: Information system for validation of an aircraft digital twin in flight tests. In: Polytransport Systems: Proceedings of the XII International Scientific and Technical Conference. Novosibirsk, -pp. 117–123 (2022)
26. Bychkov, E.V., Zagrebina, S.A., Zamyshlyaeva, A.A., Keller, A.V., Manakova, N.A., Sagadeeva, M.A., Sviridyuk, G.A.: Development of the theory of optimal dynamic measurements. Bulletin of the South Ural State University. Ser. Math. Model. Prog. **15**(3), 19–33 (2022). https://doi.org/10.14529/mmp220302
27. Larkin, E., Bogomolov, A., Gorbachev, D., Privalov, A.: About approach of the transactions flow to poisson one in robot control systems. In: Interactive Collaborative Robotics, pp. 113–122 (2017). https://doi.org/10.1007/978-3-319-66471-2_13
28. Larkin, E., Bogomolov, A., Privalov, A.: Estimation of events flow generated with ergodic semi-Markov processes. In: 2nd International Ural Conference on Measurements (UralCon), 124–129 (2017). https://doi.org/10.1109/uralcon.2017.8120698.
29. Bychkov, E., Sviridyuk, G., Bogomolov, A.: Optimal control for solutions to Sobolev stochastic equations. Electr. J. Diff. Equ. **51**, 1–11 (2021). http://ejde.math.txstate.edu
30. Golosovskiy, M.S., Bogomolov, A.V., Evtushenko, E.V.: An algorithm for setting sugeno-type fuzzy inference systems. Autom. Document. Math. Ling. **55**(3), 79–88 (2021). https://doi.org/10.3103/S000510552103002X
31. Tobin, D., Bogomolov, A., Golosovskiy, M.: Model of organization of software testing for cyber-physical systems. In: Kravets, A.G., Bolshakov, A.A., Shcherbakov, M.V. (eds.) Cyber-Physical Systems: Modelling and Industrial Application. Studies in Systems, Decision and Control, vol. 418, pp. 51–60. Springer, Cham (2022). https://doi.org/10.1007/978-3-030-95120-7_5
32. Soldatov, A., Scherbakov, I., Osipov, A.: Cyber-physical approach to building a flight experiment control system. In: Silhavy, R. (ed.) Data Science and Algorithms in Systems, vol. 597, pp. 495–507. Springer, Cham, LNNS. (2023). https://doi.org/10.1007/978-3-031-21438-7_39
33. Balyk, O., Zolotaeva, M., Bogomolov, A., Soldatov, A.: Cyber-physical test facility for certification of robotic unmanned aerial systems. In: Silhavy, R. (ed.) Software Engineering Application in Systems Design, vol. 596, pp. 385–396. Springer, Cham, LNNS (2023). https://doi.org/10.1007/978-3-031-21435-6_33

Tracked Robot Motion Control System

M. J. Almashaal, A. R. Gaiduk, and S. G. Kapustyan

Abstract The task of finding a suitable control system for nonlinear systems is a difficult task. The great reliance on robots, especially mobile robots these days, which led to an increase in interest in studying these robots. Despite the multiplicity of research and a large number of researchers in the field of robotics, the problem of finding a mathematical model for the robot in addition to finding an appropriate control system is still a problem under discussion, since robots in general and mobile robots, in particular, are considered non-linear systems. Development of the mathematical model of the small-size tracked mobile robot and also the algorithm of control system design by its movement on the trajectory set by the program in time is considered. The design of the control system is performed by the algebraic polynomial-matrix method with the use of the quasilinear model. It's shown that the use of the algebraic polynomial-matrix method helps to find an appropriate control system that makes a nonlinear system stable, this is obvious by finding a useful control system of the studied tracked robots in this chapter. The research of the designed system is conducted by its computer simulation.

Keywords Tracked robot · Mathematical model · Movement · Turn · Control system · Design

1 Introduction

The expression "mobile robots" is mainly used to refer to robots that can move their own base with some sort of locomotion mechanism. The main advantage of mobile robots over stationary manipulators is the almost unlimited working space

M. J. Almashaal (✉) · A. R. Gaiduk
Southern Federal University, 44 Nekrasovsky, Taganrog 347928, Russia
e-mail: almashaal.jalal@gmail.com

S. G. Kapustyan
Southern Scientific Center of the Russian Academy of Sciences, 41 Chekhov, Rostov-On-Don 344006, Russia

A. G. Kravets et al. (eds.), *Cyber-Physical Systems Engineering and Control*, Studies in Systems, Decision and Control 477,
https://doi.org/10.1007/978-3-031-33159-6_7

[1]. These robots have already migrated from closed research centers and innovation fairs to the reality of everyday life. Mobile robots are used to supply warehouses, autonomous mowers in home gardens, smart vacuum cleaners in homes, remote-controlled robots, or bomb-disposal robots are becoming not only a common piece of company equipment but also a product purchased by the average consumer. Ground mobile robots can be classified according to a number of criteria, and the systems they have built differ in types of locomotion. These can be wheeled, walking, tracked, or balancing robots [2]. Modern field robotics is dominated by two main mobility systems: wheels (cylindrical, spherical and omnidirectional [1, 2]) and tracks. In the field, the choice of a particular mobility system depends on the terrain where the robot must work. Alternatively, tracks are more effective on rough terrain, including soft surfaces such as mud and sand, where the robot's load is distributed over a larger area of contact with the ground [3]. Tracked robots can be classified as a compromise between wheeled and walking robots. Tracks provide mobility for the robot in difficult terrain and help it maintain a satisfactory speed of movement. This type of locomotion is recognized, researched, and used more largely than walking systems [4].

At present, intensive research is underway to create motion control systems for mobile robots. The problem is complicated by the fact that the robot is unmanned and its mathematical model is nonlinear [5–7]. The problem of control systems design for such nonlinear objects is still in the research stage. To resolve it, several approaches have been proposed: the feedback linearization method [8], the pacification method [9–11], the backstepping method [12], the method Jordan controlled form [7], the quasilinear approach [13, 14], and position-trajectory method [15, 16]. When using each of the methods, it is first necessary to transform the original nonlinear objects belonging to a certain special form [7]. For example, the method step-by-step can be applied if the control object's equations are transformed to the triangular type. With such a check of the conditions, one can construct an appropriate transformation or construct this transformation comparable in complexity [17] with the construction of the desired nonlinear control.

One of the simplest methods in this regard is the algebraic polynomial-matrix method using quasilinear models [18]. For the existence of these models, only the differentiability of the nonlinearities of the control object with respect to all arguments is necessary, and the construction of a quasilinear model consists in taking partial derivatives and integrating them with respect to an auxiliary variable [7].

The purpose of this chapter is to synthesize a non-linear motion control system for a tracked robot, by use of the algebraic polynomial-matrix (APM) method with the application of the quasi-linear model [18].

This chapter consists of 5 parts. The first is an introduction, the second one definition of the APM method, the third part is devoted to the creation of the mathematical model of the movement of the tracked robot and the design of the robot control system by the APM method, the fourth part is a computer simulation of the designed system and the last part is the conclusion.

2 Algebraic Polynomial-Matrix Method

This method is used for the design of a control system of linear or nonlinear objects of arbitrary order, which are given as quasilinear models and have the form

$$\dot{x} = A(x)x + b(x)u,\ y = cT(x)x \tag{1}$$

In this case, the desired control model is given in the form

$$u = u(g, x) = k_0(x)g - k^T(x)x = k_0(x)g - [k_1(x)x_1 + k_2(x)x_2 + \ldots + k_n(x)x_n] \tag{2}$$

where $k_i(x)$ are the coefficients of the matrix-row $k^T(x)$ calculated during the synthesis, which are the feedback coefficients on state variables x_i in a closed system, $i = \overline{1,\ n}$ $g = g(t)$ and is the setting action. Usually $g(t) = g_0 * 1(t)$ [19].

Substituting (2) into (1) shows that the mathematical model of a closed system will also have the form corresponding to the structure of a quasilinear model, i.e.

$$\dot{x} = D(x)x + k_0(x)b(x)g \tag{3}$$

where D(x) is also a functional matrix of the same dimension as A(x). The structure of this matrix, as follows from the substitution, is determined by the following expression:

$$D(x) = A(x) - b(x) \cdot kT(x) \tag{4}$$

For determining $k_i(x)$ there are polynomials that have the form

$$A(p, x) = \det(pE - A) = p^n + \alpha_{n-1}(x)p^{n-1} + \ldots + \alpha_1(x)p + \alpha_0(x) \tag{5}$$

$$V_i(p, x) = e_i \cdot \mathrm{adj}(pE - A(x)) \cdot b(x) = v_{i,n-1}p^{n-1}(x) + \ldots + v_{i,1}(x)p + v_{i,0}(x) \tag{6}$$

In these expressions $\alpha_i(x)$ are the coefficients of the characteristic polynomial of the matrix $A(x)$ of the control object (10); e_i- i -th row of the identity $n \times n$ matrix E; $\mathrm{adj}(pE - A(x))$ is the associated matrix for $(pE - A(x))$; $v_{ij}(x)$ are the coefficients of the polynomials $V_i(p, x)$, $i = \overline{1,\ n}$.

A polynomial is formed as;

$$D^*(p) = \prod_{i=1}^{n}(p - p_i^*) = p^n + \delta_{n-1}p^{n-1} + \delta_{n-2}p^{n-2} + \ldots. + \delta_1 p + \delta_0 \tag{7}$$

where δ_1 *and* δ_0 are any positive numbers on which the duration of the transient process depends. Next, the difference of polynomials $D^*(p)$ and A(p,x) and its coefficients:

$$R(p,x) = D^*(p) - A(p,x) = \overline{\delta}_{n-1}p^{n-1} + \overline{\delta}_{n-2}(x)p^{n-2} + \ldots. + \overline{\delta}_1(x)p + \overline{\delta}_0(x) \tag{8}$$

A system of algebraic equations is compiled:

$$\begin{bmatrix} v_{1,0} & v_{2,0} & \cdots & v_{n,0} \\ v_{1,1} & v_{2,1} & \cdots & v_{n,1} \\ \vdots & \vdots & \vdots & \vdots \\ v_{1,n-1} & v_{2,n-1} & \cdots & v_{n,n-1} \end{bmatrix} \cdot \begin{bmatrix} k_1 \\ k_2 \\ \vdots \\ k_n \end{bmatrix} = \begin{bmatrix} \overline{\delta}_0 \\ \overline{\delta}_1 \\ \vdots \\ \overline{\delta}_{n-1} \end{bmatrix} \tag{9}$$

The matrix D(x) is calculated by using expression (4). The coefficient $k_i(x)$ from (2) is found with condition $y^o = g_0$ by the formula

$$k_0(x^\circ) = \frac{-1}{c^T(x^\circ)D^{-1}(x^\circ)b(x^\circ)} \tag{10}$$

where $x^\circ = limx(t)_{t\to\infty}$

3 Rectilinear Movement of the Tracked Robot

Consider one tracked robot and its drive, consisting of a DC motor and a gearbox. Engine equations have the form [20]:

$$U_p = R_m I + L\dot{I} + c_e\omega_p \tag{11}$$

$$M_\partial = c_m I \tag{12}$$

where ω_p is the angular velocity of rotation of the rotor; I is motor armature winding current; c is EMF motor constant, U_p is armature voltage, R_m, L are resistance and inductance of the armature winding; M_d is torque; c_m is the moment constant.

As is known, when moving without slippage, the angular velocity of the caterpillar drive wheel is given as $\omega_k = V/r$, where V is the linear speed of the robot, and r is the radius of the caterpillar drive wheel. At the same time, the speeds of the engine rotor and the caterpillar drive wheel are related by the ratio $\omega_k = \omega_p/N$ [21], i.e., where $\omega_p = N\omega_k = NV/r$. Substituting ω_p into (11), we get:

$$L\dot{I} = U_p - R_m I - (c_e N/r)V \tag{13}$$

The equation of rectilinear motion of the robot:

$$0,5m\dot{V} = F_t - C_v^{'}|V|V \tag{14}$$

where m is the mass of the robot; F_t is the traction force developed by one caterpillar and applied to the body of the robot, $and C_v^{'}|V|V$ is the force of viscous resistance to the movement of the caterpillar on the surface without slipping. The force F_t is equal to the torque M_t applied to the drive wheel of the caterpillar, divided by the radius of this wheel, i.e., $F_t = M_t/r$. As is known, the torque on the output shaft of the gearbox $M_t = M_\partial N$, where N is the gear ratio of the gearbox. Taking into account (12), we have $F_t = c_m NI/r$. Then Eqs. (13) and (14) take the form:

$$V = a_v I - C_v|V|V, \dot{I} = L^{-1}U_p - L^{-1}R_m I - ((c_e N)/Lr)V \tag{15}$$

where $a_v = 2c_m N/mr; C_v = 2C_v^{'}/m$. Equations (15) describe all the movements of the system both in the transient and in the steady state relative to the total variables of the control object–engine-caterpillar-robot.

When solving the design problem of the motion control system of a tracked robot (15), it is supposed to use the algebraic polynomial-matrix method (APM) [19]. To apply this method, the equations are needed in deviations from the steady state of the control object, in this case, the robot. We write these equations based on (16); introducing the notation for the steady-state values: $V°$, $U°$, $I°$, we set $V = V° + x_1$,

$I = I° + x_2, U = U° + u$, where x_1, x_2, and u are the deviations of the corresponding variables. To simplify conversions, we will assume that $V° + \Delta_V > 0$. Taking into account in (15) the equalities $a_v I_0 = C_v V^{\circ 2}$ and $R_m I_0 + (c_e N/r)V° = U°$, which follow from Eqs. (15) corresponding to the steady motion of the robot (at$\dot{V} = \dot{I} = 0$), we obtain.

$$\dot{x}_1 = a_v x_2 - C_v(2V° x_1 + x_1^2),$$

$$\dot{x}_2 = -L^{-1}R_m\, x_2 - (c_e N/Lr)x_1 + L^{-1}u \tag{16}$$

In vector–matrix form, these equations have the form.

$$\dot{x} = \begin{bmatrix} -C_v(2V° + x_1) & a_v \\ -c_e N/Lr & -L^{-1}R_m \end{bmatrix} x + \begin{bmatrix} 0 \\ L^{-1} \end{bmatrix} u,\ y = x_1 \tag{17}$$

Equations (16), and (17) describe the effect of one caterpillar on the robot. The equations of thc other caterpillar have a completely similar form. In this case, the numerical values of the coefficients in Eqs. (15) are: $V° = 0, 1$ m/s; $C_v = 0, 132$ 1/m, $a_v = 0.0486$ 1/sec; $L = 0, 0123$ gn; $c_e N/Lr = 11, 106$ A/m; $L^{-1}R_m = 25, 543$ 1/sec For these values, the quasilinear model (17) has the form:

$$\dot{x} = \begin{bmatrix} -0,0264 - 0,132x_1 & 0.0486 \\ -11,10558 & -25,54288 \end{bmatrix} x + \begin{bmatrix} 0 \\ 81,3 \end{bmatrix} u,\ y = x_1 \tag{18}$$

Turning to the synthesis of the control system for the rectilinear movement of a caterpillar robot by the APM method, we note that it is applied to a quasi-linear model of control objects, which in the general case has the form (1) In this case [17]:

$$A(x) = \begin{bmatrix} a_{11}(x)\ a_{12} \\ a_{21} \quad a_{22} \end{bmatrix},\ b(x) = \begin{bmatrix} 0 \\ b_2 \end{bmatrix},\ y = \begin{bmatrix} 1\ 0 \end{bmatrix} x \tag{19}$$

Here and below, it is taken into account that in (13), (14) $n = 2$, a_{11} depends on x, and $b_1(x) = 0 c_1(x) = 1, c_2(x) = 0$.

Since in this case $n = 2$, the control action can be found using (2) by the expression:

$$u(x) = k_g(x)g - k_1(x)x_1 - k_2(x)x_2 \tag{20}$$

First of all, the controllability condition is checked $\det U(x) \neq 0$; here $U(x) = [\,b(x)\ A(x)b(x)\,]$. As applied to Eq. (18), we have $\det U(x) = -0,0486 \neq 0$. Therefore, object (18) is controllable, i.e. the design problem has a solution.

Following the APM method and taking into account that n = 2, we determine the polynomials (5) and (6) as:

$$A(p, x) = p^2 + (0.132x_1 + 25.56928)p + (3.37166x_1 + 1.21406) \tag{21}$$

$$V_1 = \begin{bmatrix} 1\ 0 \end{bmatrix} \cdot \begin{bmatrix} p - a_{22} & a_{12} \\ a_{21} & p - a_{11}(x) \end{bmatrix} \cdot \begin{bmatrix} 0 \\ b_2 \end{bmatrix} = 3,95118 \tag{22}$$

$$V_2 = \begin{bmatrix} 0\ 1 \end{bmatrix} \cdot \begin{bmatrix} p - a_{22} & a_{12} \\ a_{21} & p - a_{11}(x) \end{bmatrix} \cdot \begin{bmatrix} 0 \\ b_2 \end{bmatrix} = 81,3p + 2.14632 + 10.7316\text{x}_1 \tag{23}$$

Further, the n-order Hurwitz polynomial (7) is formed. In this case:

$$D^*(p) = \prod_{i=1}^{2} (p - p_i^*) = p^2 + \delta_1 p + \delta_0 \tag{24}$$

Then, using (21) and (24), the polynomial is determined

$$R(p, x) = [\delta_1 - 0.132x_1 - 25.56928]p + \delta_0 - (3.37166x_1 + 1.21406) \tag{25}$$

The coefficients $k_1(x)$ and $k_2(x)$ from (20) are determined by solving the system (9) by substituting Eqs. (21), (22) and (25) in (10) will have the system.

$$\begin{bmatrix} 3,95118\ 2.14632 + 10.7316_1 \\ 0 \qquad\qquad 81.3 \end{bmatrix} \begin{bmatrix} k_1 \\ k_2 \end{bmatrix}$$

$$= \begin{bmatrix} \delta_0 - (3.37166x_1 + 1.21406) \\ \delta_1 - (0.132x_1 + 25.56928) \end{bmatrix} \tag{26}$$

The solution of this system gives the expressions:

$$k_2(x) = \frac{-(\delta_1 - (0.132x_1 + 25.56928))}{81,3} \tag{27}$$

$$k_1(x) = \frac{-\delta_0 + 1.21406 + 3.37166x_1 - (2.17884 + 10.7316x_1)k_2}{-3.951} \tag{28}$$

The coefficient $k_g(x)$ from (10) in this case is determined by the expression [17]

$$k_g(x) = -[c^T(x^\circ)D^{-1}(x^\circ)b(x^\circ)]^{-1} = \delta_0/0,0486 \times 81,3 \tag{29}$$

where x° is the state vector of the system in steady state, i.e. at $\dot{x}^\circ_1(t) = 0$?

Calculation of the parameters of the control system for the rectilinear motion of a caterpillar robot (18) was performed at $\delta_1 = 8$, $\delta_0 = 15$. Substituting these values, as well as the values from (18) into expressions (21)–(29), we find the coefficients from control (20):

$$k_1(x) = \frac{293,205144 - 21,58477x_1 + 0,35843x_1^2}{81,3} \tag{30}$$

$$k_2(x) = \frac{17,5693 + 0,132x_1}{81,3} \tag{31}$$

$$k_g(x) = \frac{15}{0,0486 \cdot 81,3} \tag{32}$$

Substituting formulas (30–33) into this expression (20), will get the control system

$$u(x) = \left(\frac{15}{0,0486 \cdot 81,3}\right)g - \left(\frac{293,205144 - 21,58477x_1 + 0,35843x_1^2}{81,3}\right)x_1 - \left(\frac{17,5693 + 0,132x_1}{81,3}\right)x_2 \tag{33}$$

The matrix D (x) of the closed system (18), (20) obviously has the form

$$D(x) = \begin{bmatrix} -(0,0264 + 0,132x_1) & 0.0486 \\ -304.31072 + 21.58477x_1 - 0.35843x_1^2 & -8 - 0.132x_1 \end{bmatrix} \tag{34}$$

To make sure that the characteristic polynomial of the closed system is equal to the polynomial $D^*(p)$, we calculate the polynomial

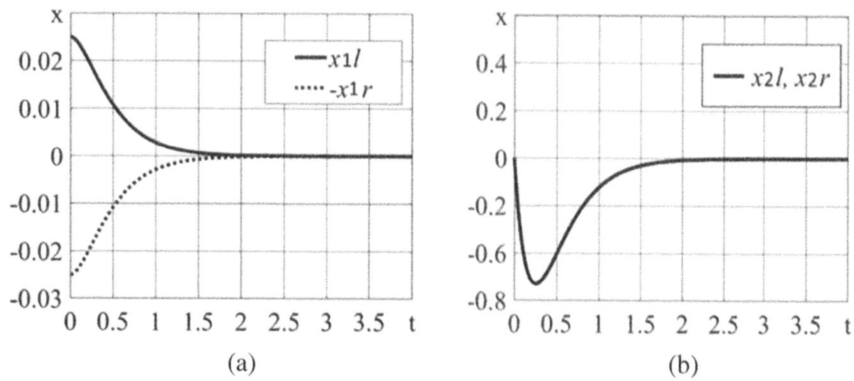

Fig. 1 Transient processes of the control system: **a** changes in speed deviation; **b** changes in motor winding currents

$$D(p,x) = det(pE - D(x)) == p^2 + (-a_{11}(x) - a_{22} + \delta_1 + a_{11}(x) + a_{22})p + (a_{11}(x)\cdot a_{22} - a_{12}a_{21} + \delta_0 - a_{11}(x)\cdot a_{22} + a_{12}\cdot a_{21}) \quad (35)$$

Substituting formulas (34) into this expression (31), taking into account (21), will have

$$D(p,x) = p^2 + \delta_1 p + \delta_0 = D^*(p) \quad (36)$$

To assess the quality of the synthesized control system (18), (19), and (33) for the rectilinear movement of the tracked robot, it was modeled in MATLAB. It's noticed that the designed control makes the system asymptotically stable with a duration of transient not more than 2 s. The results of working out by it the deviation of the speed from the given $V_0 = 0{,}05$ m/sec are shown in Fig. 1.

The given graphs testify to the stability of the rectilinear movement of the caterpillar robot. Obviously, it can be said that using APM can design an efficient enough and easy-to-find control system.

4 Rotational Movement of the Tracked Robot

Let us assume that the rotation of the tracked robot is carried out by simultaneously reducing the speed of one caterpillar and increasing the speed of the other. In this case, turning to the left (in the positive direction) can be thought of as moving the robot around a circle centered at point O, as shown in Fig. 2. Let R_n is the radius of the circle along which the center of mass of the robot moves that (in Fig. 2, its arc is shown by a dashed line). Let's say the robot passed this arc in time dt and turned during this time at an angle $d\varphi$. Then, in accordance with the figure, the left and right tracks will cover the distances:

Fig. 2 To turn the robot

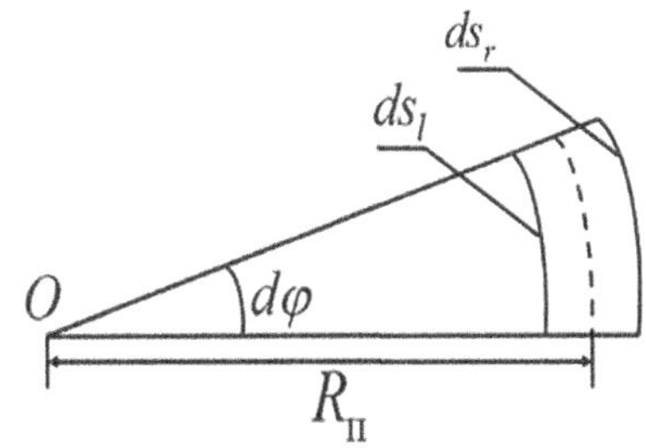

$$ds_l = (R_\pi - l)d\varphi . ds_r = (R_\pi + l)d\varphi \quad (37)$$

where l is half the distance between the tracks.

Different lengths of ds_l and ds_r are due to different caterpillar speeds, i.e.

$$ds_l = (V - \Delta_V)dt . ds_r = (V + \Delta_V)dt \quad (38)$$

where V is the rectilinear speed of the robot, before the start of the turn; Δ_V–change in the speeds of the left $(-\Delta_V)$ and right $(+\Delta_V)$ caterpillars. For $\Delta_V = \text{const}$ from expressions (37), (38) we deduce.

$$ds_r - ds_l = 2ld\varphi, ds_r - ds_l = 2\Delta_V dt \quad (39)$$

Equating the right parts of expressions (39) to each other, we obtain the following equation, which describes the turn of the caterpillar robot in the case under consideration $\dot{\varphi} = (-1)^\nu k_\varphi \Delta_V$, where $k_\varphi = 1/l$ c is the transmission coefficient; ν- the value that determines the direction of rotation: ν = 1 turns right; ν = 2 is turn left.

To work out the turn at a given angle φ^* and maintain a given direction, it is proposed to use a control system turn, the equations of which are:

$$\varepsilon_\varphi = \varphi^* - \varphi,\ u_\varphi = \begin{cases} 0, & if\ |\varepsilon_\varphi| < \delta_\varphi, \\ \Delta_V sign\, \varepsilon_\varphi, & f\ |\varepsilon_\varphi| \geq \delta_\varphi, \end{cases}\ \dot{\varphi} = k_\pi k_\varphi u_\varphi \quad (40)$$

where u_φ is the control of the rotation of the robot; k_t is the coefficient of proportionality.

In accordance with the definition of the turn angle φ, the coordinates of the center of gravity of the robot in the rectangular coordinate system XY are determined by solving the equations:

$$\dot{x}_6 = 0, 5(x_1 + x_3)\cos\varphi;\ \dot{x}_7 = 0, 5(x_1 + x_3)\sin\varphi \quad (41)$$

where $x_6 = X$, $x_7 = Y$ (see Fig. 3b).

Equations (19), (20), (30–33), (40), and (41) describe the motion control system of a tracked robot. The required trajectory of the robot's movement can be set by the

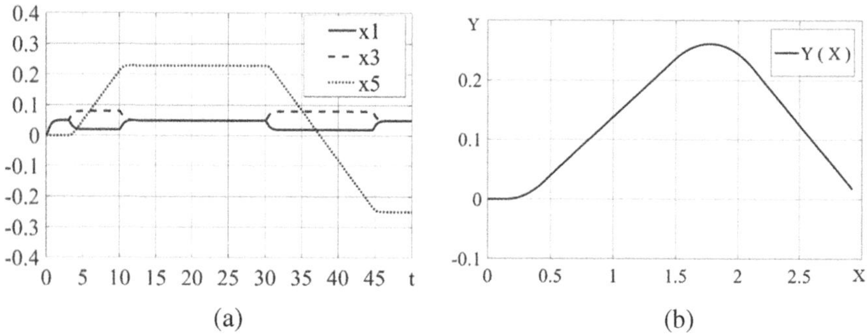

Fig. 3 Simulation of the movement of a caterpillar robot: **a** graphs of the speeds of the caterpillars and the angle of rotation; **b** the resulting trajectory of the robot

time program. The simulation results are shown in Fig. 3, where x_1 and x_3 are the speeds of the left and right tracks; x_5 angle φ.

On Fig. 3a clearly shows the relationship between the change in the speed of the robot caterpillars and the change in the angle of rotation of the robot. On Fig. 3b shows the resulting trajectory of the robot corresponding to the given one.

5 Conclusion

Based on the foregoing, it can be concluded that the obtained mathematical model of the caterpillar robot quite fully describes the movements of the caterpillar robot. The algebraic polynomial-matrix method makes it possible to synthesize non-linear control systems for non-linear objects for which a quasi-linear model has been built. The following is always possible if the nonlinearities of the control object are differentiable in all their arguments.

The publication is prepared within the implementation of State order Southern scientific center, Russian Academy of Sciences, project No. 122020100270–3.

References

1. Moothedath, S., Chaporkar, P., Belur, M.N.: Minimum cost feedback selection for arbitrary pole placement in structured systems. IEEE Trans. Automat. Control **63**(11), 3881–3888 (2018)
2. Kotenko, I., Parashchuk, I.: Determining the parameters of the mathematical model of the process of searching for harmful information. In: Kravets, A.G., Bolshakov, A.A., Shcherbakov, M.V. (eds.) Cyber-Physical Systems: Industry 4.0 Challenges, vol. 260, pp. 225–236. Springer
3. Mei, W., Friedkin, N.E., Lewis, K., Bullo, F.: Dynamic models of appraisal networks explaining collective learning. IEEE Trans. Autom. Control **63**(9), 2898–2912 (2018)
4. Manianin, S., Molev, Y., Vakhidov, U., Makarov, V., Belyakov, V., Dorofeev, R., Zonov, I., Tyurikov, P.: Experimental Research of the Effect of Track Chains Pre-static Tension on the

Mobility of a Tracked Vehicle. pp. 1–4 (2019). https://doi.org/10.1109/MILTECHS.2019.8870025
5. Kersting, S., Buss, M.: Direct and indirect model reference adaptive control for multivariable piecewise affine systems. IEEE Trans. Autom. Control **62**(11), 5634–5649 (2017)
6. Yang, G., Liberzon, D.: Feedback stabilization of switched linear systems with unknown disturbances under data-rate constraints. IEEE Trans. Autom. Control **63**(7), 2107–2122 (2018)
7. Gaiduk, A.R.: Nonlinear control systems design by transformation method//Mekhatronica. Avtomatizatsiya, Upravlenie **19**(12), 755–761 (2018)
8. Elameen, M.H., Karsiti, M.N., Ibrahim, R.: Nonlinear model feedback linearization control strategy of a pH neutralization process. In: 5th International Conference on Intelligent and Advanced Systems (ICIAS), vol. 2, pp. 1–5. IEEE (2014)
9. Madeira, D.D. S., Adamy, J.: Feedback control of nonlinear systems using passivity indices. In: Proceedings of the IEEE Conference on Control Applications, Sydney, Australia, pp. 263–268. (2015)
10. Byrnes, C., Isidori, A., Willems, J.: Passivity, feedback equivalence and the global stabilization of minimum phase nonlinear systems. IEEE Trans. Automat. Contr. **36**, 1228–1240 (1991)
11. Xia, M., Rahnama, A., Wang, S., Antsaklis, P.J.: Control design using passivation for stability and performance. IEEE Trans. Control **63**(9), 2987–2993 (2018)
12. Zhu, Y., Su, H.-Y., Krstic, M.: Adaptive backstepping control of uncertain linear systems under actuator delay. Automatica **54**, 256–265 (2015)
13. Li, I.H., Chen, Y.H., Wang, W.Y., Kao, Y.F.: Hybrid Intelligent algorithm for indoor path planning and trajectory-tracking control of wheeled mobile robot. Int. J. Fuzzy Syst. **18**(4), 595–608
14. ShiNung, C., Yongsoon, E., Pierre, T.K., Semyon, M.M.: Quasilinear control theory: an overview. IFAC Proc. **44**(1), 10958–10963
15. Tebueva, F., Pavlov, A., Satybaltina, D.: A method for planning the trajectory of a mobile robot in an unknown environment with obstacles. In: E3S Web of Conferences (2021).
16. António, M.G., Ghanbari, P., Ghanbari, M.: Piecewise approximation of contours through scale-space selection of dominant points. IEEE Trans. Image Process. **19**(6), 1442–1450 (2010)
17. Matokhina, A., Kravets, A.G., Volodina, D., et al.: Analysis of dynamics in human-exoskeleton collaborative system. In: Studies in Systems, Decision and Control, vol. 272, pp. 89–109 (2020). https://doi.org/10.1007/978-3-030-37841-7_8
18. Neydorf, R., Gamayunov, N., Gaiduk, A.: The multiplicative-isolating principle of significantly nonlinear mathematical models creation. In: Studies in Systems, Decision and Control, pp. 23–32 (2021). https://doi.org/10.1007/978-3-030-66077-2_3
19. Izadi, M., Srivastava, H.M.: Generalized bessel quasilinearization technique applied to bratu and lane–emden-type equations of arbitrary order. Fractal Fract. **5**, 179p (2021)
20. Gaiduk, A.R.: Theory and methods of analytical synthesis of automatic control systems (polynomial approach). Fizmatlit, pp. 281–310 (2012),
21. Kabanov, A.A.: Modeling and motion control of a caterpillar mobile robot. In: Proceedings of the conference "UTEOSS". St. Petersburg, State Scientific Center of the Russian Federation JSC "Concern" Central Research Institute "Elektropribor", pp. 740–743 (2012)

Software Package for Simulation Modeling of Mobile Robots

Maksim Kuprin, Igor Osipov, Arkady Klyuchikov, and Nikita Samokhin

Abstract In this chapter, the possibility of using Unity as a simulation environment for mobile robots is considered, and the possibilities of Unity Editor for use in simulation modeling are determined. The possibility of using a standard set of development tools to simulate the behavior of mechanical connections, kinematic circuits, and rigid body physics is evaluated. The architecture diagram of the developed simulation modeling system is compiled. The methods of implementation and functionality of simulating modules are described, such as ModuleBaseScript for the simulation of rigid mechanical connections, HingeBaseScript for axial connections, ServoMotorScript and MotorBaseScript for electric motors and servos, respectively, MicrocontrollerBaseScript, GPIOBaseScript, DataTransferMasterScript for microcontrollers, MotorDriverScript for electric motor drivers, RangeFinderBaseScript for rangefinders and DataTransferSlaveScript for transmission protocols data. The received package of tools was tested, and a virtual model of a mobile robot with a rangefinder was created. The shortcomings of the project have been identified and ways to solve problems have been proposed.

Keywords Simulation modeling · Computer modeling · Mobile robots · Physics engine · Unity

M. Kuprin (✉) · I. Osipov · N. Samokhin
Yuri Gagarin State Technical University of Saratov, 77 Politekhnicheskaya, Saratov 410054, Russia
e-mail: Krok9407@mail.ru

A. Klyuchikov
Biotechnology and Engineering Named After N.I. Vavilov, Saratov State University of Genetics, Theatre Sq. 1, Saratov 410012, Russia

A. G. Kravets et al. (eds.), *Cyber-Physical Systems Engineering and Control*,
Studies in Systems, Decision and Control 477,
https://doi.org/10.1007/978-3-031-33159-6_8

1 Introduction

Currently, software tools are used to design each stage in the development of robotic systems, for example, creating drawings, 3d models of parts and elements, prototyping electronic circuits of the hardware, mathematical calculations, modeling of individual elements of a robot and testing the algorithms of the software component of the robotic complex. [1, 2] In total these processes are used in simulation modeling.

Specialized programming software is used for simulation modeling process visualization. For this purpose, it is possible to use game engines. The most popular engines are Unity and Unreal Engine. Based on research findings [3, 4], it is possible to use Unity to design simplified robotic systems for educational purposes.

2 Problem Statement

In connection with the identified need for simulation in the field of designing mobile robots, a theoretical goal was set to study the possibility of using Unity as a simulation environment for mobile robots. The following tasks have been set:

- recreate close-to-realistic behavior of mechanics and electronics;
- develop microcontroller programming interface;
- develop a flexible system of settings for the elements of the simulated platform;
- provide a base of prepared elements with the possibility of creating extensions;
- develop data transfer protocols;
- develop odometry, measuring through sensors.

3 Analysis of the Possibilities of Using the Unity Game Engine for the Design of Simulation Systems

Unity is a cross-platform game engine for the development of 2D and 3D games and applications, but developers create plugins for robot modeling based on it [5]. NASA Curiosity rover simulation [6], the 3D-planning system of the 2014 Winter Olympics [7], and US Navy Virtual Training [8] have been developed and used for research purposes.

Component-based development is used in the Unity environment. Object properties on the scene are defined by components. The default Unity tool kit is not sufficient to solve the problem, but it is possible to create your own components in Unity. C# and JavaScript are used for script programming.

One of the criteria affecting the quality of simulation modeling of robotic systems is the realism and accuracy of physical calculations. Unity uses the PhysX engine to determine the interaction of three-dimensional bodies and their external forces of them [9]. PhysX has sufficient accuracy and stability to simulate simple robots [10–14]. Physical aspects are taken into account in the engine:

- rigid-body physics [15];
- collisions [16];
- joints [17];
- articulations [18].

Usage of these components, taking into account correlations and mathematical models of the system, allows for the simulation of mechanical joints of mobile robots and robot manipulators.

Standard Unity tools cannot simulate electronic parts, microcontrollers, electric motors, and data transfer protocols. It is necessary to create specialized scripts that expand the capabilities of the development environment to simulate the hardware.

4 Architecture of the Design System for Simulation Modeling of Mobile Robots

The development of the mechanical system of a robot is a time-consuming process and transferring into a development environment requires a number of additional steps. Specialized modules were developed in Unity for the automation of adding and configuring mechanical joints. This script's system allows you to simulate a mobile robotic system. The architecture scheme of the design system for simulation modeling of mobile robotic systems is shown in Fig. 1.

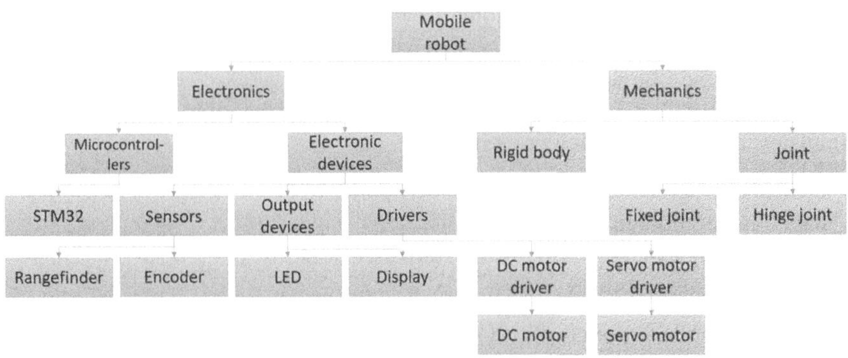

Fig. 1 Architecture scheme

5 Development of Modules for Simulating the Mechanical Part of a Robotic Complex

ModuleBaseScript and HingeBaseScript were developed as base modules to simplify the user experience.

ModuleBaseScript connects two objects by fixing their position relative to each other. Logic is developed by using the Fixed joint [19] component. If the connected elements do not have a RigidBody component, it will be added automatically by the RequireComponent [20] attribute.

HingeBaseScript connects two objects by creating an axial joint between them with one degree of freedom. Functioning is developed using the Hinge joint [21].

When the scene is launched, these elements are added to all the necessary components and set the values of the joint parameters.

Using the Hinge joint allows you to simulate the rotation of a motor or servo. The component can adjust the torque and achieve the specified angular velocity or specified angle. Hinge joint also adjusts the restrictions on the angle of rotation of one body relative to another (Fig. 2).

Hinge joint has an excessive number of parameters. It makes the setup process time-consuming. For this reason, the scripts "ServoMotorScript" and "MotorBaseScript" were developed. They simulate servo motor and electric motor.

The script "ServoMotorScript" accesses GPIOScript by using object reference to GPIO and receives port values with PortPWMId number. The resulting value is between 0 and 1 and is used to calculate the end position of the servo shaft. The required angle is the multiplication of the control signal and the maximum angle of rotation of the shaft.

MotorBaseScript sets up Motor, maximum angular speed, and torque and receives a control signal from the electric motor driver using the "motorDriver" object reference. The multiplication of the control signal and the maximum angular velocity is transferred to the "Target velocity" parameter.

A microcontroller is used to control the hardware of the robot. The module that simulates it has the following properties:

- control of external devices through GPIO ports;
- data transfer through special protocols;
- programming and interaction with external devices;
- time control using timers.

"MicrocontrollerBaseScript", "GPIOBaseScript", "DataTransferMasterScript" are used to create a simulation of a microcontroller.

MicrocontollerBaseScript is the base script for microcontroller simulation. The script contains GPIOBaseScript and DataTransferMasterScript object references, controlling GPIO ports and data bus, and also processes an algorithm according to program code.

GPIOBaseScript is used to simulate the general-purpose I/O interface. The script stores two arrays of values: digital ports and with the possibility of pulse-width

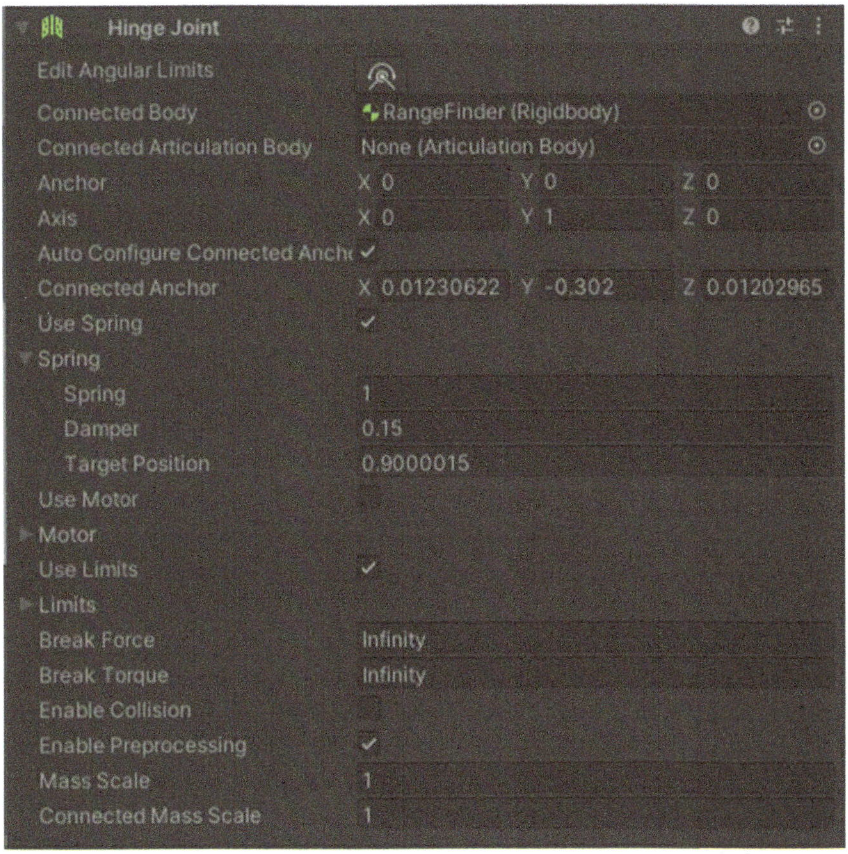

Fig. 2 Hinge joint setup

modulation. GPIO communicates and receives signals from external devices. Any device connected to GPIO can get port value and/or write it.

DataTransferMasterScript and DataTransferSlaveScript allow transferring text through a data bus. The script contains the list of devices connected to the bus. The master can request data from slave devices and send data back. DataTransferSlaveScrpit allows to receive data from the master and send data to the master on request.

Power control of motors in robots is implemented by using a set of power keys (motor driver). This module is implemented in the MotorDriverScript in the software package. It is used for generating a control signal for the electric motor. The script connects to the GPIO pins of the microcontroller and receives data from two digital and one PWM port. The data is used to determine the direction and speed of rotation of the motor shaft (Fig. 3).

Also, in the package, there is the possibility of modeling a rangefinder. The "Physics" [22] class in Unity contains the "Raycast" [23] method, which allows

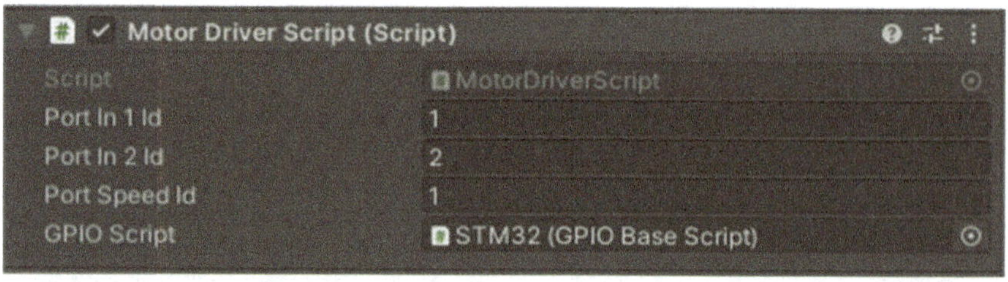

Fig. 3 Motor driver setup

one to cast a ray from a point in space in a target direction to detect obstacles and a distance to them. RangeFinderBaseScript allows the use of Raycast to measure distances and transfer the values using the data transfer protocol.

6 Approbation of the Software Package on the Example of a Mobile Robot with an Ultrasonic Rangefinder

The current version of the software package allows you to simulate simplified constructions of mobile robots and test programs for microcontrollers. A mobile robot was developed to test the package. It has a four-wheeled platform, a microcontroller, a display, a rangefinder, a servo motor, and four drivers (Fig. 4). A video [24] with a demonstration of the work of a virtual robot.

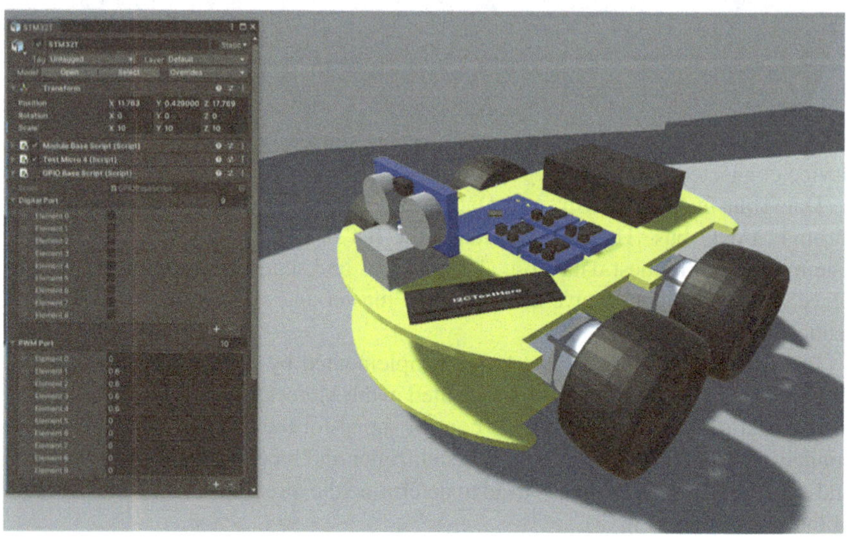

Fig. 4 Basic mobile robot

During the simulation and testing of the robot, the following disadvantages of the software package were identified:

- Unity standard tool kit doesn't provide enough information about joints during setting up. It can be difficult to determine the direction of rotation of the wheels during setting up their joint to the motor shaft;
- code for simulating the operation of the microcontroller and the robot control program are located in one file. It is necessary to copy the microcontroller script file to create a new robot program;
- it is impossible to simulate omnidirectional and "Mecanum wheel" operations.

Currently, the directions for the development of the project have been determined and ways to solve existing problems have been proposed.

It is necessary to create arrows with Unity tools that indicate the direction of the axis and the direction of rotation of the motor shaft with a positive value of the control signal to fix the first problem. This solution will provide clarity and simplify the tuning of electric motors.

The second problem can be solved by splitting the microcontroller modeling script into two files. The first defines the operation of the microcontroller, the second defines a robot's controlling algorithm. It can reduce the dependence on variables and make the system more flexible.

7 Conclusion

The developed tools have sufficient functionality for modeling robotic systems. During the tests, disadvantages, and limitations were found and ways of improving the software package were developed.

Moreover, for further improvement of the software packages, the following tasks were suggested:

- add to MotorBaseScript the ability to attach the Rigidbody component to the wheels automatically;
- develop new programming tools for microcontrollers that allow the creation of firmware without creating additional components;
- implement the possibility of hinged joints of bodies using Unity physics tools;
- develop a new component simulating a motor including encoders operations;
- develop a component simulating the functionality of stepper motors and servo drives;
- create a tool modeling the different types of wheels.

References

1. Bräunl, T.: Robot adventures in Python and C. Springer, Berlin (2020)
2. Guo, S., Li, H.-W., Ji, J.C., Ming, Z.-F.: Kinematic analysis and simulation of a new-type robot with special structure. Adv. Manuf. **2**, 295–302 (2014). https://doi.org/10.1007/s40436-014-0094-x
3. Manyuxina, A.N.: Game engine usage for robotic devices modeling systems development. In: Conference «Integraciya mirovyh nauchnyh protsessov kak osnova obshetvennogo progressa», pp. 131–137 (2015) (in Russian)
4. Zawadniak, P.V.F., Piardi, L., Brito, T., Lima, J., Costa, P., Monteiro, A. L. R., Costa, P., Pereira, A. I.: Micromouse 3D simulator with dynamics capability: a Unity environment approach. SN Appl. Sci. (2021) (Research Article)
5. Matokhina, A., Dragunov, S., Popova, S., Kravets, A.G.: Method of the exoskeleton assembly synthesis on the base of anthropometric characteristics analysis. Stud. Syst. Decis. Control **259**, 25–42 (2020). https://doi.org/10.1007/978-3-030-32579-4_3
6. Unity Powers NASA Virtual Mars Rover Experience [Online] // Unity Real-Time Development Platform|3D, 2D VR & AR Engine (2012). https://unity.com/our-company/newsroom/unity-powers-nasa-virtual-mars-rover-experience. Last accessed 5 Sept 2022
7. Olympic 3D. The thorny path from the disclosure of sudden to the solution of the tasks of national projects. [Online] // CG EVENT — Main event in computer modeling sphere. https://cgevent.ru/archives/8134. Last accessed 5 Sept 2022 (in Russian)
8. Assembly/disassembly of the Navy-Heartwood drive [Online]//Train Faster, Smarter with Interactive Simulations URL: https://hwd3d.com/videos/navy-actuator-virtual-training/. Last accessed 5 Sept 2022
9. Unity-Manual: Physics [Online]//Unity-Manual: Unity User Manual 2021.3 (LTS). https://docs.unity3d.com/2020.3/Documentation/Manual/PhysicsSection.html. Last accessed 6 Sept 2022 (in Russian)
10. Roennau, A., Sutter, F., Heppner, G., Oberlaender, J., Dillmann, R.: Evaluation of physics engines for robotic simulations with a special focus on the dynamics of walking robots. In: International Conference on Advanced Robotics (ICAR), Montevideo (2013)
11. Drumwright, E.: An evaluation of methods for modeling contact in multibody simulation. In: Drumwright, E., Shell, D. (eds) IEEE International Conference on Robotics and Automation, pp. 1695–1701 (2011)
12. Erez, T.: Simulation tools for model-based robotics: comparison of Bullet, Havok, MuJoCo, ODE and PhysX/T. Erez,
13. Hummel, J.: An evaluation of open source physics engines for use in virtual reality assembly simulations
14. Seugling, A: Evaluation of physics engines and implementation of a physics module in a 3d-authoring tool
15. Unity-Manual: Rigidbody [Online]// Unity User Manual 2021.3 (LTS). https://docs.unity3d.com/Manual/class-Rigidbody.html. Last accessed 6 Sept 2022
16. Unity-Manual: Introduction to collision [Online]//Unity User Manual 2021.3 (LTS). https://docs.unity3d.com/Manual/CollidersOverview.html. Last accessed 6 Sept 2022
17. Unity-Manual: Introduction to joints [Online]//Unity User Manual 2021.3 (LTS). https://docs.unity3d.com/Manual/Joints.html. Last accessed 6 Sept 2022
18. Unity-Manual: Introduction to physics articulations [Online]//Unity User Manual 2021.3 (LTS). https://docs.unity3d.com/Manual/physics-articulations.html. Last accessed 6 Sept 2022
19. Unity-Manual: Fixed Joint component reference [Online]// Unity User Manual 2021.3 (LTS) URL: https://docs.unity3d.com/Manual/class-FixedJoint.html. Last accessed 7 Sept 2022
20. Unity-Scripting API: Require Component [Online]//Unity Scripting API 2021.3 (LTS). https://docs.unity3d.com/ScriptReference/RequireComponent.html. Last accessed 7 Sept 2022
21. Unity-Manual: Hinge Joint component reference [Online]//Unity User Manual 2021.3 (LTS). https://docs.unity3d.com/2021.3/Documentation/Manual/class-HingeJoint.html Last accessed 7 Sept 2022

22. Unity-Scripting API: Physics [Online]//Unity Scripting API 2022.1 (LTS). https://docs.unity3d.com/2022.1/Documentation/ScriptReference/Physics.html. Last accessed 8 Sept 2022
23. Unity-Scripting API: Physics. Raycast [Online]//Unity Scripting API 2021.3 (LTS). https://docs.unity3d.com/ScriptReference/Physics.Raycast.html. Last accessed 8 Sept 2022
24. RSMA: Simple mobile robot [demonstration of the software package]//YouTube. https://www.youtube.com/watch?v=4CZjw4ji71s&t=50s. Last accessed 24 Sept 2022

New Materials Engineering for Cyber-Physical Systems

Analysis of Obtaining a Dispersed Mixture with Secondary Raw Materials for Cyber-Physical Support of Recycling

D. V. Stenko, A. B. Kapranova, D. D. Bahaeva, D. V. Fedorova, and A. E. Lebedev

Abstract The use of secondary components in the process of obtaining a dispersed mixture from polypropylene for the development of 3D printing technologies indicates the urgency of the problem of cyber-physical support for recycling. The study is devoted to stochastic modeling of the process of obtaining a dispersed mixture with secondary raw materials in the chamber of a rotary mixer at the first stage of its operation, as a preliminary stage of operation of an injection molding machine. The analysis of the results of the model is carried out from the standpoint of the known criterion of the efficiency of the mixing process of bulk components in the analysis of the most significant factors for increasing this efficiency in the formation of rarefied flows of mixed media by double-row elastic blades. The used modeling method relates to the energy method of modeling the mechanism of the behavior of mixed non-spherical particles in rarefied flows.

Keywords Cyber-physical support · Recycling · Rotary apparatus · Mixing · Mixture · Secondary component · Polypropylene · Elastic blades · Model · Parameters · Information variables

1 Introduction

The problem of improving cyber-physical support for the recycling process [1–4] within the development of digital technologies, including at various stages of 3D printing, meets the needs of a modern energy-efficient approach to environmental issues and energy saving [5, 6]. The development of technologies for the practical implementation of the 3D printing process in the direction of attracting polymer secondary raw materials [7–9] is relevant and requires solving several problems related to the preparatory stages of this operation. Note that the use of secondary products increases the requirements for the organization of the 3D printing process

D. V. Stenko (✉) · A. B. Kapranova · D. D. Bahaeva · D. V. Fedorova · A. E. Lebedev
Yaroslavl State Technical University, Moskovskiy Prospect, 88, Yaroslavl 150023, Russia
e-mail: dvs3d@yandex.ru

A. G. Kravets et al. (eds.), *Cyber-Physical Systems Engineering and Control*,
Studies in Systems, Decision and Control 477,
https://doi.org/10.1007/978-3-031-33159-6_9

itself due to the need to comply with tests not only for the color fastness of the finished product but also for its strength characteristics under various operating conditions [3, 4]. These additional factors include indicators of heat resistance, frost resistance. Changes in the temperature parameter or its multiple cyclic loads should not affect the regulatory effect of maintaining the stability and reliability of the structure obtained by 3D printing. The solution of these problems is greatly facilitated with the involvement of the cyber-physical approach [10, 11] to the support of the recycling process when it is required to take into account many sets of parameters to organize an efficient process for obtaining a high-quality 3D printing product.

One of several known schemes for the practical implementation of recycling in this area of digital technologies involves the use of a two-component polymer-dispersed mixture, consisting of secondary and primary bulk materials [12–14]. The proportional content of these solid dispersed materials in the mixture is controlled by the customer's regulations [15], and usually corresponds to a ratio of 1:9 at a given granulometric composition.

The problem of achieving the specified requirements for the homogeneity [16–18] of the specified polymer-dispersed mixture can be solved in the working volume of the rotary mixer [19] at the stage of preliminary preparation of the injection-molding machine. The purpose of this study is to analyze the most significant factors for increasing the efficiency of the operation of mixing polymer components during the formation of their rarefied flows. The latter are formed using double-row elastic blades installed tangentially to the surfaces of two mixing cylindrical drums in the working volume of the rotary mixer [19] of the injection molding machine.

Liberation of the projected process of mixing polymer components from the negative impact of the consequences of segregation [20, 21] is one of the main tasks in the development of new mixing equipment of the rotary type. The selection of sets of parameters of a given mechanical process according to various criteria (constructive, regime, physical and mechanical) and their accounting according to the system-structural analysis [22, 23] at each design stage is decisive of a decisive nature. In this case, as a methodological basis for modeling the process of obtaining a two-component polymer-dispersed mixture, consisting of secondary and primary bulk materials, a stochastic approach in the form of the energy method of Yu. L. Klimontovich was used [24, 25].

The specified energy method has been tested by the authors on examples of modeling the formation of rarefied flows of bulk components by flexible brush elements of various types [10, 16–18, 26] and elastic blades [11, 27, 28]. Known deterministic and indeterminate approaches [29, 30] to mathematical modeling of the process of mixing solid dispersed media. The second one has been developing most actively in recent years. At the same time, it is possible to conditionally divide the existing group of indeterminate mixing models into subgroups. Among them, the following descriptions of the behavior of the components being mixed are distinguished: cybernetic [31, 32], using the theory of time series [33, 34], based on information theory and statistical thermodynamics [35], Markov chains [36]. In particular, within the framework of the latter, variations [37–41] and the energy method [24, 25] with approbation [10, 16–18, 26–28] can be distinguished.

2 A Mechanical Method for the Formation of Rarefied Flows of Polymer-Dispersed Components

According to the mechanical method proposed by the authors [19], the formation of rarefied flows of polymer-dispersed components occurs in the working volume of a rotary mixer at two stages of its operation (Fig. 1). The first stage refers to the production of crossing flows of secondary and primary polymeric bulk materials after their vertical loading and dosing into the gaps of two drums with three-row elastic blades installed tangentially to their cylindrical surfaces (Fig. 2).

The second stage of mixing serves to enhance the efficiency of the projected process due to the operation of the conical bump surfaces. This study focuses on the first stage (Fig. 2) of the process of mixing polymer bulk materials in the working volume of a rotary mixer [19] at the stage of preliminary preparation of the injection molding machine (Fig. 1).

The solution to the problem of cyber-physical support of recycling when obtaining a dispersed polymer mixture in the presence of a secondary component is closely related to the issue of choosing the parameters of the mixing process as a preliminary stage in the operation of an injection molding machine (Fig. 1).

Description of information variables $z(t) \equiv \{z_\nu(t)\} = \{x(t), y(t), a_{i,ki} = const, b_{i,li} = const\}$, $\nu = \overline{1, s_0}$ the simulated process of mixing polymer materials is performed using sets of parameters: input $x(t) \equiv \{x_{i,mi}(t)\}$, $m_i = \overline{1, s_{1i}}$; output $y(t) \equiv \{y_n(t)\}$, $n = \overline{1, s_2}$; constructive $a \equiv \{a_1, a_2\} = \{a_{i,ki} = const\}$, $k_i = \overline{1, s_{3i}}$; regime $b \equiv \{b_1, b_2\} = \{b_{i,li} = const\}$, $l_i = \overline{1, s_{4i}}$. The specified sets are given by expressions

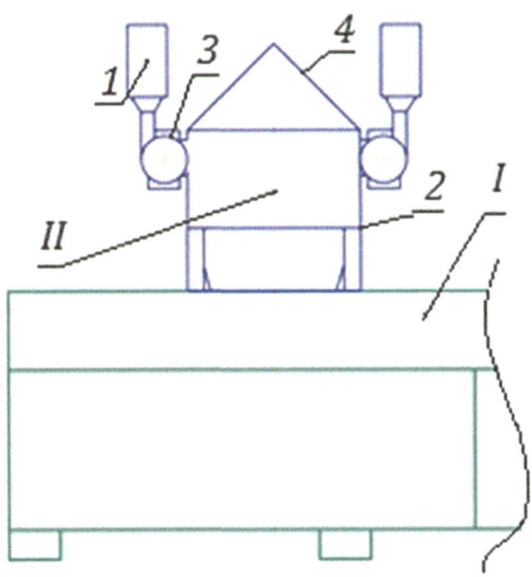

Fig. 1 A schematic diagram of the installation of a rotary mixer for secondary and primary polymeric bulk materials on an injection molding machine 1500–475: *I*—injection molding machine; *II*—rotary mixer; *1*—metering hopper; *2*—mixer body; *3*—mixing drum cover; *4*—conical bump

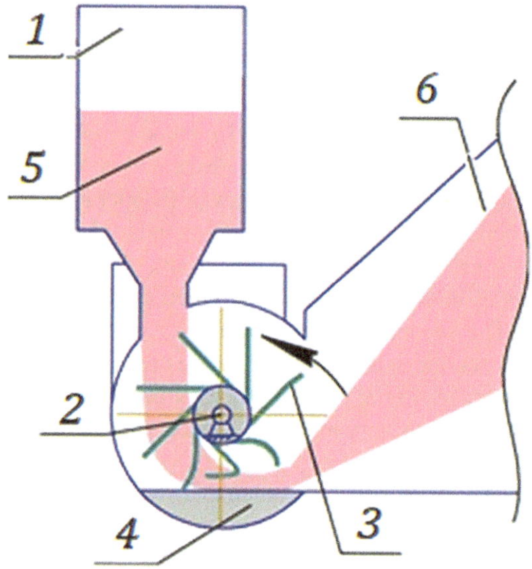

Fig. 2 Conditional scheme for the implementation of the first stage of mixing polymer bulk materials in a rotary mixer: *1*—metering hopper; *2*—mixing drum; *3*—elastic shoulder blades; *4*—fixed surface; *5*—dispersed component; *6*—rarefied flow

$$\left\{x_{i,mi}(t)\right\} = \{Q_{Vi}, C_{Vi}\},\ \left\{y_{i,ni}(t)\right\} = \left\{x_{i,mi}(t), V_C^{reg}, \Delta V_C\right\} \tag{1}$$

$$\{a_{1,k1}\} = \{r_{ci}, s_{ci}, D_{wi}, l_{bi}, h_{bi}, \eta_{bi}, \varphi_i, u_{ri}, u_{si}\},\ \left\{a_{2,k2}\right\} = \{\gamma_{bs}, L_{bs}\} \tag{2}$$

$$\left\{b_{i,li}\right\} = \{\omega_i, h_{Li}, h_{0i}\},\ q = \left\{\rho_i, \overline{d_i}, k_f, k_u\right\} \tag{3}$$

Expression (1) includes: Q_{Vi} is the volumetric flow rate of the polymer component $i = \overline{1,2}$; C_{Vi} is volume fractions of the specified components; V_C^{reg}, ΔV_C are regulatory values of the coefficient of heterogeneity and its absolute error for the finished dispersed polymer mixture. Expression (2) contains the following notation related to sets of design parameters $\{a_{1,k1}\}$, $\{a_{2,k2}\}$ for two stages of the process of mixing polymer components, according to Fig. 1, 2:

- for cylindrical drums providing a rarefied flow of the polymer component $i = \overline{1,2}$ (s_{ci} is length, r_{ci} is radius);
- bodies of the specified drums (D_{wi} is diameter);
- elastic rectangular blades installed tangentially to the cylindrical surfaces of the drums (l_{bi} is length, h_{bi} is height, η_{bi} is width, φ_i is clamping offset angle, u_{ri} is the number of rows, u_{si} is the number in one row);
- conical bump surface (γ_{bs} is the opening angle, L_{bs} is generatrix length).

A set of operating parameters $\left\{b_{i,li}\right\}$ from expression (3) for a given technological operation includes: ω_i are angular speeds of rotation of mixing drums with the formation of a rarefied flow of a component $i = \overline{1,2}$; h_{Li} is the full height of the component layer $i = \overline{1,2}$ at the entrance to the gap between the drum and the stationary support surface; h_{0i} is the height of a given gap. Thus, according to expressions (1)–(3), the total number of information variables of the studied technological operation corresponds to $s_0 = 36$. Last value s_0 indicates the relevance of the problem of cyber-physical support of recycling in the process of preparing a polymer mixture with a secondary component of polypropylene for the efficient operation of an injection molding machine (Fig. 1). In this case, an additional set of parameters of the mixing model is the set $q = \{\rho_i, \overline{d_i}, k_f, k_u\}$ from (3), relating to the characteristics of the physical and mechanical properties of substances of mixed polymer-dispersed media and elastic blades. Here is the notation: ρ_i is the true density of the polymer component $i = \overline{1, 2}$; $\overline{d_i}$ is averaged over fractions particle diameter of a given component; k_f is aspect ratio; k_u is the angular coefficient of stiffness of an elastic blade.

One of the main tasks of modeling the process of mixing polymer materials in a rotary apparatus [19] is to establish from a set of information parameters $\{z_\nu(t)\}$ or the optimal values of the optimized parameters, or the limits of their change from the standpoint of the rational flow of the specified technological operation. The practical needs of designing mixing equipment are often limited by the second condition, which is sufficient for the implementation of the factor of liberation from segregation effects. However, the analysis of optimized or rational parameters in terms of the degree of significance and impact on the efficiency of the rotary mixer [19] from the standpoint of the system-structural scheme [22, 23] is one of the priority research tasks due to the need to provide conditions for controlling the mixing process for special purposes. This control becomes possible when formulating a criterion for feedback between information parameters $\{z_\nu(t)\}$. For example, such a criterion is usually the quality indicator of the resulting polymer mixture [18] in the framework of the implementation of recycling [1–4], including when solving 3D printing problems [12–15] using an additional rotary mixer [19] installed on the injection molding machine (Fig. 1). In particular, the criterion indicator, in this case, can be the coefficient of heterogeneity of the polymer dispersed mixture V_C [11–13] at the given required values V_C^{reg} from the set (1) for the output parameters of the technological operation under study $\{y_{i,ni}(t)\}$.

3 Stochastic Modeling of the Process of Obtaining a Dispersed Mixture with Secondary Raw Materials in the Chamber of a Rotary Mixer

The development of the fundamental aspects of the process of mixing polymer components of the secondary and primary types is proposed to be carried out on the methodological basis of the stochastic approach of modeling in the form of the energy method of Yu. L. Klimontovich [24, 25].

The authors have tested mathematical models of the movement of particles of dispersed materials in rarefied flows formed by elastic blades [11, 28], installed tangentially to the cylindrical surfaces of the drum, at different values of the design parameter u_{ri} from set $\{a_{1,k1}\}$ according to (2). In particular, for $u_{ri} = 1$ there is a single-row case of attachment of the blades [11]; at $u_{ri} = 2$—double row; at $u_{ri} = 3$—three-row [28] etc. In this chapter, we consider the second case ($u_{ri} = 2$).

The features of the proposed model for mixing polymer-dispersed materials are different from the spherical shape of polypropylene particles of both secondary and primary components. According to (3), the set $q = \left\{\rho_i, \overline{d_i}, k_u\right\}$ includes the averaged over fractions particle diameter of this component $\overline{d_i} = \sum_{t=1}^{n_t} \overline{d_{fit}}/n_t$, where $\overline{d_{fit}}$ is average particle diameter in the fraction $t = \overline{1, n_t}$. Form factor introduction k_f when calculating the volume of single particles of components, they will quantitatively take into account their non-sphericity. In addition, the supposed stochastic modeling for the behavior of macrosystems of particles of the mixed polymer components is carried out outside the macroscale fluctuations of the states of these macrosystems. This fact is reflected in the assumption about the absence of collisions of particles during their scattering by elastic blades located on different rows $j = \overline{1, 2}$ one drum $i = \overline{1, 2}$. Taking into account the assumptions made, the construction of the desired distribution of the number of particles of each polymer component is carried out within the framework of a random process by A. A. Markov as homogeneous and stationary. Let the choice of the elementary phase volume $d\tau_{ij} = dv_{xij} dv_{yij} = -\omega_i^2 r_{ij} dr_{ij} d\theta_{ij}$ is determined by the phase variables in the form of the components of the velocity of the center of mass of the particle $(v_{xij}; v_{yij})$ component $i = \overline{1, 2}$ when interacting with a row blade $j = \overline{1, 2}$. Then, according to the stationary solution of the kinetic Fokker–Planck equation $\phi_i = a_{ij} \exp\left(-E_{ij}/E_{0ij}\right)$ with the specified choice $d\tau_{ij}$ we have the following exponential dependence for a decrease in the number of particles dN_{ij} in a given elementary volume and the normalization equation for the coefficient a_{ij}

$$dN_{ij} = a_{ij} \exp\left(-E_{ij}/E_{0ij}\right) d\tau_{ij} \tag{4}$$

$$\int_{\tau_{ij}} a_{ij} \exp\left(-E_{ij}/E_{0ij}\right) d\tau_{ij} = 1 \tag{5}$$

Expression (4) contains the energy of stochastic motion E_{ij} for a particle of the component $i = \overline{1, 2}$ when operating double-row elastic blades $j = \overline{1, 2}$ and parameter E_{0ij}, as the value of this energy at the moment of stochastization of macrosystems of particles of polymer components. Addiction E_{ij} on phase variables in different interpretations, for example, $(v_{xij}; v_{yij})$ or $(r_{ij}; \theta_{ij})$, is determined by the relationship between the velocity of the center of mass of the particle in polar coordinates $v_{r\theta ij}$ and coordinates of points $D_{ij}\left(r_{D_{ij}}; \theta_{D_{ij}}\right)$ deformable blades at the ends of their projections in the transverse plane of the drum. In general, the energy of stochastic motion E_{ij} for the specified particle of the macrosystem is given by the sum of three terms. Note, These terms correspond to the energies: translational motion together with the center of mass of a single particle, its rotational motion relative to a given center, and the energy of interaction of this particle with a deformed elastic blade when leaving the gap between a drum and a stationary surface (Fig. 2)

$$E_{ij}\left(r_{ij},\theta_{ij}\right)=\sum_{\nu=1}^{3}E_{ij\nu}\left(r_{ij},\theta_{ij}\right) \tag{6}$$

Explicit dependency $E_{ij}\left(r_{ij},\theta_{ij}\right)$ according to (6) is not given due to its cumbersomeness. A method is used to describe the motion of a blade deformed in the gap between the drum and the stationary surface (Fig. 2), using the Archimedes spiral equation $r_{SAij}\left(\theta_{ij}\right)=\mu_{0i}+\mu_{1i}\theta_{ij}$ in the polar coordinate system $\left(r_{ij},\theta_{ij}\right)$, proposed by the authors [11, 28]. In this case, the position for the ends of the projections of the blades on the transverse plane of the drum is determined by the equation $r_{D_{ij}}\left(\theta_{ij}\right)$ for point movement D_{ij}, corresponding to the indicated Archimedes curve $r_{SAij}\left(\theta_{ij}\right)$, through the design parameters of the working volume of the mixer $\{a_{1,k1}\}$ from expression (2)

$$\begin{aligned}r_{D_{ij}}\left(\theta_{ij}\right)&=(\{2r_{SAij}\left(\theta_{ij}\right)\cos\left[3/2\psi_i\left(\theta_{ij}\right)\right]\}^2\\&-4\{[r_{SAij}\left(\theta_{ij}\right)]^2-r_{ci}^2\})^{1/2}/2+r_{SAij}\left(\theta_{ij}\right)\cos[3/2\psi_i\left(\theta_{ij}\right)\end{aligned} \tag{7}$$

Here it is indicated: $\beta_{0\mathrm{i}}=2\pi/u_{si}$; $\psi_i\left(\theta_{ij}\right)=\mathrm{arctg}\left[\mu_{0i}/r_{SAij}\left(\theta_{ij}\right)\right]$; $\mu_{0i}=h_{0i}+(1-\cos\beta_{0\mathrm{i}})r_{ci}$; $\mu_{1i}=(\{\left[(1+\cos\beta_{0\mathrm{i}})r_{ci}\right]^2+(r_{ci}+l_{bi})^2\}^{1/2}-(\mu_{0i}+h_{0i}/3))/\gamma_{\delta i}$; $\gamma_{\delta i}=\pi+\mathrm{arctg}\{(r_{ci}+l_{bi})/[r_{ci}(1+\cos\beta_{0\mathrm{i}})]\}$.

Note that the parameter E_{0ij}, included in expressions (4), (5), is given by the energy balance equation. Taking into account the equality between the energy of particles with mass m_{ij} for macrosystems of polymer components $i=\overline{1,2}$ when capturing with double-row elastic blades $j=\overline{1,2}$ in the gap, the drum is a fixed surface, and when dissemination with these blades we have

$$\begin{aligned}&\sum_{j=1}^{2}N_{ij}m_{ij}\left\{h_{0\mathrm{i}}^2/2+\beta_{0\mathrm{i}}^{-1}\int_0^{\beta_{0\mathrm{i}}}[r_{D_{ij}}\left(\theta_{ij}\right)]^4/\left[K_{ij}\left(\theta_{ij}\right)\right]d\theta_{ij}\right\}+\\&+\sum_{j=1}^{2}a_{ij}\int_{\gamma_{0ij}}^{\gamma_{1ij}}d\theta_{ij}\int_{\mu_{0i}}^{r_{D_{ij}}\left(\theta_{ij}\right)}E_{ij}\exp(-E_{ij}/E_{0ij})r_{ij}dr_{ij}=0\end{aligned} \tag{8}$$

where $K_{ij}\left(\theta_{ij}\right)=[U_{1ij}\left(\theta_{ij}\right)]^2+u_{20ij}^2$; $U_{1ij}\left(\theta_{ij}\right)=u_{0ij}+u_{1ij}\theta_{ij}$; $u_{0ij}=\cos\left[\psi_i\left(\theta_{ij}\right)/2\right]$; $u_{1ij}=\{\mu_{1i}^2/\left[2\left(\mu_{0i}^2+\mu_{1i}^2\right)\right]\}\sin\left[\psi_i\left(\theta_{ij}\right)/2\right]$; $u_{20ij}=[u_{0ij}\left(1-u_{0ij}\right)]^{1/2}$ at values γ_{0ij} and γ_{1ij}, the correspond $\gamma_{\delta i}$.

Expression (4), taking into account the normalization (5) and the value of the parameter E_{0ij} from (8) determines the form of two differential distributions of the number of particles of macrosystems of polymer components $i=\overline{1,2}$ by dissemination angle θ_{ij}:

- functions $g_{ij}\left(\theta_{ij}\right)$ after the work of the elastic blades of the row $j=\overline{1,2}$ in the gap, the drum is a fixed surface

$$g_{ij}(\theta_{ij}) = (1/N_{ij})dN_{ij}/d\theta_{ij}, \tag{9}$$

- full function $G_i(\theta_{ij})$ after the work of double row elastic blades

$$G_i(\theta_{ij}) = \prod_{j=1}^{2} g_{ij}(\theta_{ij}). \tag{10}$$

4 Simulation Results

Consider an example of calculating the desired characteristics for the process of obtaining a two-component polymer-dispersed mixture, consisting of a secondary ($i = 1$; $\rho_1 = 0.91 \times 10^3$ kg/m^3; $\overline{d_1} = 5.490 \times 10^{-3}$ m; $k_f = 0.28$) and primary ($i = 2$; $\rho_2 = 0.93 \times 10^3$ kg/m^3; $\overline{d_2} = 4.435 \times 10^{-3}$ m; $k_f = 0.30$) polypropylene GOST 29,996–86. Table 1 contains data on the obtained parameter values E_{0ij}. Function simulation results $g_{ij}(\theta_{ij})$ and $G_i(\theta_{ij})$ depending on the angular speed of rotation of the drums ω_i, angle of dispersion of particles of macrosystems of polymer-dispersed components θ_{ij} and the angle of displacement when fixing the elastic blades φ_i are shown in Figs. 3 and 4. According to the described sets $\{a_{1,k1}\}$, $\{b_{i,li}\}$ from (2), (3) let us clarify the corresponding values of operating and design parameters or the limits of their change: $\{r_{ci} = 3.0 \times 10^{-2}$ m; $l_{bi} = 4.5 \times 10^{-2}$ m; $\varphi_i = (0 - 0.3927)$ rad; $u_{ri} = 2$; $u_{si} = 8\} \in \{a_{1,k1}\}$; $\{\omega_i = (41 - 53)$ s^{-1};$h_{0i} = 3.0 \times 10^{-2}$ m; $h_{Li} = 3.0 \times 10^{-2}$ m; $k_u = 5.4 \times 10^{-4}$ kg×m/rad $\} \in \{b_{i,li}\}$.

The obtained dependences (9), (10) for the functions $g_{ij}(\theta_{ij})$ and $G_i(\theta_{ij})$ allow analyzing to analyze the most significant factors of increasing the efficiency of the operation of mixing polymer components (Figs. 3 and 4) when forming their rarefied flows by double-row elastic blades (Fig. 2). Note that this method of forming flows of mixed components of polypropylene of two types is implemented at the first stage

Table 1 Data on the values of the parameter E_{0ij} at $u_{ri} = 2$; $\omega_i = 57, 6$ s^{-1}

Design parameter values		Parameter values E_{0ij}, × 10^{-3} J	
Offset angle φ_i, rad	The degree of deformation of the elastic blade δ_i	Secondary component GOST 29,996–86 ($i = 1$)	Primary component GOST 29,996–86 ($i = 2$)
0	0.60	1.902	1.098
	0.67	1.621	0.935
	0.70	1.484	0.857
0.3927	0.60	2.064	1.191
	0.67	1.752	1.011
	0.70	1.648	0.951

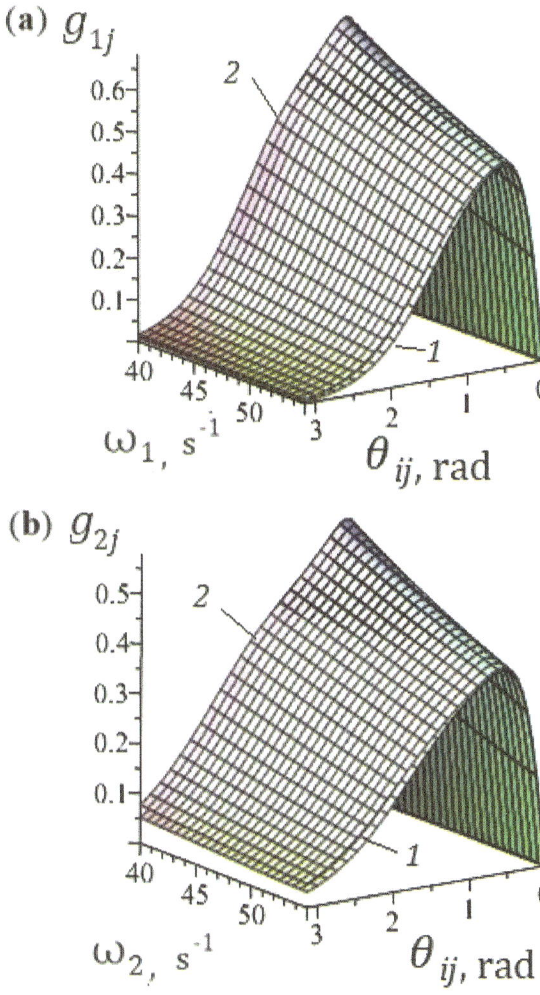

Fig. 3 Dependence $g_{ij}(\theta_{ij}, \omega_i)$: a—secondary polypropylene GOST 29,996–86 ($i = 1$); b—primary polypropylene GOST 29,996–86 ($i = 2$); $u_{si} = 8$; $u_{ri} = 2$; $\delta_i = 0, 67$; *1*—$j = 1, \varphi_i = 0$; *2*—$j = 2, \varphi_i = 0, 3927$ rad

of operation of a rotary mixer used for premixing dispersed polymer materials before loading into an injection molding machine (Fig. 1).

It was theoretically established that from the sets of parameters (design $\{a_{1,k1}\}$ and regime $\{b_{i,li}\}$) in expressions (2), (3) the following characteristics of the mixing process are of particular importance in assessing the required quality of the mixture:

- combined parameter $\delta_i = r_{ci}/l_{bi}$ (the degree of deformation of elastic rectangular blades when scattering polymer components $i = \overline{1, 2}$);
- constructive parameters: u_{ri} is the number of rows of deformable blades ($j = \overline{1, u_{ri}}$; $u_{ri} = 2$) on cylindrical surfaces of mixing drums for the formation of the rarefied component flows $i = \overline{1, 2}$; φ_i are angles of displacement when fixing

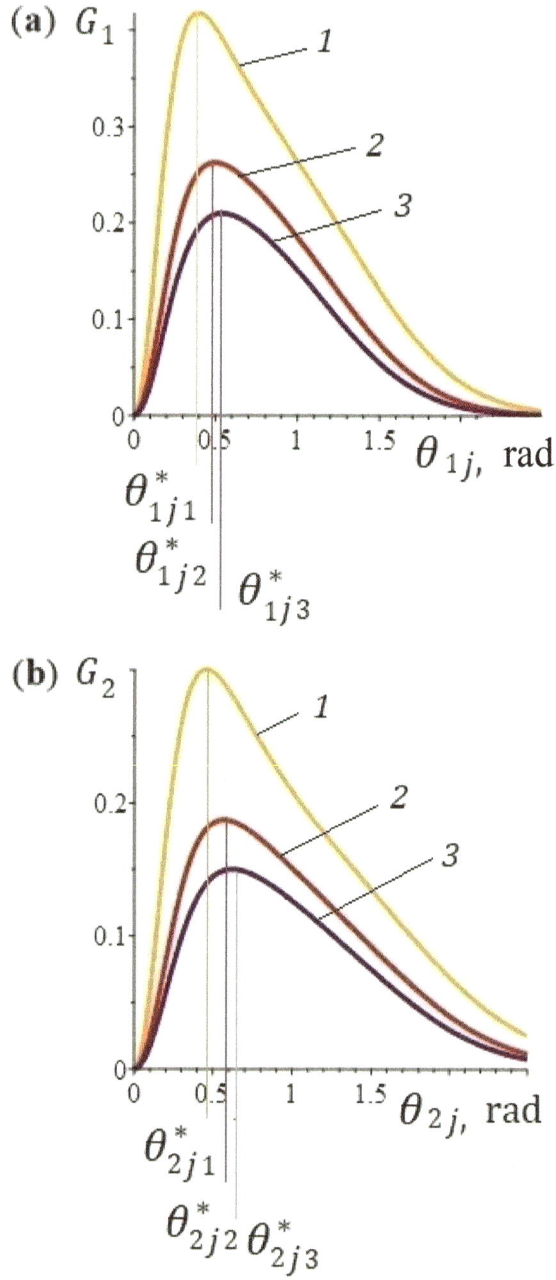

Fig. 4 Dependence $G_i(\theta_{ij}, \omega_i)$: a—secondary polypropylene GOST 29,996–86 (i = 1); b—primary polypropylene GOST 29,996–86 ($i = 2$); $u_{si} = 8$; $u_{ri} = 2$; $\omega_i = 57,6\ \mathrm{s}^{-1}$; *1*—$\delta_i = 0,60$; *2*—$\delta_i = 0,67$; *3*—$\delta_i = 0,70$

elastic blades on the surface of cylindrical drums of the specified components $i = \overline{1,2}$;

- operating parameters: ω_i are angular speeds of rotation of mixing drums with the formation of rarefied flows of components $i = \overline{1,2}$.

According to Fig. 3 surfaces for differential distribution functions $g_{ij}(\theta_{ij})$ number of particles of secondary polypropylene ($i = 1$, Fig. 3a) and primary ($i = 2$, Fig. 3b) by dissemination angle θ_{ij} at (9) with deformations of elastic blades of the row $j = \overline{1, u_{ri}}$; $u_{ri} = 2$ have one pronounced maximum. At the same time, in the case of both secondary (graphs 1 and 2, Fig. 1a) and primary (graphs 1 and 2, Fig. 1b) polypropylene, close values are observed for the indicated maxima related to different rows of elastic blades. In particular, for the ratios of the maximum values corresponding to the rows of elastic blades at $j = \overline{1,2}$, conditions are met: $g_{i1}^{\max}/g_{i2}^{\max} \approx 0.99$; $\theta_{i2}^* - \theta_{i1}^* = 0.01$ rad. Increasing dissemination θ_{ij} angle 3 times compared to extreme values θ_{i1}^*, θ_{i2}^* leads to a drop in the values of the function $g_{ij}(\theta_{ij})$ at $g_{11}/g_{12} \approx 0.88$; $g_{21}/g_{22} \approx 0.81$.

This result shows that the presence of the second row of elastic blades ($j = 2$), mounted on a cylindrical drum with angular displacement φ_i, is reflected in the increase not only in the values of the distribution functions $g_{ij}(\theta_{ij})$ (Fig. 3), but also on the bursts of the graphs for the full dissemination functions $G_i(\theta_{ij})$ (Fig. 4). Comparison of the latter $G_1(\theta_{1j})$, $G_2(\theta_{2j})$, for example, for the values of the degree of deformation of the blade $\delta_i = 0,70$ for miscible polymer components $i = 1$ (graph 3, Fig. 4a) and $i = 2$ (graph 3, Fig. 4b) gives a picture of the approach of the extreme values of the scattering angles of the polymer components θ_{1j}^*, θ_{2j}^*. In this case, the conditions for relations are fulfilled $\theta_{1j}^*/\theta_{2j}^* \approx 1.23$; $G_1^{\max}/G_2^{\max} \approx 1.28$ и выполняются условия для отношений $\theta_{1j}^*/\theta_{2j}^* \approx 1.23$; $G_1^{\max}/G_2^{\max} \approx 1.28$.

A similar trend is observed for these functions with other values of the combined parameter $\delta_i = 0,60$ (graph 1, Fig. 4a; graph 1, Fig. 4b), when $\theta_{1j}^*/\theta_{2j}^* \approx 1.22$; $G_1^{\max}/G_2^{\max} \approx 1.40$ and at $\delta_i = 0.67$ (graph 2, Fig. 4a; graph 2, Fig. 4b), when $\theta_{1j}^*/\theta_{2j}^* \approx 1.25$; $G_1^{\max}/G_2^{\max} \approx 1.40$. So, the indicated conditions for the functions $g_{ij}(\theta_{ij})$ and $G_i(\theta_{ij})$ characterize the efficiency of the process of obtaining a dispersed polymer mixture at the first stage of the operation of the rotary apparatus (Fig. 2) according to the criterion of convergence of curves for mixed components $i = \overline{1,2}$ proposed in [18]. The additional use of the second row of elastic blades allows the dispersed flows of dispersed polymer components to be displaced into a narrower range of extreme values for the angles of their dissemination θ_{1j}^*, θ_{2j}^*.

5 Conclusion

The urgent problem of recycling to improve the technical support of digital technologies, including at various stages of 3D printing, can be solved with the help of an additional process of mixing polymer components of the secondary and primary types in a rotary apparatus. At the same time, the proposed mathematical model is the theoretical basis for constructing a cyber-physical system, taking into account a set of information variables $z(t) \equiv \{z_\nu(t)\} = \{x(t), y(t), a_{i,ki} = const, b_{i,li} = const\}$, $\nu = \overline{1, s_0}$ at $s_0 = 36$ parameters. Note that the results of modeling the process of obtaining a two-component polymer-dispersed mixture, consisting of secondary and primary bulk materials, were obtained taking into account an additional set $q = \{\rho_i, \overline{d_i}, k_f, k_u\}$, relating to the characteristics of the physical and mechanical properties of dispersed media and elastic blades. These results are analyzed from the standpoint of the criterion of the effectiveness of the mixing process of bulk components proposed by the authors earlier [18].

The main conclusions and results of the study include the following.

- The assumptions were confirmed that the sets of parameters (constructive $\{a_{1,k1}\}$ and regime $\{b_{i,li}\}$) the following characteristics of the mixing process: combined parameter $\delta_i = r_{ci}/l_{bi}$; constructive parameters u_{ri}, φ_i; regime parameters ω_i, $i = \overline{1, 2}$.
- The analysis of the results of the model from the standpoint of the known criterion of the effectiveness of the mixing process of bulk components was carried out.
- It is shown that the use of the second row of elastic blades makes it possible to displace the scattered flows of polypropylene components of the secondary and primary types into a narrower range of variation of the extreme values for the angles of their scattering.

An increase in the degree of deformation of elastic blades within $\delta_i = (0, 60 - 0, 70)$ is reflected in the growth of the bursts of the graphs for the full spreading functions $G_i(\theta_{ij})$ in the following ranges of change in relations $G_1^{\max}/G_2^{\max} = (1, 28 - 1, 40)$ and $\theta_{1j}^*/\theta_{2j}^* = (1.22 - 1, 25)$.

References

1. Martikka, O., Kärki, T.: Promoting recycling of mixed waste polymers in wood-polymer composites using compatibilizers. Recycling **4**(1), 6 (2019). https://doi.org/10.3390/recycling4010006
2. Tratzi, P., Giuliani, C., Torre, M., Tomassetti, L., Petrucci, R., Iannoni, A., Torre, L., Genova, S., Paolini, V., Petracchini, F., DiCarlo, G.: Effect of hard plastic waste on the quality of recycled polypropylene blends. Recycling **6**(3), 58 (2021). https://doi.org/10.3390/recycling6030058
3. Turner, R.P., Kelly, C.A., Fox, R., Hopkins, B.: Re-formative polymer composites from plastic waste: novel infrastructural product application. Recycling **3**(4), 54 (2018). https://doi.org/10.3390/recycling3040054
4. Turku, I., Kasala, S., Kärki, T.: Characterization of polystyrene wastes as potential extruded feedstock filament for 3D printing. Recycling **3**(4), 57 (2018). https://doi.org/10.3390/recycling3040057
5. Suárez, L., Domínguez, M.: Sustainability and environmental impact of fused deposition modeling (FDM) technologies. Int. J. Adv. Manuf. Technol. **106**, 1267–1279 (2020). https://doi.org/10.1007/s00170-019-04676-0
6. Gebler, M., Schoot Uiterkamp, A.J.M., Visser, C.: A global sustainability perspective on 3D printing technologies. Energy Policy. 74, 158–167 (2014). https://doi.org/10.1016/j.enpol.2014.08.033

7. Cunico, M.W.M., Kai, D.A., Cavalheiro, P.M., de Carvalho, J.: Development and characterisation of 3D printing finishing process applying recycled plastic waste. Virtual Phys Prototyp. **14**(1), 37–52 (2019). https://doi.org/10.1080/17452759.2018.1521248
8. Kreiger, M.A., Mulder, M.L., Glover, A.G., Pearce, J.M.: Life cycle analysis of distributed recycling of post-consumer high density polyethylene for 3-D printing filament. J. Cleaner Prod. **70**, 90–96 (2014). https://doi.org/10.2139/ssrn.3331165
9. Hopewell, J., Drovak, R., Kosior, E.: Plastics recycling: challenges and opportunities. Philos. Trans. R. Soc. B **364**(1526), 2115–2126 (2009). https://doi.org/10.1098/rstb.2008.0311
10. Kapranova, A.B., Verloka, I.I., Bahaeva, D.D.: About Preparation of the Analytical Platform for Creation of a Cyber-Physical System of Industrial Mixture of Loose Components. Cyber-Physical Systems: Advances in Design & Modelling. Studies in Systems, Decision and Control. 259, 81–91 (2020). https://doi.org/10.1007/978-3-030-32579-4_7
11. Kapranova, A.B., Bahaeva, D.D., Stenko, D.V., Verloka, I.I.: A Model of Rotational Mixing of Loose Environment on the Platform of Cyber-Physical. In: Kravets A.G., Bolshakov A.A., Shcherbakov M. (eds) Cyber-Physical Systems: Design and Application for Industry 4.0. Studies in Systems, Decision and Control. Springer, Cham, 342, 167–178 (2021). https://doi.org/10.1007/978-3-030-66081-9_13
12. Pakkanen, J., Manfredi, D., Minetola, P., Iuliano, L.: About the Use of Recycled or Biodegradable Filaments for Sustainability of 3D Printing. In: Campana G., Howlett R., Setchi R., Cimatti B. (eds) Sustainable Design and Manufacturing . Smart Innovation, Systems and Technologies, 68. Springer, Cham (2017). https://doi.org/10.1007/978-3-319-57078-5_73
13. Cantrell, J.T., Rohde, S., Damiani, D., Gurnani, R., DiSandro, L., Anton, J., Young, A., Jerez, A., Steinbach, D., Kroese, C., Ifju, P.G.: Experimental characterization of the mechanical properties of 3D-printed ABS and polycarbonate parts, In: Sanichiro Yoshida, Luciano Lamberti (eds) Cesar Sciammarella Rapid Prototyping Journal, 23(4), 811–824 (2017). https://doi.org/10.1108/RPJ-03-2016-0042
14. Dong, Y., Milentis, J., Pramanik, A.: Additive manufacturing of mechanical testing samples based on virgin poly (lactic acid) (PLA) and PLA/wood fibre composites. Adv. Manuf. **6**, 71–82 (2018). https://doi.org/10.1007/s40436-018-0211-3
15. Eleni Gkartzou, Elias, P. Koumoulos and Costas A.: Charitidis Production and 3D printing processing of bio-based thermoplastic filament Manufacturing Rev. Manufacturing Rev. 4, 1 (2017). https://doi.org/10.1051/mfreview/2016020
16. Kapranova, A.B., Verloka, I.I., Bahaeva, D.D., Tarshis, M.Yu., Cherpitsky, S.N.: To the calculation of the average value of the volume fraction of the key bulk component at the intermediate stage of mixing with an inclined bump, Frontiers in Energy Research: Process and Energy Systems. 8, article 135, 1–11, (2020). https://doi.org/10.3389/fenrg.2020.00135
17. Kapranova, A.B., Bakin, M.N., Verloka, I.I.: Simulation of the Quality Criterion of a Mixture in a Drum-Belt Apparatus. Chemical and Petroleum Engineering. 54(5), 287–297 (2018). https://doi.org/10.1007/s10556-018-0477-0
18. Kapranova, A.B., Verloka, I.I.: Stochastic Description of the Formation of Flows of Particulate Components in Apparatuses with Brush Elements. Theor. Found. Chem. Eng. **52**(6), 1004–1018 (2018). https://doi.org/10.1134/S0040579518050330
19. Lebedev, A.E., Kapranova, A.B., Stenko D.V. Patent: 2749510 Russian Federation, IPC B01F 3/18, B01F 5/04, B01F 7/00, B01F 9/10. Mixer of bulk materials. Bul. No. 17. (2021)
20. Makarov, Yu. I.: Devices for mixing bulk materials. 216 p. Mechanical Engineering, Moscow (1973)
21. Verloka, I., Kapranova, A., Tarshis, M., Cherpitsky, S.: Stochastic modeling of bulk components batch mixing process in gravity apparatus. Int. J. Mech. Eng. Technol. **9**(2), 438–444 (2018)
22. Bogomolov, B., Boldyrev., V, Zubarev, A., Meshalkin, V., Men'shikov V/: Intelligent Logical Information Algorithm for Choosing Energy- and Resource-Efficient Chemical Technologies, Theoretical Foundations of Chemical Engineering, 53(5), 709–718 (2019). https://doi.org/10.1134/S0040579519050270
23. Kafarov, V.V., Dorokhov, I.N, Arutyunov, S.Yu.: System analysis of processes of chemical engineering. 440 p. Science, Moscow (1985)

24. Klimontovich, Yu. L.: Turbulent Motion and the Structure of Chaos: A New Approach to the Statistical Theory of Open Systems, 328 p. LENAND, Moscow (2014) Klimontovich, Yu. L.: Turbulent Motion and the Structure of Chaos: A New Approach to the Statistical Theory of Open Systems, 328 p. LENAND, Moscow (2014)
25. Klimontovich, Yu. L.: Turbulent Motion and the Structure of Chaos. Series: Fundamental Theories of Physics. V. 42. Springer Netherlands, pp. 401 (1991). https://doi.org/10.1007/978-94-011-3426-2
26. Kapranova, A., Verloka, I., Bahaeva, D.: Factors of increasing the performance of the bulk material mixer of the gravity type. In: Anatolijs, B., Nikolai, V., Vitalii, S. (eds) Proceedings of EECE 2019. Lecture Notes in Civil Engineering, vol. 70, pp. 337–347. Springer, Cham (2020). https://doi.org/10.1007/978-3-030-42351-3_30
27. Kapranova, A., Bahaeva, D., Stenko, D., Vatagin, A., Lebedev, A., Lichak, D.: Distribution of the components of the building mixture in the presence of secondary raw materials during rotary mixing. In: E3s Web of Conference, vol. 220, 01060 (2020). https://doi.org/10.1051/e3sconf/202022001060
28. Kapranova, A., Bahaeva, D., Stenko, D., Fedorova, D., Vatagin, A., Lebedev, A.: Stochastic analysis of impact mixing of bulk materials in a rotary apparatus. In: E3s Web of Conference, vol. 247, 01048 (2021). https://doi.org/10.1051/e3sconf/202124701048
29. Röpke, G.: Statistical Mechanics for Non-Equilibrium. German Publishing House of Sciences, Berlin (1987)
30. Bharucha-Reid, A.T.: Elements of the Theory of Markov Processes and Their Applications. McGraw-Hill, New York (1960)
31. Geng, T., Sau, L.L., Xiaochuan, Ya., Moo, S.H.: A dimensionless analysis of residence time distributions for continuous powder mixing. powder technology **315**, 332–338 (2017). https://doi.org/10.1016/j.powtec.2017.04.007
32. Lobasenko, B.A., Borodulin, D.M., Kotlyarov, R.V., Golovacheva, Y.S., Bakin, I.A.: Mathematical modeling of the membrane concentration of whey by the state space method. Int. Rev. Model. Simul. **13**(4), 194–206 (2020). https://doi.org/10.15866/iremos.v13i4.18643
33. Ghaderi, A.: Continuous mixing of particulate materials. In: Proceedings of the 4th International Conference for Conveying and Handling of particulate solids, At Budapest, Hungary, Vol. 2 (2003). https://doi.org/10.13140/2.1.3487.6801
34. Johnson, N., Kendall, M., Stuart, A.: J. R. Stat. Soc. Ser. D (The Stati.) **12**(2), 138 (1962)
35. Kordas, M., Pluskota, D., Rakoczy, R.: The characterization of the residence time distribution in a fluid mixer by means of the information entropy. In book: Practical Aspects of Chemical Engineering, pp. 201–216 (2018)
36. Leonchik, B.I., Mayakin, V.P.: Measurements in Dispersed Flows Energy, 248 p. Moscow (1985)
37. Akhmadiev, F.G., Gizzyatov, R.F.: Modeling of separation of granular materials on multiple-deck classifiers using the theory of stochastic processes. Theor. Found. Chem. Eng. **52**(3), 360–370 (2018)
38. Mizonov, V., Balagurov, I., Berthiaux, H., Gatumel, C.: Gatumel Markov chain model of mixing kinetics for ternary mixture of dissimilar particulate solids. Particuology **31**, 80–86, (2016). https://doi.org/10.1016/j.partic.2016.05.006
39. Dehling, H.G., Gottschalk, T., Hoffmann, A.C.: Stochastic Modeling in Process Technology, 279 p. Elsevier Science, London (2007)
40. Zhukov, V.P., Belyakov, A.N.: Simulation of combined heterogeneous processes based on discrete models of the Boltzmann equation. Theor. Found. Chem. Eng. **51**(1), 88–93 (2017). https://doi.org/10.1134/S0040579517010158
41. Zhuang, Y., Chen, X., Liu, D.: Stochastic bubble developing model combined with Markov process of particles for bubbling fluidized beds. Chem. Eng. J. 291, 206–214 (2016). https://doi.org/10.1016/j.cej.2016.01.095

Cyber-Physical System of a Polymer Composition Optimization Based on the Solution of a Fuzzy Programming Problem

Egor Feoktistov and Ilya Germashev

Abstract A model of a cyber-physical system is proposed for calculating the optimal concentration of active additives for the best manifestation of the properties of the polymer composition. The substantiation and relevance of the task are given. This work is based on the problem of fuzzy programming, the solution of which is based on the approach with alpha levels. A computational experiment illustrates the proposed solution method. Data visualization allows restoring the membership function and obtaining a fuzzy concentration vector for solving a fuzzy programming problem.

Keywords Cyber-physical system · Optimization methods · Fuzzy programming · Alpha levels · Polymer composition · Composite materials

1 Introduction

Polymer compositions consist of many components and are complex cyber-physical systems. In modern production, composite materials with unique properties are increasingly used to solve various technological problems. Controlling the properties of polymer compositions (PC) is a multicriteria optimization problem, which can be solved in multiple approximations by mathematical, quantum-chemical, and chemical methods. The empirical approach, in this case, requires enormous resource costs. To give not only a qualitative but also a quantitative representation of the manifestation of the list of possible properties of a cyber-physical system under the conditions of chemical interaction is an excellent and still unsolved problem. One of the most common ways to control properties is active and passive additives in PC [1]. Certain chemicals are introduced into the overall composition to increase transparency, plasticity, fire resistance, and other critical properties in production.

Composite materials are widely used in applied problems of our time; however, the synthesis of such systems is constantly faced with new requirements and demands

E. Feoktistov (✉) · I. Germashev
Volgograd State University, 100, Universitetsky Avenue, Volgograd 400062, Russia
e-mail: faa-201_193934@volsu.ru

A. G. Kravets et al. (eds.), *Cyber-Physical Systems Engineering and Control*,
Studies in Systems, Decision and Control 477,
https://doi.org/10.1007/978-3-031-33159-6_10

of industries in various fields of human activity. The control and determination of the properties of these compositions is a multicriteria problem, which is solved by multiple methods, including empirical, mathematical, and quantum-chemical ones [1]. The relevance of polymers in the current context of the main areas of science and practice is explained by a whole set of unique chemical and physical properties that are a consequence of the chain structure of molecules. Viscosity, transparency, flexibility, and fire resistance are some of the many possible properties of such materials.

These compositions form products, the use of which can be found everywhere, from industry and construction to radio-electronic technologies. The range of properties, including reliability, stability, and durability, determines the widespread use of these materials; however, there are many individual factors and implementation conditions in each specific case. It is rather challenging to consider the total variability of such factors, especially considering the filling of polymer matrices with various passive and active additives for the sake of functionality or simplification of technological processes. Therefore, the synthesis of new materials is rarely accompanied by preliminary theoretical work; much more often, such methods are used at the stage of evaluation of finished substances after empirical studies. As a result, the use of mathematical and informational methods and models can become an extremely relevant and promising tool with predictive functionality, regardless of the chosen stage of designing a polymer product.

The most well-known polymer compositions were synthesized about half a century ago. Actual tasks are concentrated more around optimizing the properties of already known polymers, although the synthesis is also carried out in new directions. One of the promising development vectors is nanocomposites, which require a more complex technological process of synthesis, but at the same time, have much higher critical properties. However, an important task remains to identify the properties of these chemical systems in the context of implementation for practical purposes. More often and technically easier to analyze polymers empirically, that is, experimentally; however, this method, as a rule, is accompanied by the use of expensive equipment and a relatively large-scale and costly technological process. It explains the prospects of using non-empirical methods based on mathematical, informational, and chemical approaches, namely their predictive operational functionality [2–4].

The most common in the scientific literature is the stage of controlling the properties of polymer compositions using passive fillers or active additives [5–9]. However, preliminary mathematical calculations based on any theories or the construction of information systems are rarely used in such procedures. Much more often, one can meet the methods of quantum chemistry with their implementation to obtain polymerization results while controlling certain of its characteristics. On the other hand, the interaction and influence of active additives are described mathematically in rare specific cases if empirically verified information is available.

The theoretical basis for describing the chemical interaction is of a point nature and, as a rule, is found empirically. A single patent dealing with a specific polymer matrix and specific additives can describe the relationship between the performance and concentrations of certain active additives. Building a general system for determining the effect of additives on properties is a rather difficult task due to the multicriteria and the variety of chemical and physical interactions of polymer matrices and additives [1, 10]. Moreover, each PC has a whole range of properties, the manifestation of which will differ significantly with varying concentrations of ingredients. Since the customers of materials demand from the manufacturer a complete set of properties that should manifest themselves relatively equally well, this task is promising in the current conditions of material synthesis.

The problem is also defined by the fact that there is no single correct method for each specific task. A wide range of properties of a considerable number of various polymers opens up endless possibilities for ongoing analysis. In this regard, the actualization of generalization in these approaches is a priority task that has a deep practical meaning in the long term.

Identifying ingredients for designing a polymer composition with the required characteristics and determining the structure and composition of a chemical system are still unsolved problems in science [1, 11–15]. Often, complex compositional systems are described using fuzzy mathematics [16–18]. This method has its advantages since the data is often presented in ranges or not at all. The task is complicated by the often polar opinions of experts on the same issue or recipe, which leads to the need to describe the variables linguistically.

Since PC compositions often do not have exact concentration values but are presented as a range of acceptable values, they are proposed to use the basics of fuzzy analysis [19]. In our previous studies, in optimizing the concentration of additives for the best manifestation of the properties of the polymer composition, the problem of fuzzy quadratic programming was used, in which only variables were undefined [20]. Therefore, in [20], the defuzzification of variables was carried out, and the problem of actual quadratic programming was obtained, which was solved. This chapter considers a fuzzy programming problem in which both variables and parameters are fuzzy, so a solution using alpha levels is proposed.

Given the fuzzy nature of the data, obtaining accurate values in the future can become an essential task within some practical issues or technological processes. The apparent method that suggests itself after calculating the alpha levels is the interpolation of the concentration vector. Sometimes customers may be critical of the exact melting temperature limit of the composition, which in these calculations will require the use of the interpolation method. However, at this stage, it is proposed to visualize the results obtained in order to determine the need for approximation or interpolation of intermediate values. This chapter offers a solution to an abstract example to prove the possibility of implementing the approach when working with fuzzy data. As a result, the membership function will be visualized for fuzzy vectors of each alpha level.

2 Problem Statement

The formal problem is that there is a polymer P with a certain set of properties $Q_1 \dots Q_m$, the manifestations of which in P will be denoted by $x_1 \dots x_m$. In addition to matrix P, PC includes ingredients $s_1 \dots s_n$ with concentration $c_1 \dots c_n$. It is necessary to find a set of values $c_1, \dots, c_n$, that provides the greatest manifestation of the Q_j for a fixed j.

2.1 *Mathematical Model*

We fix some value j and further we will consider a specific property Q_j.

To solve the problem, it is proposed to formulate the problem of mathematical programming, using fuzzy values as variables $c_1, \dots, c_n$. In this case, we denote by $\hat{p}_i^{min}$ minimum possible content of s_i, by $\hat{p}_i^{max}$ denote the maximum possible content s_i, by $\hat{p}$ denote the largest total content of all additives, i.e.

$$\sum_{i=1}^{n} \hat{c}_i \le \hat{p}, \hat{c}_i \le \hat{p}_i^{max}, \hat{c}_i \ge \hat{p}_i^{min}, i = \overline{1,n}, \tag{1}$$

where $\hat{c}_i$—fuzzy value for c_i.

These limitations make it possible to take into account such aspects of modeling as, for example, the fact that some additives, when passing a certain concentration threshold, may have an undesirable significant effect on the chemical system. When solving various applied problems, the content of a certain additive in the composition can be fundamental, as a result of which the third limitation is presented.

Denoting by the elements of the fuzzy matrix $\hat{D} = \left(\hat{d}_{i_1 i_2}\right)$ specific, per unit concentration, impact on a property Q_j pair of additives s_{i_1} and s_{i_2}, and elements of vector $\hat{l} = \left(\hat{l}_i\right)$—specific effect of the additive s_i on property's manifestation Q_j respectively. We will assume that the matrix $\hat{D}$ and vector $\hat{l}$ are given.

By changing the concentrations $\hat{c}_i$ it is possible to optimize the fuzzy manifestation $\hat{q}_j$ of the property Q_j of the composite material, which is described by the following relationship:

$$\hat{q}_j = -\frac{1}{2}\hat{c}^T D\hat{c} + l^T\hat{c}$$

Thus, we obtain the problem of fuzzy programming.

$$\hat{q}_j(\hat{c}) = -\frac{1}{2}\hat{c}^T \hat{D}\hat{c} + \widehat{l^T}\hat{c} \to max_{\hat{c}} \tag{2}$$

under constraints (1), where the matrix elements $\hat{D} = \left(\hat{d}_{i_1 i_2}\right)$ are given as follows:

$$\hat{d}_{i_1 i_2} = e^{-\frac{\left(x - d_{i_1 i_2}\right)^2}{\delta^2_{i_1 i_2}} \ln 2},$$

where the matrices D and Δ are assumed to be given:

$$D = \begin{pmatrix} 1 & 0.5 & 0.75 & 0 \\ 0.5 & 0.5 & 1.3 & 0.6 \\ 0.75 & 1.3 & 1.5 & 0.85 \\ 0 & 0.6 & 0.85 & 1.05 \end{pmatrix}, \ \Delta = \begin{pmatrix} 0.05 & 0.05 & 0.05 & 0.05 \\ 0.05 & 0.05 & 0.05 & 0.05 \\ 0.05 & 0.05 & 0.05 & 0.05 \\ 0.05 & 0.05 & 0.05 & 0.05 \end{pmatrix}$$

We assume that the total proportion of all ingredients is limited by $\hat{p} = e^{-100(x-0.7)^2 \ln 2}$. Each active additive also has boundary fuzzy values $\hat{p}_i^{max} = e^{e^{-100(x-0.5)^2 \ln 2}}$ and $\hat{p}_i^{min} = e^{-10^6(x-0.001)^2 \ln 2}$.

It is proposed to calculate the α-level for all fuzzy parameters, variables of problem (2) and constraints (1) for some fixed value $\alpha \in [0, 1]$. As a result, we obtain a problem of quadratic programming over real numbers. Solving the obtained problem, we obtain the corresponding α-levels for the optimal fuzzy solution. Next, we repeat this procedure for other values of the α-level. To do this, the designations D_α and l_α, are introduced, on the basis of which, at values of alpha levels from 0.1 to 1, with a step of 0.1, the concentration vectors $\vec{c}_\alpha$ will be calculated for each of the four additives taken in the computational experiment.

3 Computational Experiment

Let us construct α-levels for the model parameters. For example, for α-level $\alpha_1 = 0.1$ the calculated matrix and vector D_{α_1}, l_{α_1} for the left and right boundaries take the following form, shown in Fig. 1, depicting the graph for the function $\hat{d}_{11} = e^{\frac{-(x-1)^2}{0{,}05^2} \ln 2}$ and its intersection with the straight line $y = 0.1$, corresponding to the given α-level. Solving this equation, we obtain two solutions corresponding to the left boundary $d_{11\alpha_1.left} = 0.71$ and to the right boundary $d_{11\alpha_1.right} = 1.29$ of α_1-level. Performing similar calculations for the remaining parameters, we obtain:

$$D_{\alpha_{1.left}} = \begin{pmatrix} 0.71 & 0.21 & 0.46 & -0.29 \\ 0.21 & 0.21 & 1.01 & 0.31 \\ 0.46 & 1.01 & 1.21 & 0.56 \\ -0.29 & 0.31 & 0.56 & 0.76 \end{pmatrix},$$

$$l_{\alpha_{1.left}} = (-0.04; 0.96; 0.09; 0.21),$$

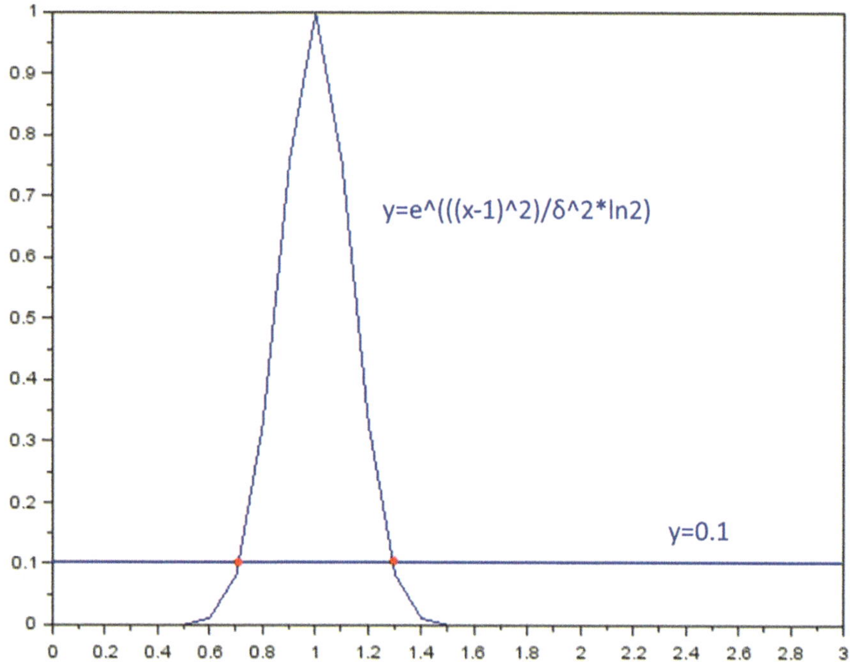

Fig. 1 Graphical representation of the equation $e^{\frac{-(x-1)^2}{0.05^2}\ln 2} = 0.1$

$$D_{\alpha_{1.right}} = \begin{pmatrix} 1.29 & 0.78 & 1.04 & 0.29 \\ 0.78 & 0.78 & 1.59 & 0.89 \\ 1.04 & 1.59 & 1.78 & 1.14 \\ 0.29 & 0.89 & 1.14 & 1.34 \end{pmatrix},$$

$$l_{\alpha_{1.right}} = (0.54; 1.54; 0.67; 0.78).$$

Then the quadratic programming problem is solved corresponding to the fuzzy programming problem (2) for the given alpha level. After these calculations, we repeat the described procedure for the remaining values of the α-level, up to 0.9. A graphical representation of the solution for one of the alpha levels is shown in Fig. 1. The resulting solutions are presented in Table 1.

Table 1 Calculation results

α-level	$\vec{c}_\alpha$ for left border	$\vec{c}_\alpha$ for right border	q_j for left// right borders
0.1	[0.356; 0; 0; 0.244]	[0.461; 0.002; 0.002; 0.335]	−0.049//0.223
0.2	[0.360; 0; 0.010; 0.230]	[0.489; 0.002; 0.004; 0.305]	−0.002//0.218
0.3	[0.374; 0; 0; 0.226]	[0.492; 0.002; 0.002; 0.204]	0.018//0.207
0.4	[0.376; 0; 0; 0.224]	[0.501; 0.002; 0.002; 0.195]	0.022//0.188
0.5	[0.379; 0; 0; 0.221]	[0.498; 0.002; 0.002; 0.298]	0.039//0.171
0.6	[0.374; 0; 0.011; 0.215]	[0.506; 0.002; 0.029; 0.263]	0.050//0.121
0.7	[0.380; 0; 0; 0.220]	[0.499; 0.002; 0.002; 0.297]	0.062//0.154
0.8	[0.380; 0; 0; 0.220]	[0.502; 0.002; 0.002; 0.294]	0.073//0.172
0.9	[0.385; 0; 0.009; 0.206]	[0.504; 0.002; 0.004; 0.316]	0.091//0.135
1.0	[0.422; 0.001; 0.001; 0.376]		0.112

4 Visualization of the Resulting Solution

The visualization of the membership function of each alpha level was carried out using the obtained discrete points from the vectors for each concentration of the active additive. This procedure was applied to the first and fourth additives since the concentration of the second and third either did not change depending on the alpha level, or the changes were insignificant. As a result of the illustration, functions and their graphs were obtained, by which it is possible to find an intermediate between known discrete values.

The most important class of such functions is the set of algebraic polynomials. Polynomials have an obvious advantage—their values are easy enough to calculate, as well as to carry out the operations of addition, multiplication, integration, or differentiation. The polynomial representation of the data, in this case, is the most appropriate method for finding intermediate values. Discrete point calculations were performed in the SciLab software environment.

As a result, a vector of fuzzy coordinates $\vec{c}_1^*$ and $\vec{c}_4^*$ was obtained for alpha levels, which are shown in Figs. 2 and 3. If in the case of the first addition, a quadratic distribution is observed behind small deviations, then in the case fuzzy concentration of the fourth additive, the scatter turned out to be quite strong.

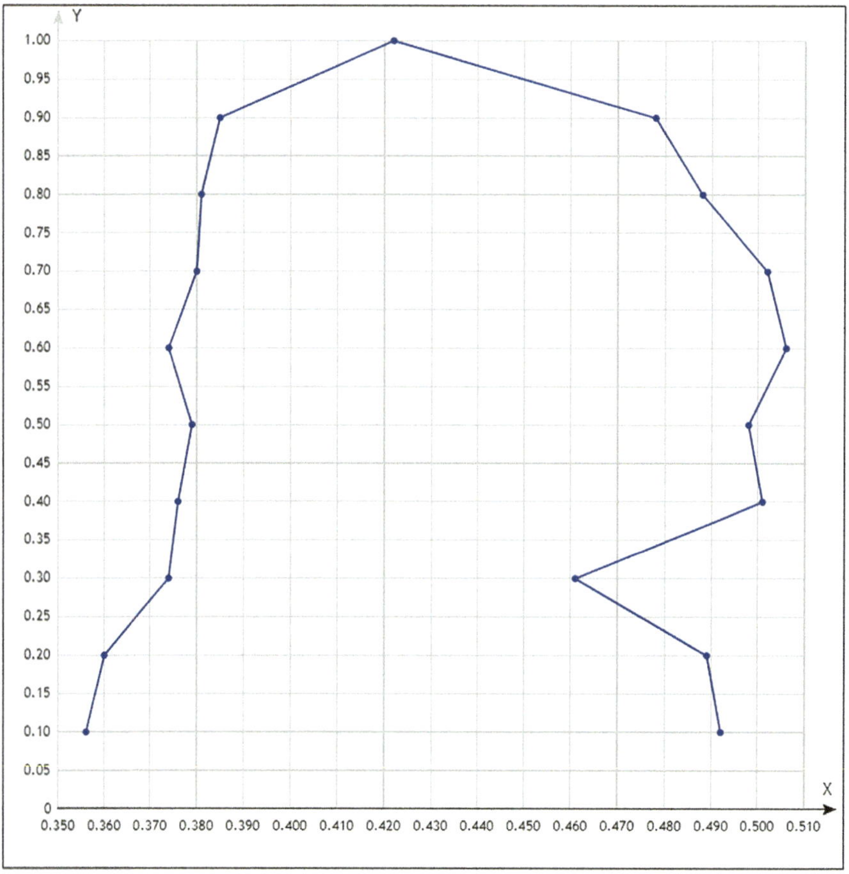

Fig. 2 Graphical representation of alpha values for $\vec{c}_1^*$

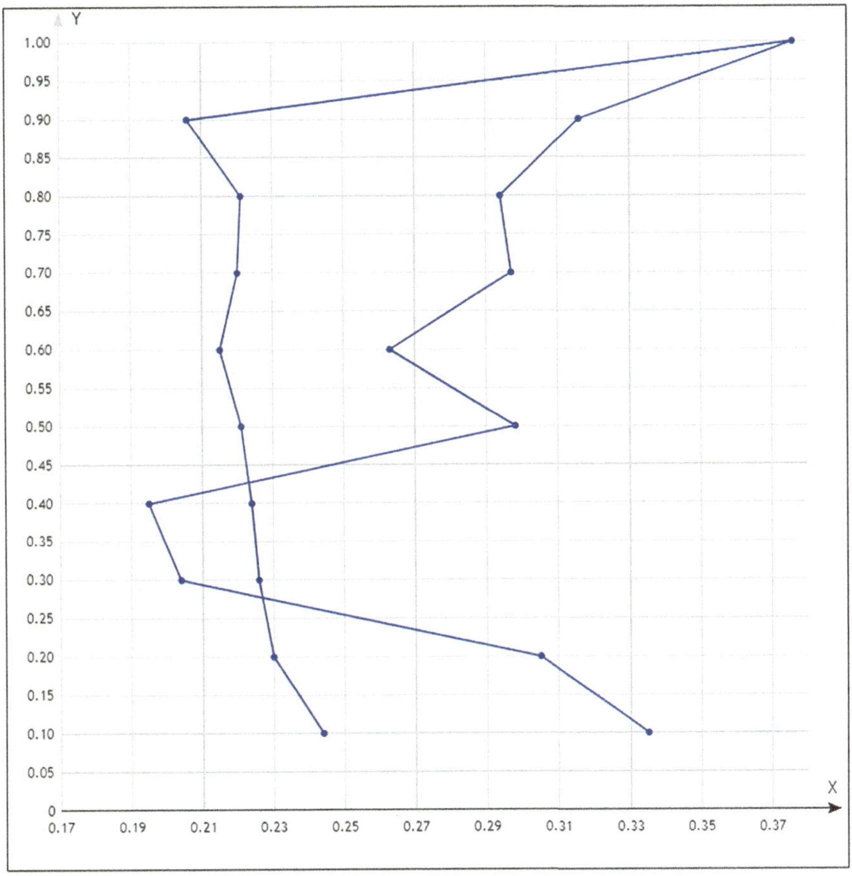

Fig. 3 Graphical representation of alpha values for $\vec{c}_4^*$

5 Conclusion

As a result of these calculations, the result was obtained in the form of alpha levels of the coordinates of the fuzzy concentration vector. The peculiarity of this model is that the concentration retains the fuzzy nature of the data at the time of the visualization and after it. A promising method for further study the obtained functions is interpolation and approximation of data to search for intermediate values from known discrete values.

Using this model, it is possible to calculate the optimal concentrations of additives while maintaining their fuzzy structure to control specific properties of a particular composition. It allows us to continue research in fuzzy mathematics while retaining the ability to carry out defuzzification at any stage and move to an accurate model. The development of the proposed model is envisaged in the future. First, it is possible to use quantum chemical methods for the practical implementation of the model with specific active additives and a polymer matrix. Secondly, obtaining the matrix D and the vector l takes place only experimentally or is also calculated using quantum chemistry methods. Empirical methods tend to be expensive in terms of time and other resources and are often challenging to implement. Future works plan to include quantum chemical methods in the developed model to refine the calculations of fuzzy concentrations and qualitatively evaluate the results. It is also possible to connect machine learning methods to analyze model parameters based on known PC recipes.

This information will later make it possible to formalize the initial data of the subject area by employing mathematical modeling, when, upon receipt of an order from a consumer, a set of requirements for a composite material is automatically generated, and then estimates of compliance with the formed requirements for specific candidates for components of a polymer composition are calculated. Then the technical side of the composition of the chemical structure is worked out, which ensures the high quality of the resulting material. The model presented here is a possible prospect for developing the applied use of mathematical modeling tools in technological production processes to improve the synthesis of new materials, optimize their composition, and, accordingly, properties. Further development will allow taking into account the economic, environmental, and many other aspects of the development of polymer composite materials.

References

1. Bobryshev, A.N., Yerofeev, V.T., Kozomazov, V.N.: Polymernie Compositsionnie Materialy: Ucheb. Posobie (Polymer composite materials: textbook). ASV, Moscow (2013) (In Russian)
2. Wang, G.: Modelling of thermal transport through a nanocellular polymer foam toward the generation of a new superinsulating material. Nanoscale **9**, 5996–6009 (2017)
3. Rentería-Baltiérrez, F.Y., Reyes-Melo, M.E., Puente-Córdova, J.G., López-Walle, B.: Correlation between the mechanical and dielectric responses in polymer films by a fractional calculus approach. Appl. Polymer **138**, (7) (2021)
4. Grigoriev, I. V. Chislennoe issledovanie processa polimerizatsii butadiena metodami matematicheskogo modelirovania (Numerical study of the butadiene polymerization process by methods of mathematical modeling). Paper presented at Differencialnie Uravnenia I Smezhnie Problemi (Differential Equations and Related Problems), Bashkir State University, Sterlitamak, 25–29 June 2018 (In Russian)
5. Akgul, Y., Ahlatci, H., Turan, M.E., Simsir, H., Erden, M.E., Sun, Y., Kilic, A.: Mechanical, tribological, and biological properties of carbon fiber/hydroxyapatite reinforced hybrid composites. Polym. Compos. **41**, 2426–2432 (2020)
6. Wu, M.C., et al.: Polymer additives for morphology control in high-performance lead-reduced perovskite solar cells. Solar RRL **4**(6), 2000093 (2020)
7. Chen, Q., et al.: Thermal management of polymer electrolyte membrane fuel cells: a review of cooling methods, material properties, and durability. Appl. Energy **286**, 116496 (2021)
8. Germashev, I.V., Derbisher, V.E., Orlova, S.A.: Evaluation of activity of the fireproofing compounds in elastomer compositions by means of fuzzy sets. Kauch Rezina **6**, 15–17 (2001). (In Russian)

9. Germashev, I.V., Derbisher, V.E., Vasil'ev, P.M.: Prediction of the activity of low-molecular organics in polymer compounds using probabilistic methods. Theor. Found. Chem. Eng. **32**(5), 514–517 (1998)
10. Derbisher, E.V., Derbisher, V.E.: Application of computational methods for the creation and selection of polymer compositions with specified properties. Matematicheskaya Fizika I Computernoe Modelirovanie **1**(22), 35–53 (2019)
11. Jozwik, J., Dziedzic, K., Barszcz, M., Pashechko, M.: Analysis and comparative assessment of basic tribological properties of selected polymer composites. Materials **13**(1), 75 (2019)
12. Yang, X., Liang, C., Ma, T., Guo, Y., Kong, J., Gu, J., Zhu, J. (2018) A review on thermally conductive polymeric composites: classification, measurement, model and equations, mechanism and fabrication methods. Adv. Compos. Hybrid Mater. **1**(2), 207–230 (2018)
13. El Moumen, A., Tarfaoui, M., Lafdi, K.: (2019) Additive manufacturing of polymer composites: processing and modeling approaches. Compos. B Eng. **171**, 166–182 (2019)
14. Zhai, S., Zhang, P., Xian, Y., Zeng, J., Shi, B.: Effective thermal conductivity of polymer composites: theoretical models and simulation models. Int. J. Heat Mass Transf. **117**, 358–374 (2018)
15. Raju, G.M., Madhu, G.M., Khan, M.A., Reddy, P.D.S.: Characterizing and modeling of mechanical properties of epoxy polymer composites reinforced with fly ash. Mater. Today Proc. **5**(14), 27998–28007 (2018)
16. Anand, G., Alagumurthi, N., Elansezhian, R., Palanikumar, K., Venkateshwaran, N.: Investigation of drilling parameters on hybrid polymer composites using grey relational analysis, regression, fuzzy logic, and ANN models. J. Braz. Soc. Mech. Sci. Eng. **40**(4), 1–20 (2018)
17. Sinha, A.K., Narang, H.K., Bhattacharya, S.: Experimental determination, modelling and prediction of sliding wear of hybrid polymer composites using RSM and fuzzy logic. Arab. J. Sci. Eng. **46**(3), 2071–2082 (2021)
18. Upputuri, H.B., Nimmagadda, V.S., Duraisamy, E.: Optimization of drilling parameters on carbon fiber reinforced polymer composites using fuzzy logic. Mater. Today Proc. **2020**(23), 528–535 (2020)
19. Germashev, I.V., Derbisher, V.E., Derbisher, E.V.: Two-component model of the interaction of the ingredients of the polymer composition. Model of paired and solitary influence of ingredients of polymer composition. Cyber-Phys. Syst. Design Appl. Ind. 4.0 205–217 (2021)
20. Germashev, I.V., Feoktistov, E.F., Derbisher, V.E., Derbisher, E.V.: Optimization of the ingredients of the polymer composition under the conditions of pair interaction of active additives. Stud. Syst. Decis. Control **418**, 311–321 (2022)

Cyber-Physical Complex for the Optimal Design of Installation for Surface Hardening

Yu. Pleshivtseva, A. Pavlushin, A. Popov, and A. Yevelev

Abstract The chapter presents the structure of a cyber-physical complex for optimal design and control of inductors for surface hardening of complex-shaped workpieces. The core of the complex is an optimization procedure based on the alternance method of parametric optimization of systems with distributed parameters. It allows solving systems of transcendental equations closed regarding all unknown process parameters, including the design and operational mode characteristics of the induction heating system. A 2D numerical ANSYS model of the heating stage of surface hardening is integrated into the optimization procedure, developed in MATLAB. The program part of the presented cyber-physical complex is applied to solve the optimal inductor design problem for surface induction hardening of steel L-shaped workpieces.

Keywords Cyber-physical complex · Optimal design · Numerical simulation · Digital twin · Surface induction hardening · ANSYS · MATLAB

1 Introduction

Efficiency improvement of electro-technological processes is an actual and complex problem, which can be solved via optimization of design or operation modes of electro-technological equipment. One of the popular trends in the field of design and optimization of complex systems is the application of nonlinear multidimensional numerical mathematical models or «digital twins», allowing to adequately imitate the behavior of real objects considering a full range of their key features.

Modern technologies of heat treatment of metals often involve using induction installations during the heating stage due to their undeniable advantages over other competitive equipment. One of the widely used heat treatment processes in metallurgy is surface induction hardening of steel [1–5], aimed at raising the processed

Yu. Pleshivtseva (✉) · A. Pavlushin · A. Popov · A. Yevelev
Samara State Technical University, 244, Molodogvardeyskaya Str., 443100 Samara, Russia
e-mail: yulia_pl@mail.ru

A. G. Kravets et al. (eds.), *Cyber-Physical Systems Engineering and Control*,
Studies in Systems, Decision and Control 477,
https://doi.org/10.1007/978-3-031-33159-6_11

material's wear resistance by achieving some specific changes in the microstructure of its surface layer.

The surface quenching process consists of two stages: heating of the billet's surface layer above the critical austenitizing temperature and rapid cooling of the heated layer below the temperature of martensite formation. Herewith, the main goal is to obtain the most uniform martensite structure in the surface layer of the workpiece, which depends on the uniformity of the temperature field along the border of the hardened layer at the end of the heating stage and thermal stress value, arising during rapid cooling [1].

To obtain these qualitative characteristics in the surface layer of the product, it is often insufficient to use typical induction heating installations, which confirms the relevance of the problem of optimization of the design and operating characteristics of the inductor [6–8]. One of the test cases of such an optimization problem is solved in the presented work.

2 General Formulation of the Optimal Design Problem

Let us consider the problem of optimal design of an inductor, aimed at implementing the heating stage for surface hardening of critical complex-shaped parts. An induction system can be represented as a control object with distributed parameters (ODP), which function of the state is described by the known system of interrelated Maxwell and Fourier equations for electromagnetic and thermal fields [9, 10]. It is suggested to use the 2D nonlinear numerical model developed in ANSYS Mechanical APDL [11] as a model of ODP. The FEM software ANSYS has been chosen due to its wide range of features for transient electromagnetic, thermal, and elastic static analysis, including the embedded possibility to simulate the thermal stress fields.

In the problem under investigation a vector $\overline{P} = (p_1, p_2, \ldots, p_n)$, including different design and operating characteristics of induction heating installation, is considered as a vector of optimized parameters. The components of the vector can represent the sizes and number of inductor turns, current and frequency of power supply, etc. In the general case the optimized parameters are limited by the following constraints:

$$p_{i\,min} \leq p_i \leq p_{imax}, i = \overline{1, n} \quad (1)$$

To obtain the uniform temperature distribution within the hardened layer at the endpoint τ^0 of a heating stage, it is reasonable to define a maximal admissible value ε of absolute deviation of final temperature distribution from the prescribed temperature T^* [9]. Due to the mentioned above, it is suggested to consider the optimization criterion in a form (2), which minimizes deviations of the final temperature $T(X, l, \tau^0, \overline{P})$ along the border of the hardened layer with a depth from the required value T^*:

$$J = \frac{max}{l\epsilon[0, L]}|T(X, l, \tau^0, \overline{P}) - T^*| \rightarrow \frac{min}{\overline{P}}, \quad (2)$$

here l is a coordinate of a point on the border of the hardening layer with the length L, $l\epsilon[0; L]$; [3].

One of the particular characteristics of the surface induction hardening process is high-speed heating of the billet's surface layer up to the temperatures above the Curie point, which can lead to local overheating or high value of thermal stresses above their limits. Thereby, it is reasonable to include the corresponding phase restrictions on the maximal admissible value of surface temperature T_{adm} and the thermal stress value σ_{adm} in a formulation of the optimization problem. These restrictions can be written in the following form:

$$T_{\max}(t) = \frac{\max}{r, l \in \Omega} T\left(r, l, t, \overline{P}\right) \leq T_{adm}; 0 \leq t \leq \tau^0,$$
$$\sigma_{\max}(t) = \frac{\max}{r, l \in \Omega} \sigma\left(r, l, t, \overline{P}\right) \leq \sigma_{adm}; 0 \leq t \leq \tau^0 \quad (3)$$

here Ω is a set of coordinates of points within the longitudinal cross-section of the complex-shaped workpiece.

According to the mentioned above, in the most general case, the optimal inductor design problem can be formulated as follows: it is required to obtain such values of parameters $\overline{P^{opt}}$, limited by the constraints (1), which provide the transfer of the object, described by nonlinear numerical 2D ANSYS model, from the initial state (corresponding to the temperature T_0) to the required final state defined by the minimal value of optimization criterion (2) within the specified process time τ^0 under conditions of phase restrictions (3) [9].

3 Numerical Simulation of the Surface Hardening Process

To solve the formulated optimal design problem, the nonlinear 2D axisymmetric model of surface induction hardening of the complex-shaped billet is developed in ANSYS Mechanical APDL [12]. Under certain circumstances (e.g., time scaling, identification of parameters using experimental data, etc.), the presented numerical model of the physical surface hardening process can be treated as a «digital twin», which is used for the optimization of both design and operating parameters.

The algorithm of numerical simulation presented in Fig. 1 can be divided into two steps.

On the first step of numerical simulation of the heating stage, the main nonlinear thermophysical properties of the steel C40 workpiece [13, 14] are set. Then the finite-element mesh is generated based on the created geometry of the induction heating system. Generated mesh is used for interrelated harmonic electromagnetic

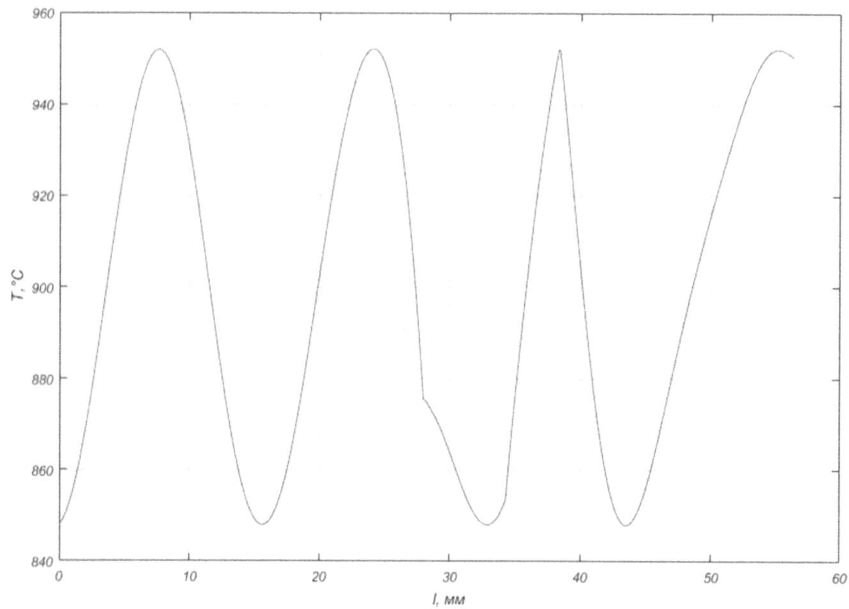

Fig. 1 Algorithm of numerical simulation of the surface hardening process

and transient thermal analyzes. The obtained time–space distributions of electromagnetic heat sources and temperatures within the heated workpiece volume are used for solving the elastic static problem considering the Huber-Mises condition.

Developing the described numerical model is accompanied by setting the proper types of finite elements from the ANSYS library corresponding to the appropriate physical problem: Plane13 (for electromagnetic problem), Plane55 (for thermal problem), Plane182 (42) (for elastic static problem). In each iteration of the numerical simulation of the heating stage, the static electromagnetic field, transient thermal field, and thermal stresses field are obtained. The simulation of the described first step ends when the current time reaches the total duration of the heating stage.

In the second step, the obtained temperature distribution and thermal stress field are used as the initial condition for numerical simulation of the spray cooling stage [15, 16]. The iterative process in this case consists of transient thermal analysis without electromagnetic heat sources. The specific feature of the cooling stage is a significantly nonlinear dependency $\alpha(T)$ of the heat transfer coefficient of the heated billet's surface versus the temperature between water and surface, as described in [17, 18]. In the general case, obtaining the particular type of such dependency is a sufficiently complex problem, that lies beyond the presented investigation. In the chapter, numerical simulation and cooling rate analysis are carried out using the equivalent nonlinear surface heat transfer coefficient $\alpha_{eq.}$ [17] as an engineering approximation of nonlinear dependency $\alpha(T)$. The results of numerical simulation of the second stage of the surface hardening process are temperature distribution and distribution of equivalent thermal stresses.

The geometry of the developed ANSYS model with generated finite-element mesh, consisting of 138,357 nodes, is presented in Fig. 2.

A test simulation on a personal computer is carried out to analyze the developed numerical model. The main PC hardware characteristics are CPU Intel Xeon E5-2680v3 2.5 GHz, 16 GB RAM. The duration of the test simulation of the 10-s heating stage with a time step of 0.5 s and a subsequent 5-s cooling stage with a time step of 0.05 s is around 3 h.

The developed formats for input, output, and data exchange between the numerical model and measuring device, and corresponding postprocessing facilities for simulation results allow integration of «digital twin» into program complexes and cyber-physical systems.

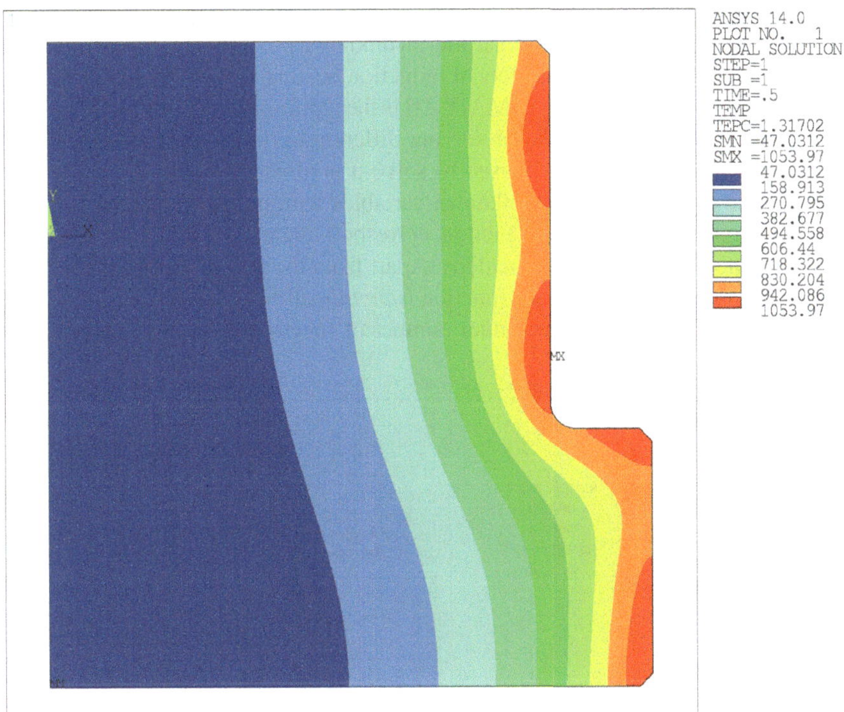

Fig. 2 2D numerical model of induction system for surface hardening, developed in ANSYS mechanical APDL

4 Structure of Cyber-Physical Complex

For optimal design of induction heating installation, the described above 2D numerical model of surface hardening is integrated into the optimization procedure, which is a core of the program part of the cyber-physical complex [19]. In cyber-physical systems, a set of process parameters are identified by field experiments on a real object, and obtained optimal values of selected operating parameters of the inductor are used as the control actions on the real-time scale.

Figure 3 shows the structure of the cyber-physical complex that provides the possibility to solve the problems of identifying of main thermophysical parameters of the digital model, optimal design of equipment, and optimal control of operating modes of the induction hardening process.

The core of the program part is the optimization procedure, developed in MATLAB software. It is based on the alternance method of parametric optimization of the system with distributed parameters, which allows solving the systems of transcendental equations closed regarding all unknown process parameters, including the design and operating parameters of induction heating system [9, 10, 20].

One of the main advantages of MATLAB is the Optimization Toolbox [21], which provides a wide range of tools for solving different optimization problems. In the paper, the function *lsqnonlin* is used to solve a nonlinear system of equations. It provides the optimal values of unknown variables within limited intervals through the minimization of a nonlinear function, corresponding to the sum of squared errors for a system of homogeneous equations. Apart from the possibility to set the limits for the optimized parameters, the function *lsqnonlin* allows customizing the solving process through setting the maximum number of function calls or iterations of the

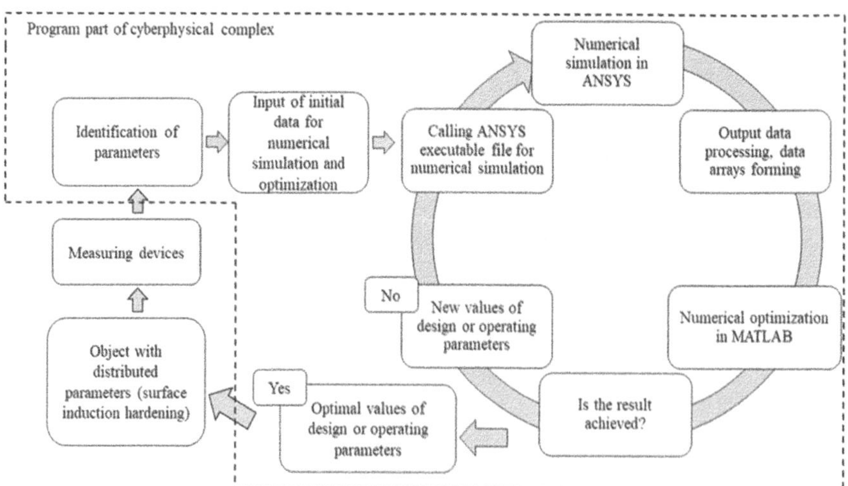

Fig. 3 Structure of cyber-physical complex

optimization procedure, minimal change of each parameter's value, and minimal change of goal function value.

Initial data for solving the optimal design problem includes required temperature value, heating time, an initial guess of optimized process parameters, and their constraints. Then in the MATLAB command window, the necessary options are set and this initial data is written in text files for its further use during numerical simulation. In the next step, the developed numerical model is run by the executable file establishing the connection between ANSYS and MATLAB software. At the end of the numerical simulation process, the obtained results are processed to form data arrays, primarily including time–space temperature distribution along the border of the hardened layer. Processed data arrays are then transferred to the numerical optimization procedure.

The special properties of the alternance method described in [9, 10, 20] allow the writing of the system of transcendental equations closed regarding all unknown parameters of the considered process. This system can be solved using standard functions of the MATLAB Optimization toolbox. If the solution of the system with prescribed accuracy is found, obtained values of optimized parameters are saved in the text files. Otherwise, the new values of optimized parameters are calculated and then again transferred into the developed numerical model for a new iteration of the optimization procedure. The iterative process ends by reaching either the prescribed accuracy of solution or iteration limit, which are set during the call of numerical solver for a system of nonlinear equations *lsqnonlin*.

The presented cyber-physical complex is a universal one that allows using either numerical models, developed in different software packages (ANSYS, Altair FLUX, etc.) or optimization procedures based on different methods (alternance method, genetic algorithms, etc.). The suggested procedure based on the alternance method allows also solving the optimization problem with an increased number of optimized parameters. In this case, the changes will affect only the syntax of the nonlinear solver call and the basic system of equations based on the alternance method, which construction is usually a non-trivial task.

5 Solving the Optimal Design Problem

In [19, 22] the application of the developed program part of the cyber-physical complex is applied to solving an optimal inductor design problem for surface hardening of steel cylindrical billets. In the paper, the presented program complex is used for solving the advanced problem of optimal inductor design for the heating stage of surface quenching of complex-shaped workpieces, which has been formulated in Sect. 2.

The geometry of the considered induction heating system is shown in Fig. 4. The vector of optimized parameters includes the following parameters: p_1—the height of turn №1, p_2—the distance between the turn №1 and №2, p_3—the maximal distance between the turn №2 and the workpiece, p_4—the distance from the point with coordinate Y1 to the top of the turn №3, p_5—the distance from the point with coordinate Y1 to the bottom of the turn №3 (the height of the turn), p_6—current of the turn №1 and p_7—current of the turn №3. The value of each listed parameter is limited by the constraint that can be written in the form (1):

$$5 \le p_1 \le 15, 2 \le p_2 \le 10, 1 \le p_3 \le 5, 2 \le p_4 \le 13, 10 \le p_5 \le 25, \\ 600 \le p_6 \le 2000, 600 \le p_7 \le 2000 \quad (4)$$

To solve the formulated problem, it is suggested to consider the spatial scan of the border ABCDE of the surface layer along the only relative axis, which is further denoted as l, and consider temperatures in points $l\epsilon[0; L]$, where $L = |AB|+|BC|+|CD| + |DE|$.

The values of optimized parameters representing the initial guess to start the optimization procedure presented in Fig. 2 are shown in Table 1.

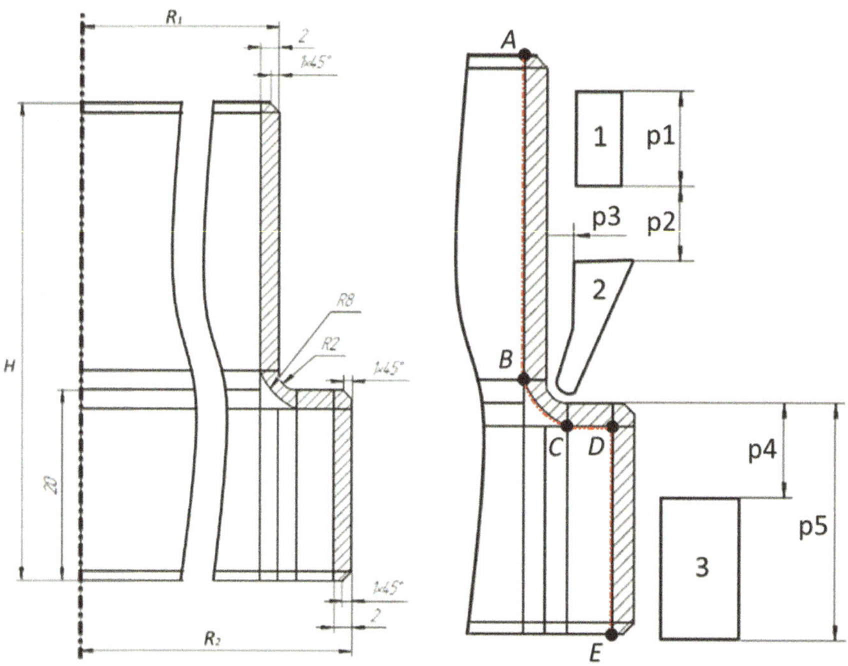

Fig. 4 The geometry of the induction heating system

Table 1 Initial guess for optimized parameters

Parameter	p1, mm	p2, mm	p3, mm	p4, mm	p5, mm	p6, A	p7, A
Value	8	6	2	8	20	1460	1443

Obtained by numerical simulation optimal values of all unknown process parameters are presented in Table 2.

The solving process of the optimal inductor design problem consists of 21 iterations with 357 goal function calls with a total duration of the numerical optimization process of around 65 h. In general, the alternance method leads to a dramatic decreasing the number of iterations of the optimization procedure. Considering a quite long simulation time, it provides a significant reduction of the total duration of solving process.

Final temperature distributions along the border of the hardened layer (a) and within the longitudinal cross-section of the treated workpiece (b) at the end of heating in the inductor of optimal design are shown in Fig. 5.

Analysis shows that the maximum temperature deviation along the border of the treated surface layer from the prescribed temperature value of 900 °C is not more than 52 °C. To improve the value of heating accuracy, the number of optimized parameters can be increased. In this case, the optimal design problem with a new number of parameters can be solved using the described above scheme.

Table 2 Optimal values of parameters

Parameter	p1, mm	p2, mm	p3, mm	p4, mm	p5, mm	p6, A	p7, A
Value	6.37	8.3	1.52	8.27	19.7	1444.8	1432.5

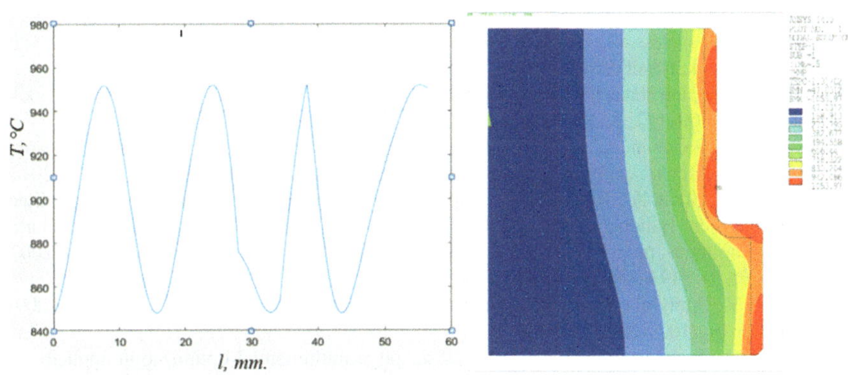

a) b)

Fig. 5 Final optimal temperature distributions: along the border of the hardened layer (**a**) and within the longitudinal cross-section of the workpiece (**b**)

6 Conclusions

In the presented paper the cyber-physical program complex for optimization of design and operating parameters of induction hardening process of complex-shaped billets is presented. The optimal inductor design problem for the heating stage of the specified quenching process is formulated. The nonlinear 2D numerical ANSYS model, which is considered an object with distributed parameters, is integrated into the optimization procedure based on the alternance method. The solution of the test case of a formulated problem is carried out using the program part of the presented cyber-physical complex. Analysis of solution results shows that the value of temperature deviation along the border of the treated surface layer of the workpiece is less than 52 °C. If it does not satisfy the induction hardening requirements, the value of heating accuracy can be improved by solving the problem with the increased number of optimized parameters.

Acknowledgements The work is supported by grant of Russian Scientific Foundation №22-29-00180, https://rscf.ru/project/22-29-00180, State Federal Budgetary Institution of Higher Education «Samara State Technical University».

References

1. Rudnev, V., Loveless, D., Cook, R.: Handbook of induction heating, 2nd edn. Marcel Dekker Inc., New York, Basel (2017)
2. Fett, G.: Induction Hardening of Axle Shafts. In: Rudnev, V., Totten G. (eds.) ASM Handbook, vol. 4C, pp. 160–172. ASM International, Ohio (2014)
3. Rudnev, V., Goodwin, A., Fillip, S., West, W., Schwab, J., St. Pierre, S.: Keys to long-lasting hardening inductors: experience, materials, and precision.. Adv. Mater. Process. 48–52 (2015)
4. Tensi, H.M., Stich, A., Totten, G.E.: Chapter 9-Quenching and Quenching Technology, Steel Heat Treatment Handbook, 2nd edn. In: Totten, G. E. (ed) Steel Heat Treatment: Metallurgy and Technologies, pp. 539–606. CRC Press Taylor & Francis Group, Boca Raton, FL. (2007) ISBN: 978–0–8493–8455–4
5. Rudnev, V.: Spray quenching in induction hardening applications. ASTM JAI. **6**(4). (2009). https://doi.org/10.1520/JAI101928
6. Baldan, M., Steinberg, T., Baake, E.: Self-adaptive multimethod optimization applied to a tailored heating forging process. IOP Conf. Ser.: Mater. Sci. Eng. **355**(012016) (2018)
7. Nacke, B., Nikanorov, A., Baldan, M.: A novel multi-surrogate multi-objective decision-making optimization algorithm in induction heating. COMPEL: Int J Comput. Maths. Electr. Electron. Eng. **39**(1), 144–157(14) (2020). https://doi.org/10.1108/COMPEL-05-2019-0222
8. Pleshivtseva, Y.U., Forzan, M., Sieni, E., Popov, A.: Multi-objective optimization of induction surface hardening process. In: Proceedings of the 44th Annual Conference of the IEEE Industrial Electronics Society (IECON'18), pp.3515–3520 (2018)
9. Rapoport, E., Pleshivtseva, Y.U.: Semi-infinite optimization methods in applied optimal control problems of systems with distributed parameters *[Metodi polubeskonechnoy optimization v prikladnikh zadachakh upravleniya sistemami s raspredelennimi parametrami]*. Nauka, Moscow (2021)
10. Rapoport, E.: Alternance method in applied optimization problems [*Alternansniy method v prikladnikh zadachakh optimizatsii*]. Nauka, Moscow (2000)

11. ANSYS manual. [Online]. Available: https://www.ansys.com/support.
12. Mannanov, E., et al.: Numerical simulation, investigation, and development of induction systems for heating of disks with complex profile. In: 2019 IEEE Conference of Russian Young Researchers in Electrical and Electronic Engineering (EIConRus), IEEE, pp. 606–608 (2019)
13. Zubchenko, A.S., Koloskov, M.M., Kashirskiy, Y.V., et al.: Grade guide of steels and alloys [*Marochnik staley i splavov 2-e izdaniye*], 2nd edn. Mashinostroyeniye, Moscow (2003)
14. Blanter, M.E.: Metal science and heat treatment [*Metallovedenie i termicheskaya obrabotka*]. Mashgiz, Moscow (1963)
15. Jescher, R., Specht, E., Köhler, Chr.: Chapter 6–Heat Transfer during Cooling of Heated Metals with Vaporizable Liquids. In: Liščić, B., Tensi, H.M., Canale, L.C.F., Totten, G.E. (eds.) Quenching Theory and Technology, 2nd edn, pp. 159–178. CRC Press, Boca Raton, FL (2010). ISBN: 978–0–8493–9279–5
16. Woche, H., Fang, Y., Specht, E.: Heat transfer analysis during metal cooling with sprays and jets. Heat Process., Vulkan-Verl. GmbH **1**, 41–47 (2018)
17. Krause, C., et al.: Wärmeübergangs-und Tropfencharakteristik für eine Spraykühlung im Temperaturbereich von 900–100° C. Forschung im Ingenieurwesen. 72(3), 163–173 (2008)
18. Stark, P., Schuettenberg, S., Fritsching, U.: Spray quenching of specimen for ring heat treatment. WIT Trans. Eng. Sci. **70**, 201–212 (2011)
19. Pleshivtseva, Y., Popov, A., Pavlushin, A.: Software solution for the optimal design of inductor for surface hardening. In: 2021 3rd International Conference on Control Systems, Mathematical Modeling, Automation, and Energy Efficiency (SUMMA), pp. 646–649 (2021)
20. Rapoport, E.: Optimal control of systems with distributed parameters [*Optimalnoe upravlenie sistemami s respredelennimi parameterami*]. Visshaya shkola, Moscow (2009)
21. MATLAB documentation. [Online]. Available: https://www.mathworks.com/help/matlab/
22. Popov, A.: Optimization of the heating stage for induction hardening of cylindrical billets. In: 2019 XXI International Conference Complex Systems: Control and Modeling Problems (CSCMP), pp. 237–241 (2019)

Contact Problems of the Theory of Roller Squeezing of Leather

Sh. Khurramov, F. Kurbanova, and K. Aliboyev

Abstract The chapter deals with the main contact problems of the theory of roller pressing of leather. Mathematical models of the shape of the roll contact curves and the distribution patterns of contact stresses along the roll contact curves were developed. A formula was derived that describes the model of friction stresses. It was revealed that the shapes of the roll contact curves of the two-roll module do not depend on the change in the moisture content of hides in the contact zone of the rollers. It was established that the model of friction stresses in a two-roll module for pressing leather does not depend on the inclination of the material layer fed to the center line and on the inclination of the upper roll relative to the vertical line. It was determined that with a decrease in the radius of the roll, with other parameters being the same, the maximum normal and shear stresses increase.

Keywords Roll squeezing of leather · Contact problems · Roll contact curves · Contact stresses

1 Introduction

The main working body of roller machines for squeezing out leather semi-finished products is a pair of working rollers with an elastic coating (a cloth). A pair of working rollers (a roll pair) and a layer of semi-finished leather after tanning (a layer of leather) together create a two-roll module for pressing leather.

In the process of roller pressing of a leather semi-finished product, two phenomena occur simultaneously—contact interaction and moisture filtration. This, in turn, requires the joint solution of two types of problems (in the theory of roller pressing of leather)—contact problems and hydraulic problems.

The basic contact problem of the theory of roller pressing of leather is mathematical modeling of the shape of the roll contact curves and the distribution patterns

Sh. Khurramov (✉) · F. Kurbanova · K. Aliboyev
Tashkent Institute of Architecture and Civil Engineering, 13, Navoi, Uzbekistan 100011
e-mail: shavkat-xurramov59@mail.ru

A. G. Kravets et al. (eds.), *Cyber-Physical Systems Engineering and Control*, Studies in Systems, Decision and Control 477,
https://doi.org/10.1007/978-3-031-33159-6_12

of contact stresses along the contact curves of rolls in a two-roll module of roller pressing of leather.

The main hydraulic tasks of the theory of roll squeezing of wet materials are mathematical modeling of the patterns of distribution of hydraulic pressure and residual moisture content of the pressed material. To solve these problems, it is necessary to know the patterns of change in the filtration rate in the squeezing area.

In [1–12], the equations of roll contact curves and patterns of contact stress distribution in two-roll modules are analytically described, without considering the moisture filtration from the wet material during the squeezing process.

An analysis of the publications devoted to the study of hydraulic problems of roll squeezing of wet materials [13–19] showed that the existing patterns of change in the filtration rate in the squeezing area were obtained with the use of models of roll equipment and materials that did not correspond to the real physical phenomena of wet materials under roll squeezing. Consequently, they do not provide an opportunity to solve problems that allow the disclosure of the hydraulic phenomenon of roll squeezing of wet materials, including leather.

This study is devoted to modeling the contact curves of two-roll modules of leather pressing. The filtration of moisture from the leather semi-finished product during the pressing process and the deformation properties of the semi-finished leather product are taken into account.

2 Solutions to the Problems Posed

The analysis of the roller machines for hides pressing [4] showed that, basically, the two-roll modules for pressing leather have a symmetrical form.

Consider a two-roll module for pressing leather, in which rollers of radii R have an elastic cloth covering of a thickness of H, the leather layer thickness is δ_1, the distance between the rolls is h, both rolls are driven (Fig. 1).

Since the two-roll module under consideration is symmetric, we will investigate the contact interaction of leather with any roller, for example, with the lower roller. The roll contact curve consists of compression and recovery zones (Fig. 1). The points of the roll compression zone are determined by the polar coordinates r_1 and θ_1, and the points of the recovery zone are determined by r_2 and θ_2, where

$$-\varphi_1 \leq \theta_1 \leq 0,\ 0 \leq \theta_2 \leq \varphi_2 \tag{1}$$

here φ_1, φ_2—are the contact angles.

From Fig. 1 it follows that.

$$2R\cos\varphi_1 + \delta_1 = 2R + h$$

Hence, we determine

Fig. 1 Scheme of interaction in a two-roll module for pressing leather

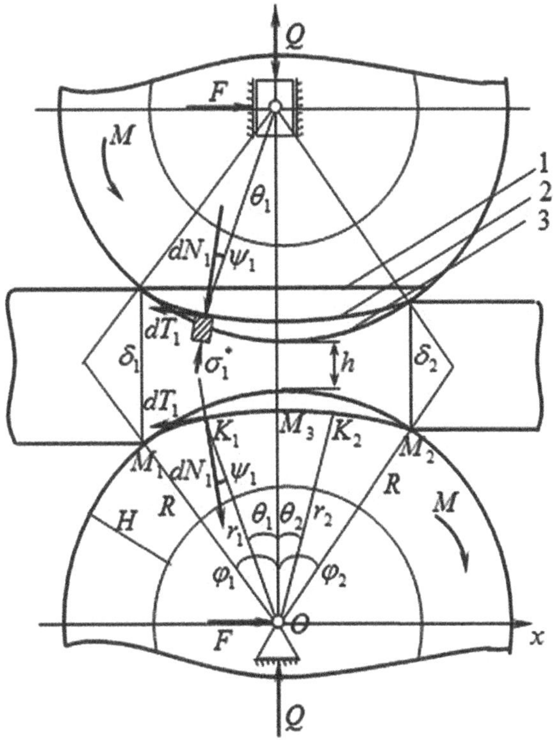

$$4\sin^2\frac{\phi_1}{2} = \frac{\delta_1 - h}{R}$$

or

$$\phi_1 = \sqrt{\frac{\delta_1 - h}{R}} \tag{2}$$

Likewise (2), we obtain

$$\phi_2 = \sqrt{\frac{\delta_2 - h}{R}}, \tag{3}$$

where δ_2- is the finite thickness of the leather layer.

According to [20], the deformation properties of leather (of the semi-finished leather product after chrome tanning) under compression and recovery are described by power-law dependences:

"Stress—relative deformation" dependence:

$$\sigma = 43,39\varepsilon_C^{3,23};\ \sigma = 85,93\varepsilon_E^{3,03},\ \sigma = 85,93\varepsilon_E^{3,03} \tag{4}$$

"Humidity—relative deformation" dependence:

$$W = 1,13\varepsilon_c^{0,2};\ W = 1,24\varepsilon_B^{0,24} \tag{5}$$

where ε_c, ε_B—are the relative deformation of compression and recovery.

According to expressions (4) and (5), the deformation properties of leather under compression are described by the following dependencies

$$\sigma_1 * = \alpha_1 \varepsilon_1^{*S_1},\ W_1^* = W_H^* - \beta_1 \varepsilon_1^{*l_1} \tag{6}$$

W_1^*—is the initial moisture of hides; $\alpha_1 = 43,39;\ \beta_1 = 1,13;\ s_1 = 3,23; l_1 = 0,2$.

From dependencies (6), we have

$$\frac{\sigma_1^*}{W^* - W_1^*} = \frac{\alpha_1}{\beta_1} \varepsilon_1^{*(s_1 - l_1)},$$

or

$$\sigma_1^* = B_1 \varepsilon_1^{*m_1^*}\left(W_H^* - W_1^*\right) \tag{7}$$

where $B_1 = \frac{\alpha_1}{\beta_1}, m_1 = s_1 - l_1 (B_1 = 37,51; m_1 = 3,03)$.

From the analysis of the research results [21–24], it follows that the deformation of the roller coating (a cloth) under compression and recovery can be described by dependencies of the form

$$\sigma_j = A_j \varepsilon_j^{m_j}(W_j - W_j),\ j = 1,2, \tag{8}$$

where W_{jH}—is the moisture content of the cloth at the initial point of the $j-$ th zone of the roll contact curve.

At each point of the contact curve, deformation of leather and roller coating occurs along the normal line $n-n$. Therefore, at each point of the compression zone $M_1 M_3$, the stress equivalence is fulfilled (according to Newton's law), that is, $\frac{\sigma_1}{\cos\psi_1} = \frac{\sigma_1^*}{\cos\psi_1}$, where ψ_1- is the angle between the radius r_1 and the line $n - n$.

Hence $\sigma_1 = \sigma_1^*$.

Therefore, it can be considered that the deformation of leather and roller coating in the two-roll module occurs in the radial direction to the axis of the rollers.

Then, according to expressions (7) and (8), we have

$$A_1 \varepsilon_1^{m_1}(W_1 - W_{1H}) = B_1 \varepsilon_1^{*m_1 *}\left(W_H^* - W_1^*\right) \tag{9}$$

where $\varepsilon_1, A_1, m_1,\ \varepsilon_1^*, B_1, m_1^*-$ are the deformation, the strain and hardening coefficients of the cloth and leather under compression, respectively.

During the squeezing process, the moisture removed from the hides to point K_1 is transferred to the roller coating along the contact curve M_1M_2. The amount of moisture removed from hides to the point

$$K_1 - W^*_{1\text{уд}} = W^*_{\text{н}} - W^*_1$$

is equal to the amount of moisture that has passed to the roller coating up to the point $W_1 = (W_1 - W_1)$.

Thus, for each point of the compression zone M_1M_3 of the lower roller, the following equivalence holds

$$(W_1 - W_{1H}) = (W^*_H - W^*_1) \tag{10}$$

Then from equivalence (9), we have

$$A_1\varepsilon_1^{m_1} = B_1\varepsilon_1^{*m_1^*}. \tag{11}$$

Differentiating both sides of Eq. (11), we determine

$$m_1A_1\varepsilon_1^{m_1-1}d\varepsilon_1 = m_1^*B_1\varepsilon_1^{*m_1^*-1}d\varepsilon_1^*.$$

Hence

$$\frac{\varepsilon_1}{\varepsilon_1^*} = \frac{m_1}{m_1^*}\cdot\frac{d\varepsilon_1}{d\varepsilon_1^*} = \frac{m_1}{m_1^*}\cdot\frac{\frac{d\varepsilon_1}{dt}}{\frac{d\varepsilon_1^*}{dt}}. \tag{12}$$

We assume that

$$\frac{\frac{d\varepsilon_1}{dt}}{\frac{d\varepsilon_1^*}{dt}} = \lambda_1. \tag{13}$$

where λ_1—is the ratio of the relative deformations of the elastic roll cover and the leather layer under compression.

From equivalence of (12) and (13), we have

$$\frac{\varepsilon_1}{\varepsilon_1^*} = \frac{m_1}{m_1^*}\lambda_1. \tag{14}$$

According to Fig. 1, the relative deformations of the contacting bodies have the form

$$\varepsilon_1 = \frac{R - r_1}{H},\ \varepsilon_1^* = \frac{r_1 - R\frac{\cos\varphi_1}{\cos\theta_1}}{\frac{\delta_1}{2\cos\varphi_1}} \tag{15}$$

From Eq. (14), taking into account expression (15), we determine

$$R - r_1 = k_1\lambda_1\left(r_1 - R\frac{\cos\varphi_1}{\cos\theta_1}\right) \tag{16}$$

where $k_1 = \frac{2m_1 H\cos\varphi_1}{m_1^*\delta_1}$.

Having solved Eq. (16) with respect to r_1, we find the equation of the contact curve of the compression zone

$$r_1 = \frac{R}{1 + k_1\lambda_1}\left(1 + k_1\lambda_1\frac{\cos\varphi_1}{\cos\theta_1}\right). \tag{17}$$

Likewise formula (17), we find the equation for the contact curve of the recovery zone

$$r_2 = \frac{R}{1 + k_2\lambda_2}\left(1 + k_2\lambda_2\frac{\cos\varphi_2}{\cos\theta_2}\right), \tag{18}$$

where $k_2 = \frac{m_2 H\cos\varphi_2}{m_2^*\delta_2}$. λ_2—is the ratio of the rates of relative deformations of the elastic roll coating and the leather layer under recovery.

Generalizing Eqs. (17) and (18), we obtain

$$\begin{cases} r_1 = \frac{R}{1+k_1\lambda_1}\left(1 + k_1\lambda_1\frac{\cos\varphi_1}{\cos\theta_1}\right), -\varphi_1 \le \theta_1 \le 0, \\ r_2 = \frac{R}{1+k_2\lambda_2}\left(1 + k_2\lambda_2\frac{\cos\varphi_2}{\cos\theta_2}\right), 0 \le \theta_2 \le \varphi_2. \end{cases} \tag{19}$$

The system of Eq. (19) describes the roll contact curves of the two-roll module for pressing leather.

If in a two-roll module, the rollers have a non-deformable coating, then $\lambda_j = 0$. It follows from system (19) that $r = R$, $-\varphi \le \theta \le \varphi$.

If in the two-roll module the leather layer is non-deformable, then $\lambda_j = \infty$. It follows from system (19) that $r = \frac{R\cos\varphi}{\cos\theta}$, $-\varphi \le \theta \le \varphi$.

Thus, any graph of the roll contact curve of the two-roll module of the roller squeezing of leather (curve 2) lies between the graphs of curves $r = \frac{R\cos\varphi}{\cos\theta}$ (curve 1) and $r = R$ (curve 3) (Fig. 1).

From the system (19), it was found that the shapes of the roll contact curves of the two-roll module do not depend on the change in the moisture content of hides in the contact zone of the rollers.

Figure 2 shows the graphs of the shape of the roll contact curves for various changing parameters.

Let us analyze the stress state in a two-roll module for leather squeezing.

In the steady-state process of interaction, the roll is acted upon by the force of the pressing devices Q, the horizontal response of the roll supports F, the moment of resistance forces M, elementary forces of normal pressure and friction acting along

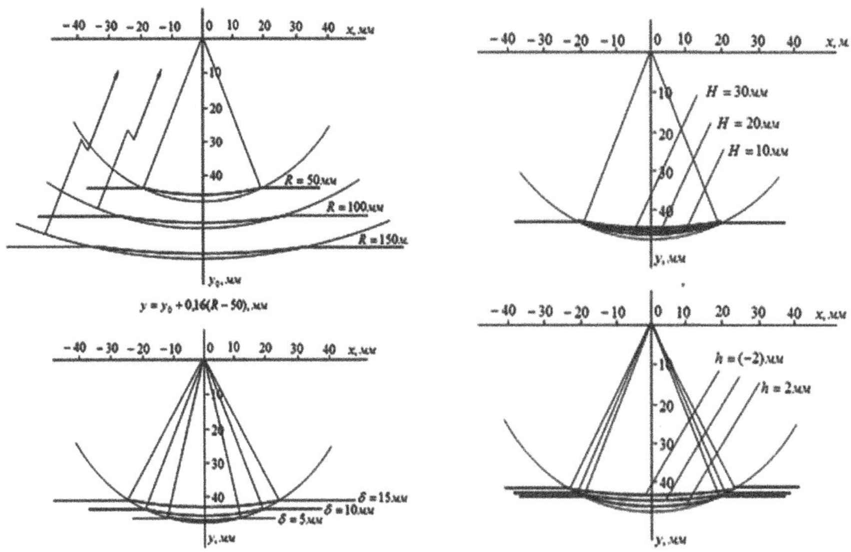

Fig. 2 The nature of the influence of factors on the shape of the roll contact curve

the entire roll contact curve. Elementary forces in compression (N_1, T_1) and recovery (N_2, T_2) zones will be presented separately.

Considering the roll in balance under the action of the applied forces, we obtain

$$\begin{cases} N_{1x} + T_{1x} + F_{1x} + Q_{1x} = 0, \\ N_{1y} + T_{1y} + F_{1y} + Q_{1y} = 0 \end{cases}$$

or

$$\begin{cases} dN_{1x} + dT_{1x} + dF_{1x} + dQ_{1x} = 0, \\ dN_{1y} + dT_{1y} + dF_{1y} + dQ_{1y} = 0, \end{cases} \tag{20}$$

where N_{1x}, N_{1y}, T_{1x}, T_{1y}—are the projections of the main normal and shear forces of the compression zone on the x and y axes.

From the diagram of the forces of the compression zone (Fig. 1), we have

$$dN_{1x} = dN_1 \sin(\theta_1 - \psi_1),$$

$$dN_{1y} = -dN_1 \cos(\theta_1 - \psi_1),$$

$$dT_{1x} = -dT_1 \cos(\theta_1 - \psi_1),$$

$$dT_{1y} = -dT_1 \cos(\theta_1 - \psi_1), dF_{1x} = dF_1, dF_{1y} = 0, dQ_{1x} = 0,$$

$$dQ_{1y} = dQ_1, \tag{21}$$

where ψ_1—is the angle between force dN_1 and radius r_1.

From system (20), considering Eq. (21), for the compression zone, we have

$$\frac{dF_1}{dQ_1} = \frac{dT_1 \cos(\theta_1 - \psi_1) - dN_1 \sin(\theta_1 - \varphi_1)}{dT_1 \sin(\theta_1 - \psi_1) + dN_1 \cos(\theta_1 - \psi_1)}. \tag{22}$$

Since we are considering a steady-state process, we can assume that $\frac{F_1}{Q_1} = C_1$, where C_1—is a constant value. Hence, $\frac{dF_1}{dQ_1} = \frac{F_1}{Q_1} = C_1$.

Then from Eq. (22) we obtain

$$\frac{dT_1}{dN_1} = \frac{\sin(\theta_1 - \psi_1) + C_1 \cos(\theta_1 - \varphi_1)}{\cos(\theta_1 - \psi_1) - C_1 \sin(\theta_1 - \psi_1)}. \tag{23}$$

The following relations relate elementary forces to contact stresses [7]:

$$dN_1 = n_1\sqrt{r_{1^2} + r'_{2^1}} d\theta_1, \; dT_1 = t_1\sqrt{r_{1^2} + r'_{2^1}} d\theta_1 \tag{24}$$

where $n_1 = n_1(\theta_1)$, $t_1 = t_1(\theta_1)$, are the normal and shear stresses distributed over the compression zone of the roll contact curve, respectively.

We substitute expressions (24) into Eq. (23), then transform according to the expressions $\cos\psi_1 = \frac{r_1}{\sqrt{r_{12} + r'_{12}}}$, $\sin\psi_1 = \frac{r'_1}{\sqrt{r_{12} + r'_{12}}}$ and obtain the dependences connecting shear and normal stresses at the points of the compression zone of the roll contact curve

$$t_1 = \frac{(\sin\theta_1 + C_1 \cos\theta_1) r_1 - (\cos\theta_1 - C_1 \sin\theta_1) r'_1}{(\cos\theta_1 - C_1 \sin\theta_1) r_1 + (\sin\theta_1 + C_1 \cos\theta_1) r'_1} n_{11} \tag{25}$$

Formula (25), which connects shear and normal stresses, is also called a friction stress model.

The friction stress model of the recovery zone is obtained likewise

$$t_2 = \frac{(\sin\theta_2 + C_2 \cos\theta_2) r_2 - (\cos\theta_2 - C_2 \sin\theta_2) r'_2}{(\cos\theta_2 - C_2 \sin\theta_2) r_2 + (\sin\theta_2 + C_2 \cos\theta_2) r'_2} n_2 \tag{26}$$

where $C_2 = \frac{F_2}{Q_2}$.

Note that at the point of the contact curve lying on the line of centers, the following boundary conditions are satisfied.

$$t_1(0) = t_2(0),\ n_1(0) = n_2(0),\ r_1(0) = r_2(0),\ r_1'(0) = r_2'(0) = 0$$

These conditions lead to equivalence $C_1 = C_2$. Then, $C = C_1 = C_2 = \frac{F}{Q}$. The coefficient $C = \frac{F}{Q}$ is called a dynamic factor of the roll [7].

Therefore, generalizing Eqs. (25) and (26), we obtain

$$\begin{cases} t_1 = \frac{(Q\sin\theta_1 + F\cos\theta_1)r_1 - (Q\cos\theta_1 - F\sin\theta_1)r'_1}{(Q\cos\theta_1 - F\sin\theta_1)r_1 + (Q\sin\theta_1 + F\cos\theta_1)r'_1} n_1,\ -\varphi_1 \le \theta_1 \le 0, \\ t_2 = \frac{(Q\sin\theta_2 + F\cos\theta_2)r_2 - (Q\cos\theta_2 - F\sin\theta_2)r'_2}{(Q\cos\theta_2 - F\sin\theta_2)r_2 + (Q\sin\theta_2 + F\cos\theta_2)r'_2} n_2,\ 0 \le \theta_2 \le \varphi_2. \end{cases} \quad (27)$$

The system of Eq. (27) determines the model of friction stresses in the two-roll module for pressing leather. It shows that the model of friction stresses in a two-roll module for pressing leather does not depend on the inclination of the material layer fed to the center line and on the inclination of the upper roller relative to the vertical line.

Assuming that $tg\xi = C$, we transform Eq. (25) to the form

$$t_1 = \frac{\sin(\theta_1 + \xi)r_1 - \cos(\theta_1 + \xi)r_1'}{\cos(\theta_1 + \xi)r_1 + \sin(\theta_1 + \xi)r_1'} n_1. \quad (28)$$

Considering now that $tg\psi_1 = \frac{r_1'}{r_1}$, let us proceed to the following expression.

$$t_1 = tg(\theta_1 - \psi_1 + \xi_1)n_1 \quad (29)$$

Likewise, we transform the models of friction stresses t_2 :

$$t_2 = tg(\theta_1 - \psi_1 + \xi_1)n_1 \quad (30)$$

Generalizing Eqs. (28) and (29), we obtain

$$\begin{cases} t_1 = tg(\theta_1 - \psi_1 + \xi)n_1,\ -\varphi_1 \le \theta_1 \le 0, \\ t_2 = tg(\theta_2 - \psi_2 + \xi)n_2,\ 0 \le \theta_2 \le \varphi_2, \end{cases} \quad (31)$$

The system of Eqs. (31) defines, in a simplified form, the models of friction stresses in a two-roll module for pressing leather.

In the deformation zone under compression, we select an element of a length of dl_1 directed along the $n - n$ line since the deformation of the leather layer occurs in this direction (Fig. 1). Elementary normal dN_1 and shear dT_1 forces and the response of the cutoff parts of the leather layer act on the selected element of the leather layer from the side of the lower roller. We ignore the forces of inertia and gravity acting on the selected layer element due to their smallness.

The force components dN_1 and dT_1 in the $n - n$ direction are balanced by the force $\sigma_1^* dl_{11}$ (Fig. 1):

$$\sigma_1^* dl_1 - dN_1 \cos 0^o - dT_1 \sin 0^0 = 0$$

or

$$\sigma_1^* = n_1 \tag{32}$$

where σ_1^*—is the stress of the leather layer pressing in the $n - n$ direction.

Then, taking into account Eqs. (6) and (7), we obtain

$$n_1 = \alpha_1 \varepsilon_1^{*s_j}. \tag{33}$$

It follows from Eq. (15) that.
$\varepsilon_1^* = \frac{2\cos\varphi_1}{\delta_1}\left(r_1 - R_1 \frac{\cos\varphi_1}{\cos\theta_1}\right).$
or after substituting the expression for r_1 from Eq. (17), we obtain

$$\varepsilon_1^* = \frac{2R\cos\varphi_1}{\delta_1(1 + k_1\lambda_1)}\left(1 - \frac{\cos\varphi_1}{\cos\theta_1}\right) \tag{34}$$

Substituting expression ε_1^* into Eq. (33), we find the distribution patterns of the normal stress in the compression zone of the roll contact curve

$$n_1 = A_1\left(1 - \frac{\cos\varphi_1}{\cos\theta_1}\right)^{s_1}, \tag{35}$$

where $A_1 = \alpha_1\left(\frac{2R\cos\varphi_1}{\delta_1(1+k_1\lambda_1)}\right)^{s_1}$.

The distribution patterns of normal stresses in the recovery zone of the roll contact curve are determined likewise

$$n_2 = A_2\left(1 - \frac{\cos\varphi_2}{\cos\theta_2}\right)^{s_2}, \tag{36}$$

where $A_2 = \alpha_2\left(\frac{2R\cos\varphi_2}{\delta_2(1+k_2\lambda_2)}\right)^{s_2}$.

From Eqs. (35) and (36), taking into account the system of Eq. (31), we find the distribution patterns of shear stresses along the roll contact curve of the two-roll module for pressing leather. They have the following form

$$t_1 = A_1\left(1 - \frac{\cos\varphi_1}{\cos\theta_1}\right)^{s_1} tg(\theta_1 - \psi_1 + \xi), \tag{37}$$

$$t_2 = A_2\left(1 - \frac{\cos\varphi_2}{\cos\theta_2}\right)^{s_1} tg(\theta_2 - \psi_2 + \xi). \tag{38}$$

To analyze the distribution pattern of shear stresses along the roll contact curve, the point at which the shear stress is zero, that is, the neutral point, is of particular importance.

Let the neutral point of the roll contact curve be determined by the angle $(-\varphi_3)$.

Then, from formula (37), it follows that the following equations are true for the drive roller

$$tg(-\varphi_3 - \psi_1(-\varphi_3) + \xi) = 0$$

or

$$tg(-\varphi_3 - \psi_1(-\varphi_3) + \xi) \approx tg(-\varphi_3 - \psi_1(-\varphi_3)) + tg\xi = -\frac{Q\varphi_3 - F(1+k_1\lambda_1\cos\varphi_1)}{Q(1+k_1\lambda_1\cos\varphi_1)} = 0.$$

Hence, we have

$$\varphi_3 = \frac{F(1 + k_1\lambda_1\cos\varphi_1)}{Q}. \tag{39}$$

The analysis of the mathematical models obtained shows that the patterns of distribution of contact stresses in a two-roll module depend on the coefficients of friction of the material being processed with the contact surface of the rolls, on geometric, kinematic and deformation parameters of the contacting bodies, and on the forces acting on the rolls (Figs. 3 and 4).

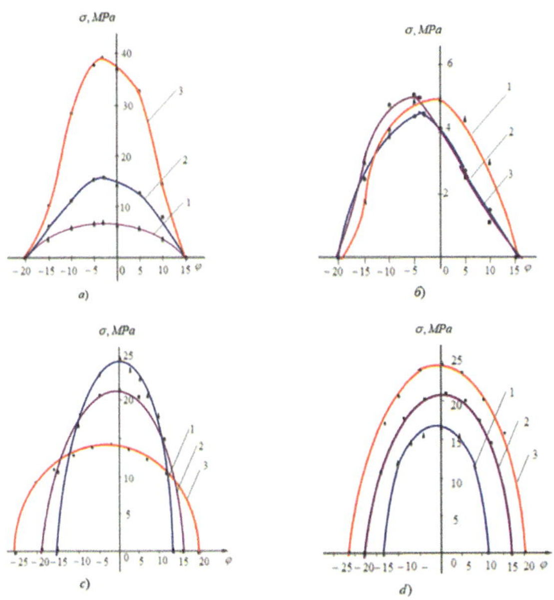

Fig. 3 Diagrams of distribution of normal stresses depending on: **a** hardening coefficient; **b** the angle of inclination of the material relative to the *OY* axes; **c** roll radius; **d** thickness of the material

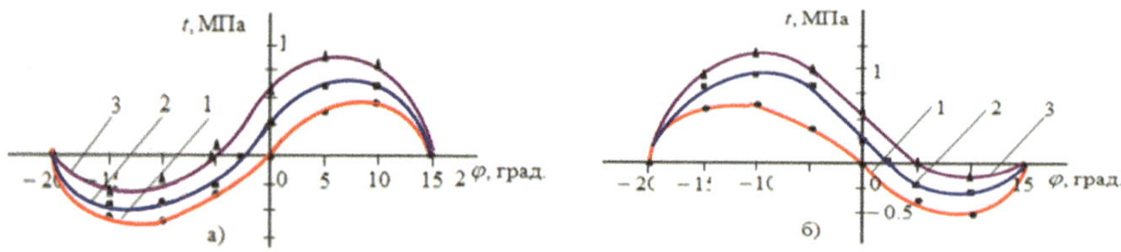

Fig. 4 Shear stress distribution diagrams: **a** in the drive roll; **b** in a free roll

3 Conclusion

Formulas that describe the shapes of roll contact curves in a generalized two-roll module are derived. All special cases of the considered two-roll module and the corresponding special forms of the obtained formulas are analyzed. It was revealed that the obtained formulas describing the shapes of the roll contact curves are general in the sense that they describe all special cases of the interaction of the processed material with roll pairs of two-roll modules.

New models of friction stresses for the lower and upper rolls of a generalized two-roll module were developed. They establish a relationship between the normal and shear stresses distributed along the contact curve of the respective roll, both in the slip and no-slip zones and with the shapes of the roll contact curves.

Using the models obtained for the shape of contact curves and friction stresses of rolls, models that describe the distribution of contact stresses of a generalized two-roll module are developed. These models take into account all the main parameters of two-roll modules and describe the distribution of contact stresses for all particular cases of interaction between the material being processed and roll pairs.

Normal and shear contact stresses are non-uniformly distributed over the contact surface of the rolls:

- Normal contact stresses vary from zero at the beginning and end of the roll contact zone to a maximum at a point lying on the center line;
- Shear contact stresses change their signs at the neutral point, which in the driving roll is located at the inlet of the material layer into the contact zone of the rolls.

References

1. Alexa V., Ratiu S., Kiss I.: IOP Conf. Series: Materials Science and Engineering (2016)
2. V.Alexa, S.A.Ratiu, I Kiss and G. Ciota. J IOP Conf. Series: Materials Science and Engineering, 106012019 (2017)
3. Khurramov, Sh.R., Abdukarimov, A., Khalturaev, F.S., Kurbanova, F.Z.: IOP Conf. Series: Materials Science and Engineering, 200 012038 (2020)
4. Khurramov, Sh.R.: Izvestiya Vysshkch Uchebnikh Zavedenii, Seriya Texnologiya Tekstilnoi promyshlennost, 4(394), 916 012051 (2021)

5. Khurramov, Sh.R., Khalturaev, F.S., Kurbanova, F.Z.: Izvestiya Vysshkch Uchebnikh Zavedenii, Seriya Texnologiya Tekstilnoi promyshlennost, 4(394) (2021)
6. Khurramov, Sh.R., Abdukarimov, A., Khalturaev, F.S., Kurbanova, F.Z.: J. of Physics: Conf. Series **1789**, 012008 (2021)
7. Alexa, V., Nussabon, A.J.: A-J. Annals of the faculty of engineering, Hunedoara, 4, pp. 7–10 (2006)
8. Khalturaev, F.S., Musirov, M.U.: J. of Physics: Conf. Series **1889**, 042024 (2021)
9. Xiawei, F., Xiochen, W., Quan, Y., Juquan, S.: Advances in Mechanical Engineering, pp. 11–18 (2019)
10. Djuraev, A., Kuliev, T.M.: International journal of Advanced Science and Technology **29**(8), 453–460 (2020)
11. Tolcha, M.A., Altenbach, H.: Metals (MDPI), 9226 (2019) doi: https://doi.org/10.3390/met9020226
12. Maximenko, O.P., Nikulin, A.A., Romanyuk, R.Ya.: News of higher educational institutions. Ferrous metallurgy (2008)
13. Gudur, P., Saiunkhe, M.A., Dixit, U.S.: Journal of Mechanical Sciences, no. 50, pp. 315–327 (2008)
14. Farhatnia, F.: Salimi Journal of Engineering. Science and Technology, **3**(4), 147–163 (2011)
15. Novikov, N.E.: Pressing paper web. Forest industry, Moscow (1972)
16. Kerekes, R.J., McDonal, J.D.: Equlibrium moisture content in wet pressing of paper. Tappi J. **19**(7), 333–340 (2020)
17. McDonald, D., Kerekes, R.J.: Estimating limits of wet pressing on paper machines. Tappi J. **16**(2), 81–87 (2017)
18. McDonald, D., Kerekes, R.J.: J. Pragmatic mathematical models of wet pressing in papermaking. BioResources, 12, pp. 9520–9537 (2017)
19. McDonald, D., Kerekes, R.J., Zhao, J.: J. Perspectives on deriving mathematical models in pulp and paper science. BioResources, 15, pp.7319–7329 (2020)
20. Kryuchkov, V. Ya.: Study of the process of pressing textile materials with roller machines. Dis. Phd. Tech. Sci. Leningrad
21. Khurramov, Sh.R., Khalturaev, F.S., Kurbanova, F.Z.: Cyber-Physical Systems: Design and Application for Industry 4.0. Studies in Systems, Decision and Control, vol. 342
22. Kukin, G.N., Soloviev, A.N., Koblyakov, A.I.: Textile materials science. Legpromizdat, Moscow (1992)
23. Marinin, A.N., Fomin, Yu.G., Sviridov, I.Kh.: Izvestiya Vysshkch Uchebnikh Zavedenii, Seriya Texnologiya Tekstilnoi promyshlennost, 7 (328) (2010)
24. Kolychev, M.V., Kokushin, N.N: Cellulose. Paper. Cardboard, 6 (2015)

Parameters Optimization of Roller Squeezing of Leather

K. Turgunov, N. Annaev, and K. Aliboev

Abstract The work is devoted to the optimization of the parameters of the roller squeezing of leather. Regression and empirical models of the residual moisture content of leather and the force of the squeezing process are determined depending on the intensity of the load, the radius and the speed of the rollers. The methods to calculate the optimal parameters of roller pressing of semi-finished leather products, taking into account the specific pressure, were developed; these methods make it possible to scientifically formulate the parameters of roller squeezing of leather. The patterns were established that allow for increasing the efficiency of leather squeezing by rollers. It was determined that the best effect of load intensity is observed in squeezing machines with low velocities and small diameters. It was found that with an increase in the specific pressure, the residual moisture content in the leather semi-finished product decreases and asymptotically approaches a certain value.

Keywords Roller squeezing of leather · Residual moisture content of leather · Force of squeezing process · Specific pressure

1 Introduction

Technological machines, along with functional improvement, must be economical both in production and in operation. Not all machines used in tannery fully comply with this requirement, including machines for squeezing leather semi-finished products (leather). This is largely due to the lack of a methodology that allows optimizing and calculating the parameters of the machine working bodies in the design of roller machines for squeezing leather.

K. Turgunov (✉) · N. Annaev · K. Aliboev
Tashkent Institute of Architecture and Civil Engineering, 13 Navoi Street, Tashkent 100011, Uzbekistan
e-mail: komilturgunov@mail.ru

A. G. Kravets et al. (eds.), *Cyber-Physical Systems Engineering and Control*,
Studies in Systems, Decision and Control 477,
https://doi.org/10.1007/978-3-031-33159-6_13

Due to the lack of such methods, the parameters of roller machines for pressing out leather semi-finished products are arbitrarily chosen by the designers, guided mainly by strength calculations and their own experience.

Incomplete knowledge of the relationship between design and technological parameters becomes the reason for the creation of machines with either unreasonably overestimated indices not used in the process or underestimated indices, which leads to non-compliance with the technological regime. Such machines are not economically viable.

In the process of roller squeezing of wet materials, including semi-finished leather, two phenomena occur simultaneously—contact interaction and moisture filtration. In accordance with two phenomena, the theory of roller squeezing wet materials represents a joint solution to two types of problems: the first is a contact interaction in two-roll modules (a contact problem); the second is the filtration of moisture in a deformable nonhomogeneous porous medium a hydraulic problem.

The Refs. [1–13] are devoted to the study of contact problems of the theory of roller pressing of wet materials. They present solutions to such contact problems as modeling the shape of the roll contact curves, modeling the frictional stresses in roll pairs, and an analytical description of the distribution patterns of contact stresses.

The main hydraulic problems of the theory of roller pressing of wet materials are the analytical description of the hydraulic pressure distribution patterns and modeling the residual moisture content of the material squeezed [14].

The studies in [14–19] are devoted to hydraulic problems of the theory of roller squeezing of wet materials.

In [17], an analytical dependence of the residual moisture content of a semi-finished leather product on the parameters of roller pressing of the leather product was obtained. However, in these models, many parameters are included implicitly, for example, the thickness of the hide in the angles of the contact, which leads to some uncertainties in the identified dependencies. Therefore, the establishment of specific relationships between the parameters and the residual moisture content of leather products requires experimental research.

This study is devoted to experimental research on the effect of the parameters of leather squeezing machines on the residual moisture of leather. The aim of the study is to obtain the dependence of residual moisture content on the main parameters, which could be used to select the optimal technological and design parameters for leather squeezing machines.

The choice of optimal parameters for leather squeezing machines depends on the setting of an extremum problem. An extreme problem to determine the diameter of the rollers, providing a minimum deflection of their working part, the content of residual moisture required during the pressing process, and the least power required to perform the pressing process is set in the chapter. To solve this problem, regression dependences of the residual moisture content of leather and the intensity of the squeezing process on the main parameters of pressing machines are required to be determined.

2 Solutions of the Problems Posed

The experiments were conducted on a setup (Fig. 1) consisting of the following main units: a frame, working rollers, a spring-screw system, and a drive. The design of the setup allows the replacement of the rollers, changing the speed of the rollers and the force of their pressing.

Fig. 1 Experimental setup

A set of rollers with diameters of 0.100, 0.210 and 0.320 m was provided (after winding the cloth, their radii were 0.058. 0.114 and 0.170 m respectively) the required pressing force was applied by a spring-screw system. The speed of the rollers was regulated by a rheostat. The length of the rollers in all cases was 0.12 m.

A set of rollers with diameters of 0.100, 0.210 and 0.320 m was provided (after winding the cloth, their radii were 0.058. 0.114 and 0.170 m respectively) the required pressing force was applied by a spring-screw system. The speed of the rollers was regulated by a rheostat. The length of the rollers in all cases was 0.12 m.

The study was conducted on semi-finished leather samples cut from bovine hide, after chrome tanning (of leather). Samples were cut out 0.06 m wide and of the required length, depending on the type of experiment. After that, they were assembled into groups.

Power consumption was determined by the formula In this case, the torque was measured by the strain gauge method [20].

Experimental studies were conducted by the methods of D optimal planning of the second-order using the K. Kano planning matrix for a three-factor experiment [21].

As a result of research in the field of the contact area of the rollers [1, 10] and the experience of operating leather squeezing machines, it was found that the main parameters affecting the residual moisture content of leather are the load intensity, the diameters of the rollers and the speed of the rollers. Therefore the residual moisture content and power were taken as response functions, and as factors of load intensity of roller radius and roller speed, respectively.

The working matrix was compiled according to the K. Kano planning matrix for a three-factor experiment.

Levels and intervals of factor variations are given in Table 1.

After the implementation of the working matrix, the arithmetic mean values $\bar{W}_{OCT}$ and $\bar{N}$ were obtained (Table 2).

Target function is approximated by the following polynomial

$$y_0 = b_0 + \sum_{i=1}^{k} b_i x_i + \sum_{i,j=1}^{k} b_{ij} x_i x_j + \sum_{i=1}^{k} b_{ii} x_i^2$$

where b_0, b_i, b_{ij}, b_{ii}—are the regression coefficients.

Table 1 Levels and intervals of factor variations

Factor	Variation level			Variation interval
	–	0	+	
Q, kH/m	15	40	65	25
R, m	0.058	0.114	0.170	0.056
V, m/s	0.10	0.22	0.34	0.12

Table 2 Obtained arithmetic mean values

Working matrix			Residual moisture content, %		Power, W	
Q	R	V	$\bar{W}_{OCT}$	$\hat{W}_{OCT}$	$\bar{N}$	$\hat{N}$
40	0.114	0.22	56.83	58.03	65.8	63.4
65	0.170	0.34	57.74	59.23	126.6	124.3
65	0.058	0.34	55.12	54.01	119.0	118.5
15	0.058	0.34	59.44	59.58	99.7	98.3
15	0.170	0.34	61.09	60.57	104.4	104.1
65	0.170	0.10	51.27	50.86	35.2	35.4
65	0.058	0.10	42.42	45.64	34.1	33.8
15	0.058	0.10	57.37	58.03	24.8	24.5
15	0.170	0.10	59.14	59.02	26.7	26.1
65	0.114	0.34	56.08	56.62	112.4	115.2
40	0.058	0.34	58.63	57.51	107.9	108.4
15	0.114	0.34	59.80	60.08	92.2	95.0
65	0.058	0.22	52.83	51.27	73.6	75.1
15	0.058	0.22	60.60	60.25	59.2	60.2
15	0.170	0.22	60.17	61.24	63.2	64.1
65	0.170	0.22	57.42	56.49	77.8	78.8
40	0.170	0.34	59.26	60.62	113.8	114.2
65	0.114	0.10	51.03	48.25	29.9	28.4
40	0.058	0.10	53.50	52.55	27.4	29.1
15	0.114	0.10	58.80	58.52	18.8	19.1
40	0.170	0.10	56.69	55.56	29.2	30.8

The homogeneity of variances was estimated by the Fisher criterion [21] at confidence level $\alpha = 0.95$:

$$F_{расч} = \frac{S^2_{max}}{S^2_{min}},$$

$$W_{ост}\colon\ S^2_{max}\ \ f_{17} = 5;\ \ S^2_{min}\ \ f_6 = 4;\ \ f_{расч} = 5.92;$$

$$N\colon\ S^2_{max}\ \ f_{10} = 5;\ \ S^2_{min}\ \ f_{12} = 4;\ \ f_{расч} = 5.86.$$

The tabular values of the Fisher criterion for $f_{max} = 5$ and $f_{min} = 4$ are $f_{табл} = 6.26$ [21]. Thus, all dispersions for W and N at $\alpha = 0.95$ can be considered homogeneous since $f_{расч} < f_{табл}$.

The hypothesis on the adequacy of the found equations was checked using the Fisher criterion [21] at confidence level $\alpha = 0{,}95$:

$$F_{расч} = \frac{S^2_{ад}}{S^2_{\{y\}}},\quad S^2_{ад} = \frac{\sum_1^N n(\bar{y}_u - \hat{y}_u)^2}{N - \frac{(k+1)(k+2)}{2}};\qquad S^2_{\{y\}} = \frac{\sum_1^N S_i^2 f_i,}{\sum_1^N S_i},$$

$$S^2_{\{y\}} = 11.434;\quad F_{расч} = 1.45;\quad f_{ад} = 11;\quad f_{\{y\}} = 91;$$

$$S^2_{\{y\}} = 10.832;\quad F_{расч} = 1.77;\quad f_{ад} = 11;\quad f_{\{y\}} = 80;$$

The tabular values of the Fisher criterion for $f_{\max} = 90$ and $f_{\min} = 91$ and $f_{\min} = 11$ are [21]. Thus, the found equations can be considered suitable with 95% confidence probability ($F_{расч} < F_{табл}$).

The significance of the coefficients of the found equations is checked by the following formulas

$$\Delta b_0 = \pm 2S_{\{b_0\}} = \pm 2\sqrt{\frac{S^2_{\{\bar{y}\}}}{21}},\quad \Delta b_i = \pm 2S_{\{b_i\}} = \pm 2\sqrt{\frac{S^2_{\{\bar{y}\}}}{16}}$$

$$\Delta b_{ii} = \pm 2S_{\{bii\}} = \pm 2\sqrt{\frac{S^2_{\{\bar{y}\}}}{21}},\quad \Delta b_{ij} = \pm 2S_{\{bij\}} = \pm 2\sqrt{\frac{S^2_{\{\bar{y}\}}}{212}}$$

where $S^2_{\{\bar{y}\}} = \frac{S^2_{\{y\}}}{\bar{n}}$.

Eliminating insignificant coefficients, we have

$$\hat{W}_{oct} = 58.03 - 0.718x_1^2 - 1.442x_3^2 + 1.058x_1x_2 + 1.705x_1x_3 - 3.431x_1 + 1.553x_2 + 2.481x_3 \quad (1)$$

$$\hat{N} = 63.4 + 6.176x_2^2 + 1.051x_3^2 + 2.716x_1x_3 + 1.035x_2x_3 + 7.377x_1 + 1.859x_2 + 40.68x_3 \quad (2)$$

After decoding Eqs. (1) and (2), we obtain

$$W_o = 57.57 - 0.0011Q^2 - 100.4V^2 + 0.754QR + 0.568QV - 0.256Q + 25R + 42V \quad (3)$$

$$N = 14.19 + 1959.4R^2 + 73V^2 + 0.905QV + 154.02RV + 0.96Q + 449.7R + 253.11V \quad (4)$$

Figures 2, 3 and 4 show graphical interpretations of dependences (3), (4).

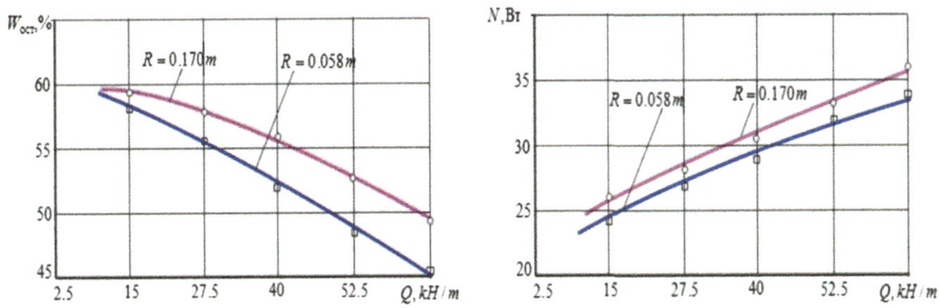

Fig. 2 Dependence of the change in residual moisture content and power on the intensity of the load Q, kN/m for V = 0,10 m/s

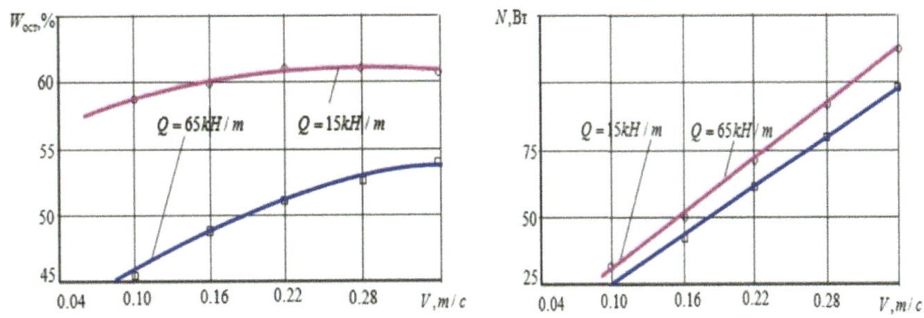

Fig. 3 Dependence of the change in residual moisture content and power on roller speed V, m/s for R = 0.058 m

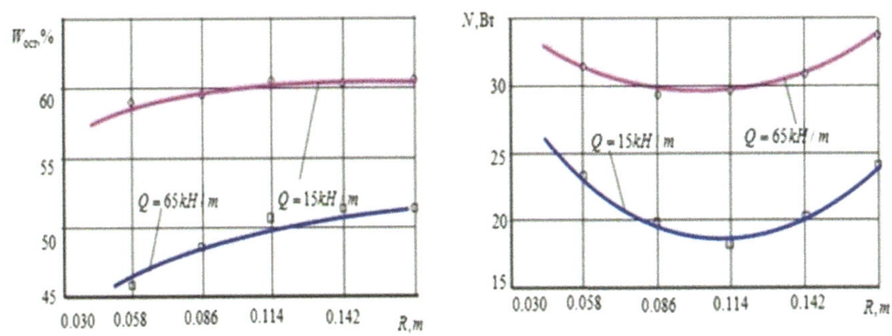

Fig. 4 Dependence of change of residual moisture content and power on the roller radii R, m for V = 0.10, m/s

We will solve a compromise problem—the task of determining the diameter of the rollers that provide the minimum deflection of their working part, the amount of residual moisture required for the pressing process and the least power required to perform the pressing process.

When calculating the optimal diameter of the rollers, we will take into account the conditions for the minimum deflection of their working part $D \geq 0.1B$ [22] or:

$$x_2 \geq \frac{0.05B + \lambda - 0.114}{0.056}, \quad (5)$$

where λ—is the thickness of the elastic coating of the roller.

We solve the extreme problem according to the following calculation scheme:

(1) Change x_3 with step 0.0625, within –1 ÷ 1;
(2) Change x_1 with step 0.05, within –1 ÷ 1;
(3) for each combination of x_1 and x_3 with the accepted values W_{oct} ($W_0 = 58$, 59, 60%) according to expression (3), x_2 is calculated;

If for values B = 1.8, 2.5, 3.2 m and $\lambda = 0.0016$ m factor x_2 satisfies condition (5), then N is calculated by expression (4);

(4) the following arrays of values N are calculated for different values x_3:

for values of $W_o = 58$, $B = 1.8$ and $x_3 = -1$;
for values of $W_o = 58$, $B = 1.8$ and $x_3 = -0.9375$;
for values of $W_o = 58$, $B = 1.8$ and $x_3 = 1$;
for values of $W_o = 59$, $B = 1.8$ and $x_3 = -0.9375$;
for values of $W_0 = 60$, $B = 1.8$ and $x_3 = 1$;

(5) the minimum value of N_0 is selected from each array;
(6) factors x_1, x_2, x_3 at given value of $N_0 = N_{min}$ provide for set values of W_o and B, and decode the natural values Q, R, V;
(7) *Wo*, *B*, *N*, *Q*, *R*, *V* are printed.

As a result of solving the extremum problem, it was revealed that, for $B = 1.8\,\text{m}$, $V = 0.16 \div 0.22\,\text{m/s}$, in order to squeeze to a moisture content of $W_{ocm} = 58 \div 60\%$, the lowest power is achieved for $R = 0.101 \div 0.111\,m$.

Roller squeezing machines used in the tannery are equipped with conveying mechanisms for feeding leather into the working area of the squeezing rollers.

The speed of the conveying mechanism is determined based on the capabilities of an operator who feeds and straightens the leather folds on the conveying mechanism. Considering this, and the effect of the roller speed on the productivity and energy costs of the machine, the values of $B = 1.8$ m–$V = 0.190$ m/s are taken as the rational speed of the rollers.

Thus, a regression model of residual moisture was obtained depending on the load intensity, roller radius and roller speed. However, as the authors of [14] testify, the residual moisture during roller pressing is determined not by linear pressure (load intensity), but by specific pressure, defined as the ratio of linear pressure to the length of the roll contact curve. Therefore, to determine the residual moisture content of leather, the dependence of the length of the roll contact curve on the main parameters of the roller squeezing of leather is also required.

In this regard, we are conducting an additional study to develop an experimental model of the length of the roll contact curve.

To measure the roll contact curve, a thin and narrow tape impregnated with a colored liquid agent was prepared (Fig. 1). The tape was attached to the front side of the leather sample. When the sample with the tape passed through the tip, a colored imprint (trace) was formed on the felt from the upper roll, the measurement of which determined the length of the roll contact curve.

When designing roller squeezing machines, engineering calculations require simpler dependences compared to regression ones. Based on this, the task was set to obtain a model of the length of the roll contact curve, as an empirical function of the main parameters of the roller squeezing of leather.

The working matrix was compiled according to the K. Kano planning matrix for a three-factor experiment (Table 1).

After the implementation of the working matrix, the arithmetic mean values L_0 were obtained (Table 3).

Table 3 Values after working matrix implementation

Working matrix			Length		ε_i^2
Q_1	R_1	V_1	L	$\bar{L}$	
1.000	1.000	1.000	0.0502	0.0460	0.0070
1.625	1.491	1.546	0.0676	0.0704	0.0017
1.625	0.509	1.546	0.0326	0.0373	0.0208
0.375	0.509	1.546	0.0179	0.0197	0.0104
0.375	1.491	1.546	0.0431	0.0437	0.0002
1.625	1.491	0.454	0.0665	0.0675	0.0002
1.625	0.509	0.454	0.0321	0.0305	0,0026
0.375	0.509	0.454	0.0234	0.0220	0.0034
0.375	1.491	0.454	0.0574	0.0588	0.0006
1.625	1.000	1.546	0.0539	0.0550	0.0004
1.000	0.509	1.546	0.0295	0.0345	0.0287
0.375	1.000	1.546	0.0355	0.0375	0.0032
1.625	0.509	1.000	0.0307	0.0344	0.0145
0.375	0.509	1.000	0.0222	0.0205	0.0058
0.375	1.491	1.000	0.0549	0.0554	0.0001
1.625	1.491	1.000	0.0637	0.0648	0.0003
1.000	1.491	1.546	0.0600	0.0643	0.0051
1.625	1.000	0.454	0.0530	0.0502	0.0028
1.000	0.509	0.454	0.0314	0.0374	0.0365
0.375	1.000	0.454	0.0467	0.0423	0.0089
1.000	1.491	0.454	0.0639	0.0676	0.0034
$\sum$			0.9362	0.9598	0.1566

The solution to the problem was based on the technique described in [12], according to which it takes the following form

$$L = C{Q_1}^{\alpha}{D_1}^{\beta}{V_1}^{\gamma} \tag{6}$$

where

$$Q_1 = \frac{Q}{Q_{cp}},\ R_1 = \frac{R}{R_{cp}},\ V_1 = \frac{V}{V_{cp}} \tag{7}$$

Taking the logarithm (6), we obtain

$$lnL = lnC + \alpha lnQ_1 + \beta lnR_1 + \gamma lnV_1 \tag{8}$$

or

$$Y = \alpha X_1 + \beta X_2 + \gamma X_3 + \eta. \tag{9}$$

where $lnW = Y$, $lnC = \eta$, $lnQ_1 = X_1$, $lnR_1 = X_2$ and $lnV_1 = X_3$.

Parameters α, β, γ and η are found by the least squares method, according to which the sum of the squares of deviations ε_i of experimental points from the straight line should be the least, that is $F(\alpha,\beta,\gamma,\eta) = \sum_{i=1}^{21}\varepsilon_i^2 = \sum_{i=1}^{21}(Y_i - (\alpha X_{1i} + \beta X_{2i} + \gamma X_{3i} + \eta))^2$—should be minimum.

Then we have

$$\frac{\partial F}{\partial \alpha} = \sum_{i=1}^{21} 2(Y_i - (\alpha X_{1i} + \beta X_{2i} + \gamma X_{3i} + \eta))(-X_{1i}) =$$

$$= 2\sum_{i=1}^{21}\left(\alpha X_{1i}^2 + \beta X_{1i}X_{2i} + \gamma X_{1i}X_{3i} + \eta X_{1i} - X_{1i}Y_i\right) = 0.$$

Hence, we have

$$\alpha\sum_{i=1}^{21} X_{1i}^2 + \beta\sum_{i=1}^{21} X_{1i}X_{2i} + \gamma\sum_{i=1}^{21} X_{1i}X_{3i} + \eta\sum_{i=1}^{21} X_{1i} = \sum_{i=1}^{21} X_{1i}Y_i$$

Determining likewise $\frac{\partial F}{\partial \beta}$, $\frac{\partial F}{\partial \gamma}$ and $\frac{\partial F}{\partial \eta}$, we obtain the following system of equations

$$\begin{cases} \alpha\sum_{i=1}^{21} X_{1i}^2 + \beta\sum_{i=1}^{21} X_{1i}X_{2i} + \gamma\sum_{i=1}^{21} X_{1i}X_{3i} + \eta\sum_{i=1}^{21} X_{1i} = \sum_{i=1}^{21} X_{1i}Y_i, \\ \alpha\sum_{i=1}^{21} X_{1i}X_{2i} + \beta\sum_{i=1}^{21} X_{2i}^2 + \gamma\sum_{i=1}^{21} X_{2i}X_{3i} + \eta\sum_{i=1}^{21} X_{2i} = \sum_{i=1}^{21} X_{2i}Y_i, \\ \alpha\sum_{i=1}^{21} X_{1i}X_{3i} + \beta\sum_{i=1}^{21} X_{2i}X_{3i} + \gamma\sum_{i=1}^{21} X_{3i}^2 + \eta\sum_{i=1}^{21} X_{3i} = \sum_{i=1}^{21} X_{3i}Y_i, \\ \alpha\sum_{i=1}^{21} X_{1i} + \beta\sum_{i=1}^{21} X_{2i} + \gamma\sum_{i=1}^{21} X_{3i} + 21\eta = \sum_{i=1}^{21} Y_i. \end{cases}$$

After substituting all the sums according to Table 3, we have

$$\begin{cases} 9,582\alpha + 0,410\beta + 0,526\gamma - 3,963\eta = 14,549, \\ 0,410\alpha + 4,925\beta + 0,293\gamma - 2,207\eta = 10,489, \\ 0,526\alpha + 0,293\beta + 6,507\gamma - 2,832\eta = 8,446, \\ -3,963\alpha - 2,207\beta - 2,832\gamma + 21\eta = -66,794. \end{cases} \tag{10}$$

System (10) has a solution: $\alpha = 0,221$, $\beta = 0,740$, $\gamma = -0,09$, $\eta = -3,074$. Hence, we have $C = e^{\eta} = e^{-3,074} = 0,046$.

Then, dependence (6) takes the following form

$$L = 0,046Q_1^{0,221}R_1^{0,74}V_1^{-0,09} \tag{11}$$

The values of the length of the roll contact curve L, calculated by formula (11), are shown in Table 3.

Let us now estimate the accuracy of approximating the experimental data by the formula obtained. For each experimental value of L_i, we determine the deviation from the mean and the square of this deviation by the formula.

$\varepsilon_i = \left(\frac{\bar{L}-L}{\bar{L}}\right)$, writing the estimates in column 6 of Table 3.

We determine the standard deviation and coefficient of variation.

$\sigma = \sqrt{\frac{\sum_{i=1}^{n}\varepsilon_i^2}{n-2}} == \sqrt{\frac{0,037}{19}} = 0,044$, $\upsilon = \frac{\sigma}{\sqrt{n}} = \frac{0,044}{\sqrt{21}} = 0,01$.

As can be seen, the accuracy of approximating the experimental data by formula (7) is more than satisfactory.

From column 5 of Table 3, it follows that the sum of the value of the length of the roll contact curve in the process of experimental research is 0.9598. Hence, $L_{cp} = 0,00457\,m$. With this in mind, from expression (11), we can obtain the following dependences

$$L_1 = 0,6Q_1^{0,221}D_1^{0,74}V_1^{-0,09} \tag{12}$$

$$L = 0,032Q^{0,221}D^{0,74}V^{-0,09} \tag{13}$$

It is known from [14], that

$$P = \frac{Q}{L}, \tag{14}$$

where $P-$ is the specific pressure.

Then, taking into account expression (13), we obtain

$$P = 11,23Q^{0,779}R^{-0,74}V^{0,09} \tag{15}$$

$$P = 56603,32L^{3,52}R^{-3,35}V^{0,41}. \tag{16}$$

Similar to formulas (12) and (13), we obtain the empirical formulas

$$W_1 = 0,97Q_1^{-0,077}D_1^{0,046}V_1^{0,074} \tag{17}$$

$$W = 92,58Q^{-0,077}D^{0,046}V^{0,074} \tag{18}$$

where $W_o = 57,25\%$;

$$N_1 = 0,98Q_1^{0,171}D_1^{0,034}V_1^{1,116} \tag{19}$$

$$N = 189,77Q^{0,171}D^{0,034}V^{1,116} \tag{20}$$

where $N = 68,7$ Вт.

We substitute the value of L from Eq. (13) into formula (14) and after simple transformations we have

$$Q = 0,0126P^{1,28}D^{0,96}V^{-0,116} \tag{21}$$

Hence

$$P = 31,12Q^{0,779}D^{-0,74}V^{0,09} \tag{22}$$

We transform formula (18) taking into account Eq. (21)

$$W = 130,97P^{-0,1}D^{-0,028}V^{0,083} \tag{23}$$

Hence

$$P = 1,37 \cdot 10^{21}W^{-10}D^{-0,28}V^{0,83} \tag{24}$$

From (23) for $D = 0.22$ m and $\upsilon = 0.19$ m/s, we have

$$W = 119,05P^{-0,1}. \tag{25}$$

From dependence (25), it follows that with an increase in specific pressure, the residual moisture content of the leather semi-finished product decreases and asymptotically approaches a certain value (for example, for R = 0.22 m, v = 0.18 m/s, it approaches the value of W = 57%).

We continue to optimize the parameters of the roller pressing of leather, taking into account the specific pressure. To do this, we solve the main extreme problem—the problem of determining the parameters of roller machines for squeezing leather, providing the amount of residual moisture required in the terms of reference.

Let the given value of the residual moisture content of the leather semi-finished product be W_T.

The calculation of the optimal parameters of roller machines for pressing the leather semi-finished product is conducted according to the following algorithm.

1. The length of the working part of the rollers B is determined depending on the width of the leather semi-finished product H. This value should be $10 \div 12$ sm greater than H, that is

$$B = H + (10 \div 12) \tag{26}$$

2. The diameter of the roller with an elastic coating (cloth) is found by solving a compromise problem.
3. We find the speed of the roll taking into account the speed of the conveying mechanism and the speed leatherV_k, determined by the following formula

$$V_k = \frac{\Pi}{60aBK} \tag{27}$$

where Π-is the specified performance of the squeezing machine; a–is the weight of $1m^2$ of the leather semi-finished product; K–is the coefficient that accounts for the feed of the leather semi-finished product, taken within $0,85 \div 0,90$ [22].

4. The average specific pressure is found by the following formula

$$P = 1,37 \cdot 10^{21} W_T^{-10} R^{-0,28} V^{0,083} \tag{28}$$

5. The intensity of the load is determined by the formula

$$Q = 0,038 P^{1,28} R^{0,96} V^{-0,116} \tag{29}$$

6. We calculate the total load applied to the rollers by the following formula

$$G = QB.$$

7. The power of the process is determined by the formula

$$N = 189.77 Q^{0.171} R^{0.034} V^{1.116}$$

8. The motor power of the roller continuous-feed machine is determined by the formula [19] $N_д = (1,15 \div 1,25) \frac{M\omega}{975 \eta_{обш} i_{обш}}$ Where $i_{общ}$ is the total gear ratio of the drive mechanism; $\eta_{обш}$ is the overall efficiency of the drive mechanism from the roller motor to the working roller.
9. Residual moisture content is determined by the following formula

$$W_p = 92.58 Q^{-0.077} R^{0.046} V^{0.074} \tag{30}$$

Table 4 Parameters of roller continuous-feed machines

Parameters	Parameter values
Width of working feed, mm	1800
Roller speed, m/s	0.18
Roller diameters (coated), m	0.220
Electric motor power, kW (for gear transmission)	4,2
Load intensity, kN/m	36

10. We compare the values of the set residual moisture content W_T and the calculated residual moisture content W_p.

If $W_P > W_T$, then it is necessary, either to reduce the load *G*, or increase the diameter *D*, checking the tensile strength of the leather semi-finished product and the deflection of the rollers.

If $W_P > W_T$, then it is necessary, either to increase the load *G* or reduce the diameter of the rollers *D*, checking the tensile strength of the leather semi-finished product and the deflection of the rollers.

If the difference of W_p and W_T does not differ by the amount exceeding the allowable value, then it is considered that the choice of parameters was performed correctly.

Calculation of the parameters of roller continuous-feed machines for squeezing the leather semi-finished product using this method will allow avoiding unreasonable overestimation or underestimatison of their indices.

After the calculations of the parameters according to the specified algorithm, it was revealed that to achieve the residual moisture content W = 57% of the leather semi-finished product, the roller continuous-feed squeezing machine must have the following basic parameters (Table 4).

3 Results

Using the methods of planning the experiment, regression formulas were determined that describe the dependence of residual moisture content of leather and power on the intensity of load, speeds and radii of the rollers.

An extremum problem was solved to determine the diameter of the rollers that provide the minimum deflection of their working part, the content of residual moisture required during the pressing process and the least power required in realizing the squeezing process.

A method was developed for calculating the optimal parameters of the roller pressing of semi-finished leather taking into account the specific pressure. This makes it possible to scientifically substantiate the parameters of the roller squeezing of semi-finished leather and opens up new possibilities for improving the existing roller squeezing machines.

4 Conclusions

An experimental study stated that with a decrease in the radius, speed of the rollers, and.

- with an increase in the intensity of the load, the quality of the roller pressing of leather improves;
- with an increase in load intensity, residual moisture content decreases; all other conditions being equal, the function of residual moisture content on the radius of the roller $W_o(R)$ can be considered a linearly increasing one;
- the pattern of change in residual moisture content from the speed of the roller $W_o(V)$ does not depend on the radius of the rollers;
- the graph of power dependence on the radius of the roller $N = N(R)$ has an extremum point, that is, there is such a radius of the rollers in which N is minimal;
- with an increase in the specific pressure, the residual moisture content of the leather semi-finished product decreases and asymptotically approaches a certain value (for example, for R = 0,22 m v = 0,18 m/s it approaches the value of $W = 57\%$).

References

1. Khurramov, S.R., Abdukarimov, A., Khalturayev, F.S., Kurbanova, F.Z.: Moduling of frrection forces in an asymmetric two-roll module. IOP Conf. Ser.: Mater. Sci. Eng. **916**, 012051 (2020). https://doi.org/10.1088/1557-899X/916/1/012051
2. Alexa, V., Ratiu, S., Kiss, I.: Metall rolling–asymmetrical rolling process. IOP Conf. Ser.: Mater. Sci. Eng. **106**, 012019 (2016)
3. Alexa, V., Ratiu, S.A., Kiss, I, Ciota, G.: Modelling pressure rolling of asymmetric rolling Process. IOP Conf. Ser.: Mater. Sci. Eng. **200**, 012038 (2017)
4. Khurramov, S.R., Abdukarimov, A., Khalturayev, F.S., Kurbanova, F.Z.: Modeling of friction stress in twin roll modules. J. Phys: Conf. Ser. **1789**, 012008 (2021). https://doi.org/10.1088/1742-6596/1789/1/012008
5. Khurramov, S.R., Bahadirov G.A.: To the solution problems of contact interaction in a two-roll module. J Phys: Conf Ser. **1889,** 042029 (2021). https://doi.org/10.1088/1742-6596/1889/1/042029
6. Tieu, A.K., Liu, Y.J.: Friction variation in the cold rolling process. **37**, 177–183 (2004)
7. Khurramov, S.R.: On the issues modeling the roll contact curves. J. Phys.: Conf. Series. **1889,** 042036 (2021). https://doi.org/10.1088/1742-6596/1889/1/042036.
8. Khurramov, S.R.: Filtration rates in roller pressing of fibrous materials. AIP Conf. Proc. **2402,** 030042 (2021). https://doi.org/10.1063/5.0071266

9. Tolcha, M.A., Altenbach, H.: Numerical modeling rolling contact problem and elasticity deformation of rolling die under hot milling. Metals (MDPI). **9**, 226 (2019). https://doi.org/10.3390/met9020226, www.mdpi.com/journal/metals
10. Khurramov, S.R., Khalturaev, F.S., Buriev, E.S.: Residual moisture content in semi-finished leather under roller pressing. AIP Conf. Proc. **2402**, 030038 (2021). https://doi.org/10.1063/5.0071265.
11. Khurramov, S.R.: Analytical description of the shape of roll contact curves in a two-roll module Izvestiya Vysshikh Uchebnykh Zavedenii. Tekhnologiya tekstilnoyi promyshlennosti. **4**, 153–158 (2021)
12. Khurramov, S.R., Khalturaev, F.S., Kurbanova, F.Z.: Theoretical analysis of capture conditions in asymmetric two-roll module. Izvestiya Vysshikh Uchebnykh Zavedenii. Tekhnologiya tekstilnoyi promyshlennosti. **4**(394), 159–163 (2021).
13. Gudur, P.P., Saiunkhe, M.A.: U.S. Dixit. A Theoretical study on the application of asymmetric rolling for the estimation of friction. Int. J. Mech. Schiences. **50,** 315–327.E 2007. Part J: J. Enginieering Tribol. **221** Novikov N.E. Pressing paper web. Moscow: Forest industry, 1972 (2008)
14. Kerekes, R.J., McDonal, J.D.: Equlibrium moisture content in wet pressing of paper. Tappi J. (7), 333–340 (2020)
15. McDonald, D., Kerekes, R.J.: Estimating limits of wet pressing on paper machines. Tappi J. (2), 81–87 (2017)
16. Khurramov, S.R., Bahadirov, G.A., Buriev, E.S., Abdukhalikova, D.T.: Modeling of the roller pressing of fibrous materials. E3S Web Conf. **264,** 01019 (2021). https://doi.org/10.1051/e3sconf/202126401019.
17. McDonald, D., Kerekes, R.J.: Pragmatic mathematical models of wet pressing in papermaking. BioResources. **12**, 9520–9537 (2017)
18. McDonald, D., Kerekes, R.J., Zhao, J.: Perspectives on deriving mathematical models in pulp and paper science. BioResources. **15**, 7319–7329 (2020)
19. Raevsky, N.P.: Methods for experimental study of mechanical parameters machines. Moscow (1952)
20. Tikhomirov, V.B.: Planning and analysis of the experiment (conducting research in the light and textile industries) Moscow: Light industry.
21. Kryuchkov, V.Y.: Study of the process of pressing textile materials with roller machines Dis... Phd. Tech. Sci. Leningrad.

Parameters of Roll Contact Curves of Two-Roll Modules

K. Turgunov, N. Annaev, and K. Aliboev

Abstract Analytical dependencies were obtained to calculate the lengths of roll contact curves, the lengths of the slip and no-slip zones, and angles that determine the position of the separation points of these zones. It was revealed that the slip zones asymptotically decrease to zero with an increase in the angle of inclination of the material to the line of centers. It was determined that an increase in the radii of the rolls and the projection of external forces on the axis, and a decrease in the projection of external forces on the axis led to an increase in the slip zone and a decrease in the no-slip zone.

Keywords Two-roll module · Roll contact zones · Slip zone · No-slip zone

1 Introduction

One of the most economical and versatile types of machines used in many industries for various technological processes is a roller machine. Any technological process in roller machines is performed in their main working body—in a two-roll module, as a result of the contact interaction of the processed material with roller pairs.

In this regard, the problems of improving existing and developing new roller machines and the technological processes implemented in them are closely related to the phenomena of contact interaction of the processed material with roller pairs.

In two-roll modules, the contact interaction of the processed material with roller pairs occurs along the roll contact curves. The shape of the contact curves of each roll of a two-roll module determines the zone of contact of this roll with the material being processed.

Various models of the shape of roll contact curves were developed in [1–13].

In two-roll modules of technological machines, the roll diameters, linear pressure, machine speed, hardness and thickness of the roll coating, the thickness of the

K. Turgunov (✉) · N. Annaev · K. Aliboev
Tashkent Institute of Architecture and Civil Engineering, Navoi, 13, Tashkent 100011, Uzbekistan
e-mail: komilturgunov@mail.ru

A. G. Kravets et al. (eds.), *Cyber-Physical Systems Engineering and Control*, Studies in Systems, Decision and Control 477,
https://doi.org/10.1007/978-3-031-33159-6_14

processed material, and other indices affect, first of all, the width of the roll contact area (the length of the roll contact curves) and the specific pressure, and through them, affect the properties of the technological parameters. Moreover, the specific pressure is characterized by the value of the linear pressure related to the length of the roll contact curves [14]. Therefore, in the theory of contact interaction in two-roll modules, both the shape and the length of the contact line are of particular importance.

There are many publications [14–18] devoted to the experimental study of the width of the roll contact zone, on the basis of which experimental and theoretical–experimental empirical formulas for the width of the roll contact zone were obtained. An analysis of the literature showed that there is no theoretical model for the width of the roll contact zone.

The analysis showed that the determination of these quantities was conducted only by experimental methods.

There are generally three zones on the contact curves of each roll of a two-roll module, that differ in kinematics—slip lagging, no-slip, and slip advance zones. The no-slip zone is separated from the slip zones by two points. The values of the lengths of the slip and no-slip zones; and the angles that determine the position of the points of separation of the no-slip zone from the slip zones are important when solving problems of contact interaction in two-roll modules.

Based on the above, this chapter considers the problem of determining the length of the roll contact curves, the lengths of the slip and no-slip zones, and the angles that determine the position of points that separate these zones.

2 Solutions to the Problems Posed

We consider a two-roll module, in which the rollers are located relative to the vertical line with an inclination to the right by angle β; they have unequal radii ($R_1 \neq R_2$) and elastic coatings made of materials with different stiffness, thickness ($H_1 \neq H_2$) and friction coefficients ($f_1 \neq f_2$), the material being processed has a uniform thickness δ_1, and is fed to the rollers with a downward incline at angle γ_1 to the Oy axis (to the line of centers) (Fig. 1).

Consider the interaction of the layer of material with the lower roll. The contact curve of the lower roll ($i = 1$) consists of compression ($j = 1$) and recovery ($j = 2$) areas (Fig. 1). The points of the compression area of the lower roll are determined by polar coordinates r_{11} and θ_{11}, and the points of the recovery area are determined by r_{12} and θ_{12}.

According to [3], the shape of the lower roll contact curve is determined by the following system of equations:

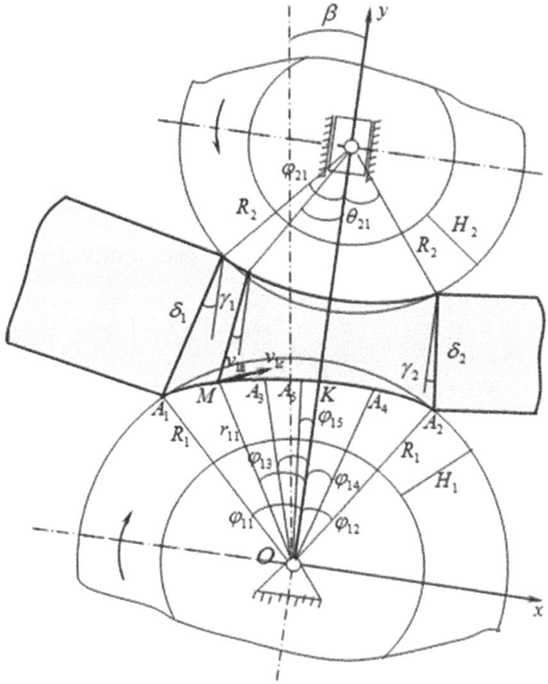

Fig. 1 Scheme of interaction in a two-roll module

$$\begin{cases} r_{11} = \frac{R_1}{1+k_{11}\lambda_{11}}\left(1 + k_{11}\lambda_{11}\frac{\cos(\phi_{11}+\gamma_1)}{\cos(\theta_{11}+\gamma)}\right), & -(\phi_{11}+\gamma_1) \le \theta_{11}+\gamma \le 0,\ \gamma = \frac{\gamma_1\theta_{11}}{\phi_{11}} \\ r_{12} = \frac{R_1}{1+k_{12}\lambda_{12}}\left(1 + k_{12}\lambda_{12}\frac{\cos(\phi_{12}+\gamma_2)}{\cos(\theta_{12}+\gamma)}\right), & 0 \le \theta_{12}+\gamma \le \phi_{12}, + \gamma_2,\ \gamma = \frac{\gamma_2\theta_{12}}{\phi_{12}} \end{cases} \quad (1)$$

where $k_{11} = \frac{n_{11}H_1\sin(\phi_{11}+\phi_{21})}{m_1\delta_1\sin(\phi_{21}-\gamma_1)}$, $k_{12} = \frac{n_{12}H_1\sin(\phi_{12}+\phi_{22})}{m_2\delta_2\sin(\phi_{22}-\gamma_2)}$, $\lambda_{11}, \lambda_{12}$ are the ratios of the rates of relative deformations of the lower roll elastic coating and the material being processed during compression and recovery, respectively; n_{11}, n_{12}—are the coefficients of strengthening of the elastic coating of the lower roll under compression and recovery, respectively; m_1, m_2—are the coefficients of strengthening of the processed material under compression and recovery, respectively.

As follows from the system of Eqs. (1), the shape of the roll contact curve mainly depends on the contact angles. Contact angles are defined according to the geometric conditions of the deformation state in a two-roll module and are evaluated by the gripping conditions [19].

For the considered two-roll module, they have the following form [19–21]:

1. when two rolls are driven:

$$\phi_{11} = \frac{1}{R_1 + R_2}(R_2(\nu_{11} + \nu_{21}) + \delta_1\gamma_1), \ \phi_{21} = \frac{1}{R_1 + R_2}(R_1(\nu_{11} + \nu_{21}) - \delta_1\gamma_1),$$
$$\phi_{12} = \frac{1}{R_1 + R_2}(R_2(\nu_{12} + \nu_{22}) + \delta_2\gamma_2), \ \phi_{22} = \frac{1}{R_1 + R_2}(R_1(\nu_{12} + \nu_{22}) - \delta_2\gamma_2) \tag{2}$$

2. when one of the rolls (a lower one) is driven:

$$\begin{aligned}\phi_{11} &= \frac{1}{R_1 + R_2}\left(R_2\left(\nu_{11} - \nu_{\mathrm{III}}\frac{r_{\mathrm{III}}}{R_2}\right) + \delta_1\gamma_1\right), \ \phi_{21} = \frac{1}{R_1 + R_2}\left(R_1\left(\nu_{11} - \nu_{\mathrm{III}}\frac{r_{\mathrm{III}}}{R_2}\right) - \delta_1\gamma_1\right),\\ \phi_{12} &= \frac{1}{R_1 + R_2}\left(R_2\left(\nu_{12} - \nu_{\mathrm{III}}\frac{r_{\mathrm{III}}}{R_2}\right) + \delta_2\gamma_1\right), \ \phi_{22} = \frac{1}{R_1 + R_2}\left(R_1\left(\nu_{12} - \nu_{\mathrm{III}}\frac{r_{\mathrm{III}}}{R_2}\right) - \delta_2\gamma_1\right).\end{aligned} \tag{3}$$

The length of the contact line of the lower roll, located in the compression zone is determined by the following formula

$$l_{11} = \int\limits_{-(\phi_{11}+\gamma_1)}^{0} \sqrt{r_{11}^2 + r_{11}'^2} d(\theta_{11} + \gamma). \tag{4}$$

Let us transform formula (4) taking into account expression $\frac{r'_{11}}{r_{11}} = tg\psi_{11}$

$$l_{11} = \int\limits_{-(\phi_{11}+\gamma_1)}^{0} \frac{r'_{11}}{\sin\psi_{11}} d(\theta_{11} + \gamma). \tag{5}$$

In formula (5), angle ψ_{11} determines the yielding of the lower roll coating and the layer of material. Depending on the yielding of the contacting bodies, ψ_{11} can be changed within $[0; \theta_{11} + \gamma_1]$, that is

$$0 \le \psi_{11} \le \theta_{11} + \gamma_1.$$

The two limit positions of angle ψ_{11} correspond to the two limit positions of the roll contact line. For $\psi_{11} = 0$, the roll coating is not deformed (Fig. 1). Therefore, the contact line is the part of a circle with radius R_1 and is determined by equation $r_{11} = R_1$. For $\psi_{11} = \theta_{11} + \gamma_1$, the layer of material is not deformed (Fig. 2). In this case, the contact line of the lower roll is the part of a straight line, which is a continuation of the lower surface of the layer and is determined by the following equation $r_{11} = \frac{R_i \cos(\phi_{11}+\gamma_1)}{\cos(\theta_{11}+\gamma_1)}$.

$$\text{We assume that } \psi_{11} = c_{11}(\theta_{11} + \gamma), \tag{6}$$

where $0 < m_{11} < 1$.

Differentiating the first equation of system (1), we have

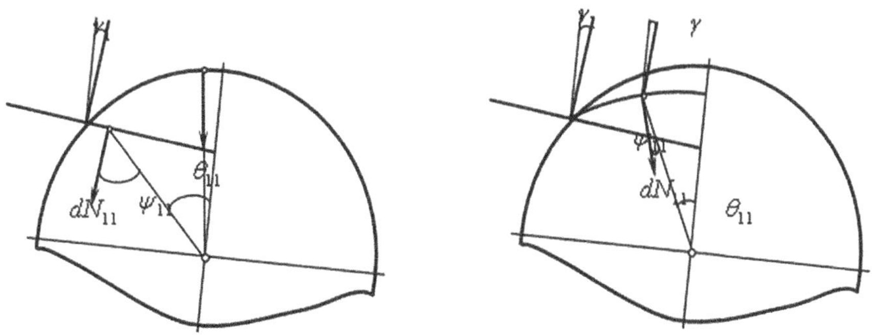

Fig. 2 Scheme of arrangement of angle ψ_{11}

$$r'_{11} = \frac{dr_{11}}{d(\theta_{11} + \gamma)} = \frac{k_{11}\lambda_{11}R_1}{1 + k_{11}\lambda_{11}} \cdot \frac{\cos(\phi_{11} + \gamma_1)}{\cos\phi(\theta_{11} + \gamma)} tg(\theta_{11} + \gamma). \tag{7}$$

From expressions (1) and (7) we find

$$tg\psi_{11} = \frac{r'_{11}}{r_{11}} = \frac{k_{11}\lambda_{11}\cos(\phi_{11} + \gamma_1)}{\cos(\theta_{11} + \gamma) + k_{11}\lambda_{11}\cos(\phi_{11} + \gamma_1)} tg(\theta_{11} + \gamma). \tag{8}$$

From Eq. (6) we have $tg\psi_{11} \approx c_{11} tg(\theta_{11} + \gamma)$.
Then, from Eq. (4), we find

$$c_{11} = \frac{k_{11}\lambda_{11}\cos(\phi_{11} + \gamma_1)}{\cos(\theta_{11} + \gamma) + k_{11}\lambda_{11}\cos(\phi_{11} + \gamma_1)} \tag{9}$$

With formulas (7) and (8), we obtain

$$\frac{k_{11}\lambda_{11}R_1\cos(\phi_{11} + \gamma_1)}{(1 + k_{11}\lambda_{11})\cos^2(\theta_{11} + \gamma)} \cdot \frac{\sin(\theta_{11} + \gamma)}{\sin(c_{11}(\theta_{11} + \gamma))}.$$

Hence, taking into account $\sin m_{11}(\theta_{11} + \gamma_1) \approx m_{11}\sin(\theta_{11} + \gamma)$, we obtain

$$\frac{k_{11}\lambda_{11}R_1\cos(\phi_{11} + \gamma_1)}{c_{11}(1 + k_{11}\lambda_{11})\cos^2(\theta_{11} + \gamma)}$$

or after substitution of c_{11} from Eq. (9) we obtain

$$\frac{R_1}{1 + k_{11}\lambda_{11}}\left(k_{11}\lambda_{11}\cos(\phi_{11} + \gamma_1)\frac{1}{\cos^2(\theta_{11} + \gamma)} + \frac{1}{\cos(\theta_{11} + \gamma)}\right). \tag{10}$$

After substituting this expression into Eqs. (5) and integrating it, we obtain a part of the length of the contact line of the lower roll located in the compression zone

$$l_{11} = \frac{R_1}{1 + k_{11}\lambda_{11}} \left(k_{11}\lambda_{11} \sin(\phi_{11} + \gamma_1) + \frac{1}{2} \ln\left(\frac{1 + \sin(\phi_{11} + \gamma_1)}{1 - \sin(\phi_{11} + \gamma_1)} \right) \right).$$

We expand the logarithmic function into a series, and considering $\sin(\phi_{11} + \gamma_1) < 1$, we restrict ourselves to its first two terms.

Then we have

$$l_{11} = \frac{R_1}{3(1 + k_{11}\lambda_{11})} \sin(\phi_{11} + \gamma_1)\big(3(1 + k_{11}\lambda_{11}) + \sin^2(\phi_{11} + \gamma_1)\big). \tag{11}$$

Likewise, we find the length of the contact line of the lower roll in the recovery zone:

$$l_{12} = \frac{R_1}{3(1 + k_{12}\lambda_{12})} \sin(\phi_{12} + \gamma_2)\big(3(1 + k_{12}\lambda_{12}) + \sin^2(\phi_{12} + \gamma_2)\big). \tag{12}$$

Thus, the total length of the contact line of the lower roll is

$$\begin{aligned} l_1 = {} & \frac{R_1}{3(1 + k_{11}\lambda_{11})} \sin(\phi_{11} + \gamma_1)\big(3(1 + k_{11}\lambda_{11}) + \sin^2(\phi_{11} + \gamma_1)\big) + \\ & + \frac{R_1}{3(1 + k_{12}\lambda_{12})} \sin(\phi_{12} + \gamma_1)\big(3(1 + k_{12}\lambda_{12}) + \sin^2(\phi_{12} + \gamma_1)\big). \end{aligned} \tag{13}$$

Likewise, we determine the length of the contact line of the upper roll

$$\begin{aligned} l_2 = {} & \frac{R_2}{3(1 + k_{21}\lambda_{21})} \sin(\phi_{21} - \gamma_1)\big(3(1 + k_{21}\lambda_{21}) + \sin^2(\phi_{21} - \gamma_1)\big) + \\ & + \frac{R_1}{3(1 + k_{22}\lambda_{22})} \sin(\phi_{22} - \gamma_1)\big(3(1 + k_{22}\lambda_{22}) + \sin^2(\phi_{12} - \gamma_1)\big). \end{aligned} \tag{14}$$

Let both rollers of the considered two-roll module be driven. Then in the contact curve of the lower roll, there are slip lagging zone A_1A_3, no-slip zone A_3A_4 and slip advance zone A_4A_2. The no-slip zone consists of two sections, where section A_3K is located in the compression zone, and section KA_4—is located in the recovery zone (Fig. 1).

Let points M_1 and M_2 simultaneously come into contact at the starting point A_1, and M_1 belongs to the coating of the lower roll, M_2—belongs to the material being processed. Entering the contact zone of the rolls, point M_2 begins to lag behind the point M_1, by the lag angle $\eta_{11}(\theta_{11} + \gamma)$.

At the end of the lagging zone, it equals to $\zeta_{11} = \eta_{11}(-(\phi_{13} + \gamma_3))$, where ϕ_{13}—is the angle separating the slip lagging zone from the no-slip zone. Then point M_2 enters the no-slip zone with a lag by angle ζ_{11}, and retains it until the exit. The no-slip zone ends at point $A_4(\phi_{14} + \gamma_4)$, where ϕ_{14}—is the angle that separates the no-slip zone from the slip advance zone. Entering the slip advance zone with lag ζ_{11}, point M_2 begins to advance point M_1 by the advance angle $\eta_{12}(\theta_{12} + \gamma)$. At the end of the contact zone of the lower roll, the angle defining the distance between the points M_1 and M_2 on the roll contact curve is $|\zeta_{12} - \zeta_{11}|$, where $\zeta_{12} = \zeta_{12}(\phi_{12} + \gamma_2)$.

Let the absolute velocities of points M_1 and M_2 are $\vec{v}_{11B}$ and $\vec{v}_{11o}$, respectively.

The value of the relative velocity of the lagging of processed material along the contact surface of the lower roll can be written as [22]:

$$v_{\mathrm{OTH}} = \omega_1\sqrt{r_{11}^2 + r_{11}^{'2}} - v_m \cos(\theta_{11} + \gamma - \psi_{11}). \tag{15}$$

Let us substitute the expressions r_{11}, ψ_{11} and $r_{11}^{'}$ from Eqs. (1), (6), (7) into this equation, and after transformations we have.

$v_{1abc} = \frac{\omega_1 R_1}{1+k_{11}\lambda_{11}}\left(k_{11}\lambda_{11}\cos(\phi_{11} + \gamma_1)\frac{1}{\cos^2(\theta_{11}+\gamma)} + \frac{1}{\cos(\theta_{11}+\gamma)}\right) - v_m \cos((1 - m_{11})(\theta_{11} + \gamma))$.

Substituting the value of c_{11} from expression (9) into this equation and introducing the following assumption $\cos((1 - m_{11})(\theta_{11} + \gamma)) \approx 1 - \frac{(1-m_{11})^2(\theta_{11}+\gamma)^2}{2}$, we have

$$v =_{1abc} \frac{R_1}{1 + k_{11}\lambda_{11}}\left(k_{11}\lambda_{11}\cos(\phi_{11} + \gamma_1)\frac{1}{\cos^2(\theta_{11} + \gamma)} + \frac{1}{\cos(\theta_{11} + \gamma)}\right) - \\ -v_m\left(1 - \frac{(\theta_{11} + \gamma)^2}{2a_{11}^2(a_{11}^2 - (\theta_{11} + \gamma)^2)}\right) \tag{16}$$

where $a_{11}^2 = 1 + k_{11}\lambda_{11}\cos(\phi_{11} + \gamma_1)$.

The relative displacement of the point of the layer of material along the contact surface of the slip lagging zone can be determined by the following formula [22]:

$S_{1\mathrm{OTC}} = \int\limits_{-(\phi_{11}+\gamma_1)}^{-(\phi_{13}+\gamma)} \frac{v_{11}}{\omega_1} d(\theta_{11} + \gamma)$.

Introducing expression v_{abc} from Eq. (16) into it, we integrate at $v_m = \mathrm{const}$, and after substituting the limits, we obtain the expression which determines the amount of lag

$$S_{1\mathrm{OTC}} = \left(\frac{R_1(1 + k_{11}\lambda_{11}\cos(\phi_{11} + \gamma_1))}{1 + k_{11}\lambda_{11}} - \frac{v_m}{\omega_1}\right)\left(\frac{\phi_{11} + \gamma_1}{\phi_{11}}\right)(\phi_{11} - \phi_{13}). \tag{17}$$

On the other hand, it follows from Fig. 1 that

$$S_{1\mathrm{OTH}} = \left|\int\limits_{-(\phi_{13}+\gamma_3)}^{-(\phi_{13}+\gamma_3)+\zeta_{11}} \sqrt{r_{11}^2 + r_{11}^{'2}}\right| d\theta_{11} = \frac{R_1(1 + k_{11}\lambda_{11}\cos(\phi_{11} + \gamma_1))}{1 + k_{11}\lambda_{11}}\zeta_{11} \tag{18}$$

Comparing two expressions of $S_{1\mathrm{OTC}}$, we determine the lag angle in the lower roll

$$\zeta_{11} = \left(1 - \frac{v_m(1 + k_{11}\lambda_{11})}{\omega_1 R_1(1 + k_{11}\lambda_{11}\cos(\phi_{11} + \gamma_1))}\right)(\phi_{11} - \phi_{13}). \tag{19}$$

The advance angle in the lower roll is determined likewise.

$$S_{1OP} = \left(\frac{R_2(1 + k_{12}\lambda_{12}\cos(\phi_{12} + \gamma_2))}{1 + k_{12}\lambda_{12}} - \frac{v_m}{\omega_1}\right)\left(\frac{\phi_{12} + \gamma_2}{\phi_{12}}\right)(\phi_{12} - \phi_{14}), \quad (20)$$

$$\zeta_{12} = \left(1 - \frac{v_m(1 + k_{12}\lambda_{12})}{\omega_1 R_1(1 + k_{12}\lambda_{12}\cos(\phi_{12} + \gamma_2))}\right)(\phi_{12} - \phi_{14}). \quad (21)$$

Thus, at the end of the lagging zone, the point of the layer of material is located to the point of the lower roll cover by angle ζ_{11}, determined by formula (19), and enters the no-slip zone with a lag by angle ζ_{11}, and retains it until exit. In the advance zone, this point, advancing the point of the lower roll cover, leaves the roll contact zone, and the distance between the points corresponding to the points of the layer of material and the lower roll cover is $|\zeta_{12} - \zeta_{11}|$, where angle ζ_{12} is determined by formula (21).

The values of angles ϕ_{13} and ϕ_{14} depend on the location of the neutral point, where the shear stress is zero.

It is known [18] that shear stresses for the lower roll of the considered two-roll module is determined by the following equations:

$$t_{11} = B_{11}\left(1 - \frac{\cos(\phi_{11} + \gamma_1)}{\cos(\theta_{11} + \gamma)}\right)^{m_1^*} \cos\psi_{11} tg(\theta_{11} + \gamma - \psi_{11} + \xi_1), -(\phi_{11} + \gamma_1) \le \theta_{11} + \gamma \le 0, \quad (22)$$

$$t_{12} = B_{12}\left(1 - \frac{\cos(\phi_{12} + \gamma_2)}{\cos(\theta_{12} + \gamma)}\right)^{m_2^*} \cos\psi_{12} tg(\theta_{12} + \gamma - \psi_{12} + \xi_1), \ 0 \le \theta_{12} + \gamma \le \phi_{12} + \gamma_2, \quad (23)$$

where $\xi_1 = arctg\frac{F_1}{Q_1}$; F_1—are the projections of the horizontal reaction and the gravity force of the lower roll on the Ox axis, Q_1—are the projections of the pressure force of the pressing device and the gravity force of the lower roll on the Oy axis.

In the driven roll, the neutral point is located in the compression area of the no-slip zone [2].

Then from Eq. (22), we determine

$$\phi_{15} = \frac{(1 + k_{11}\lambda_{11}\cos(\phi_{11} + \gamma_1))}{\phi_{11} + \gamma_1}\frac{F_1}{Q_1} \quad (24)$$

Angles ϕ_{13} and ϕ_{14} can be taken as $\phi_{13} = \frac{\phi_{11}+\phi_{15}}{2}$, $\phi_{14} = \frac{\phi_{12}+\phi_{15}}{2}$ [2], expressions (17), (18), (20) and (21) take the following form

$$S_{1OTC} = \left(\frac{R_1(1 + k_{11}\lambda_{11}\cos(\phi_{11} + \gamma_1))}{1 + k_{11}\lambda_{11}} - \frac{v_m}{\omega_1}\right)\left(\frac{\phi_{11} + \gamma_1}{\phi_{11}}\right)\frac{(\phi_{11} - \phi_{15})}{2} \quad (25)$$

$$S_{1OP} = \left(\frac{R_2(1 + k_{12}\lambda_{12}\cos(\phi_{12} + \gamma_2))}{1 + k_{12}\lambda_{12}} - \frac{v_m}{\omega_1}\right)\left(\frac{\phi_{12} + \gamma_2}{\phi_{12}}\right)\frac{(\phi_{12} - \phi_{15})}{2}, \quad (26)$$

$$\zeta_{11} = \left(1 - \frac{v_m(1 + k_{11}\lambda_{11})}{\omega_1 R_1(1 + k_{11}\lambda_{11}\cos(\phi_{11} + \gamma_1))}\right)\frac{(\phi_{11} - \phi_{15})}{2}, \quad (27)$$

$$\zeta_{21} = \left(1 - \frac{v_m(1 + k_{12}\lambda_{12})}{\omega_1 R_1(1 + k_{12}\lambda_{12}\cos(\phi_{12} + \gamma_2))}\right)\frac{(\phi_{12} - \phi_{15})}{2}. \quad (28)$$

According to Fig. 1, the length of the no-slip zone is determined by the following formula

$$l_{1\text{PR}} = l_1 - (S_{1\text{OTC}} + S_{1\text{OP}}), \quad (29)$$

here the values of l_1, $S_{1\text{OTC}}$, $S_{2\text{OP}}$ are determined by formulas (11), (25), and (26).

Similar to formulas (25)–(28), we determine the values and the lag and advance angles in the upper roll

$$S_{2\text{OTC}} = \left(\frac{R_2(1 + k_{21}\lambda_{21}\cos(\phi_{21} - \gamma_1))}{1 + k_{21}\lambda_{21}} - \frac{v_m}{\omega_2}\right)\left(\frac{\phi_{21} - \gamma_1}{\phi_{21}}\right)\frac{(\phi_{21} - \phi_{25})}{2}, \quad (30)$$

$$S_{2\text{OP}} = \left(\frac{R_2(1 + k_{22}\lambda_{22}\cos(\phi_{22} - \gamma_2))}{1 + k_{22}\lambda_{22}} - \frac{v_m}{\omega_2}\right)\left(\frac{\phi_{22} - \gamma_2}{\phi_{22}}\right)\frac{(\phi_{22} - \phi_{25})}{2} \quad (31)$$

$$l_{2\text{PR}} = l_2 - (S_{2\text{OTC}} + S_{2\text{OP}}), \quad (32)$$

$$\zeta_{21} = \left(1 - \frac{v_m(1 + k_{21}\lambda_{21})}{\omega_2 R_2(1 + k_{21}\lambda_{21}\cos(\phi_{21} - \gamma_2))}\right)\frac{(\phi_{21} - \phi_{25})}{2}, \quad (33)$$

$$\zeta_{22} = \left(1 - \frac{v_m(1 + k_{22}\lambda_{22})}{\omega_2 R_2(1 + k_{22}\lambda_{22}\cos(\phi_{22} - \gamma_2))}\right)\frac{(\phi_{22} - \phi_{25})}{2}, \quad (34)$$

where $\phi_{25} = \frac{(1+k_{22}\lambda_{22}\cos(\phi_{22}+\gamma_2))}{\phi_{22}\gamma_2} \frac{F_2}{Q_2}$, and F_2—are the projections of the horizontal reaction and the gravity force of the upper roll on the Ox axis, Q_2—are the projections of the pressure force of the pressing device and the gravity force of the upper roll on the Oy axis.

3 Results

Analytical dependencies were obtained for calculating the length of the roll contact curves, the lengths of the slip and no-slip zones and the angles that determine the position of the separation points of these zones.

It was revealed that at the end of the lagging zone, the point of the layer of material is separated from the point of the lower roll cover by angle ζ_{11}, determined by formula (19), and enters the no-slip zone with a lag by angle ζ_{11}, and retains it until exit. In the advance zone, this point, advancing the point of the lower roll cover, leaves the roll contact zone, and the distance between the points corresponding to the points of the layer of material and the lower roll cover is $|\zeta_{12} - \zeta_{11}|$, where angle ζ_{12} is determined by formula (21).

4 Conclusions

Analysis of calculations of the lengths of slip zones S_{OTC}, S_{OP} and l_{PR} by formulas (25), (26) and (29) showed that:

- the slip zone asymptotically decreases to zero with an increase in the angle of inclination of the material to the line of centers;
- an increase in the radii of the rolls and the projection of external forces on the Oy axis lead to an increase in the slip zone and a decrease in the no-slip zone;
- a decrease in the projection of external forces on the Ox axis leads to an increase in the slip zone and a decrease in the no-slip zone.

References

1. Khurramov, S.R.: Analytical description of the shape of roll contact curves in a two-roll module, pp. 153–158. Proceedings of Higher Educational Institutions, Technology of textile industry", Ivanovo, No. 4 (394) (2021)
2. Khurramov, S.R., Abdukarimov, A., Khalturayev, F.S., Kurbanova, F.Z.: Modeling of friction forces in an asymmetric two-roll module. IOP Conf. Series: Materials Science and Engineering. 916 (2020) 012051 doi:https://doi.org/10.1088/1557-899X/916/1/012051.
3. Khurramov, S.R.: On the issues of modeling the roll contact curves. J. Phys: Conf. Ser. **1889**, 042036 (2021). https://doi.org/10.1088/1742-6596/1889/1/042036
4. Khurramov, Sh.R.: Filtration rates in roller pressing of fibrous materials AIP Conference Proceedings. **2402**, 030042 (2021). https://doi.org/10.1063/5.0071266
5. Kuznetsov, G.K. Research and methods for roller squeezing units in textile machines: Dis. … Doc. Tech. Sci. Kostroma, 1970. - 287 p.
6. Eidman A.A.: Increasing the drivability of a four-wheel drive vehicle through the implementation of the maximum traction force of the wheel propeller with the hydrostatic power of wheel drive: Dis. Cand. Tech. Sci., 165 p.. Moscow (2006)
7. Khurramov, Sh.R., Abdukarimov, A., Khalturayev, F.S., Kurbanova, F.Z.: Modeling of friction stress in two-roll modules. J. Phys: Conf. Ser. **1789**, 012008 (2021). https://doi.org/10.1088/1742-6596/1789/1/012008
8. Khurramov, S.R., Bahadirov, G.A., Buriev, E.S., Abduxalikova, D.N.: Modeling of the roller pressing of fibraus materials. In: E3S Web of Conference, p. 01019, vol. 264(2021). https://doi.org/10.1051/e3sconf/202126401019
9. Pelevin, L.E., Abrashkevich, Yu.D., Balaka, M.N., Arzhaev, G.A.: Modeling of the interaction process of an elastic wheel with a deformable support surface. Min. Equip. Electromechani. - **7**, 10–16 (2013)
10. Soloviev, V.I., Shukhman, S.B., Kapralova, M.A.: Form of contact of an elastic wheel with deformable soil and calculation of the rut depth. Proceedings of MSTU "MAMI", pp. 348–355. Moscow, No. 2, Vol. 1 (2012)

11. Khurramov, S.R.: To the development of the theory of contact in two-roll modules. J. Phys: Conf. Ser. **1901**, 012115 (2021). https://doi.org/10.1088/1742-6596/1901/1/012115
12. Kim, Y.A., Bobrovich, V.A., Voitekhovsky, B.V., Isachenkov, V.S.: Influence of air pressure value in the wheels tires on the geometric parameters of the contact patch when interacting with the supporting surface. Proceedings of the Belarusian State Technical University (2018) series 1, No. 2, pp. 308–312
13. Khurramov, S.R., Bahadirov, G.A.: To the solution of problems of contact interaction in a two-rollmodule. J. Phys.: Conf. Ser. **1889** (2021), 042029. https://doi.org/10.1088/1742-6596/1889/1/ 042029
14. Novikov, N.E.: Pressing paper web, 242 p. Forest Industry, Moscow (1972)
15. Koyaidarov, B.A.: Improvement of roller mechanisms of pass-through machines forrolling of hard leathers. Dis. Cand. Tech. Sci., 168 p. Kyiv: KTI (1984)
16. Likhtzer, E.I.: Study of the spinning process for continuous processing setups for straight linen: Dis. Cand. Tech. Sci. , 170 p. Academy of Public Services, Moscow (1970)
17. Khurramov, S.R., Khalturaev, F.S., Buriev, E.S.: Residual moisture content in the semi-finishedleather under roller pressing, p. 030038. In: AIP Conference Proceedings, no. 2402(2021) https://doi.org/10.1063/5.0071265.
18. Kurbanova, F., Turgunov, K., Abdukhalikova, D.: Influence of the parameters of the contact area **2402**, 030041 (2021). https://doi.org/10.1063/5.0074272
19. Khurramov, S.R., Khalturaev, F.S., Kurbanova, F.Z.: Theoretical analysis of capture conditions in asymmetric two-roll module, pp. 159–163. News of Higher Educational Institutions, Technology in textile industry" - Ivanovo, No. 4 (394) (2021)
20. Maksimenko, O.P., Nikulin, A.A. Romanyuk, R.Y.: Theoretical analysis of exciting capacity of rolls in steady state rolling . News of Higher Educational Institutions, pp. 32–34. Ferrous metallurgy, Moscow, No. 10 (2008)
21. Tolcha, M.A., Altenbach, H.: Numerical modeling rolling contact problem and elasticity deformation of rolling die under hot milling. Metals (MDPI) **9** (2019), 226. https://doi.org/10.3390/met9020226, www.mdpi.com/journal/metals
22. Kryuchkov, V.Y.: Study of the process of pressing textile materials with rollers machines. Dis. Cand. Tech. Sci. Leningrad: LITLP, 166 p. (1970)

Contact Angles in an Asymmetric Two-Roll Module

Farkhad Khalturayev, Shukhrat Khurramov, and Kakhromon Aliboev

Abstract The problems of the contact interaction of the layer of material with the roller pairs depend mainly on the value of the angles of contact since they determine the boundary conditions of these problems. To date, there are no generally accepted models for the angles of initial contact and angles of exit in asymmetric two-roll modules. Contact angles and conditions for their evaluation are determined in the steady-state process of a generalized two-roll module, which takes into account all possible asymmetries of the interaction of a layer of material with pairs of rolls of technological machines. It is established that the sum of the capture angles during the steady-state process does not depend on the slope of the material layer feed relative to the line of centers and on the slope of the upper roll relative to the vertical.

Keywords Two-roll module · Contact problems · Contact angles · Drive roll · Free roll

1 Introduction

Roller machines are an integral part of most technological equipment for light, textile, cotton ginning, pulp and paper, metallurgy, food, chemical, and other industries. Technological processes in two-roll modules are performed as a result of contact interaction of the layer of material with roller pairs. In this regard, the problems of improving existing and developing new technological processes for the mechanical treatment of materials are closely related to the phenomena of contact interaction of a layer of material with roller pairs.

In [1–12], the contact problems of the interaction of the layer of material with the roller pair of two-roll modules of technological machines were solved.

F. Khalturayev (✉) · S. Khurramov · K. Aliboev
Tashkent Institute of Architecture and Civil Engineering, Navoi, 13, 100011 Tashkent, Uzbekistan
e-mail: fedya-spets@mail.ru

A. G. Kravets et al. (eds.), *Cyber-Physical Systems Engineering and Control*, Studies in Systems, Decision and Control 477,
https://doi.org/10.1007/978-3-031-33159-6_15

The problems of contact interaction in two-roll modules depend mainly on the values of the angles of contact of the lower and upper rollers since they determine the boundary conditions of these problems [13].

Despite numerous publications [14–20] devoted to modeling and estimation of the angles of contact in two-roll modules, there are currently no generally accepted models of the angles of initial contact and exit angles in asymmetric two-roll modules. Some of the existing modules deal with only one type of asymmetry of a two-roll module, others consider two types of asymmetry together, but there is no model that simultaneously takes into account all possible types of asymmetry. That is why the solution to the problems of contact interaction (for example, modeling the distribution patterns of contact stresses in an asymmetric two-roll module) is based, basically, on incomplete reflection of the process of contact interaction and incorrect formulation of boundary conditions.

2 Solutions to the Problems Posed

Consider a generalized two-roll module [4], in which the rollers are located relative to the vertical line with an inclination to the right at an angle of β, have unequal diameters ($D_1 \neq D_2$) and elastic coatings made of materials of different stiffness and friction coefficients ($f_1 \neq f_2$), both rollers are driven. The layer of material has a uniform thickness δ_1 and is fed with the inclination downward relative to the line of centers (the line connecting the centers of rotation of the rollers) at an angle of γ_1, the distance between the rollers is h_1.

The angles of contact of two-roll modules depend on the type and location of the guiding movable roller, which is designed mainly according to two schemes [5]: The guide of the movable roller is a slider that moves with the roller along the guide relative to the bed frame;

1. The guide of the movable roller is a lever that connects the axis of the movable roller to the bed frame.
2. The guide of the movable roller is a lever that connects the axis of the movable roller to the bed frame.

In two-roll machines, roller pairs are mainly used, where the slider of the movable roller moves along a vertical line and along a center line. Roller pairs, where the slider moves vertically, are generally used in multi-roller machines.

In [21], the angles of contact were determined in two-roll modules, where the slider of the movable roller moves along a vertical line. This study is devoted to the determination and estimation of the angles of contact where the slider of the movable roller moves along the center line.

The angles of contact in two-roll modules also depend on the way the motion is transmitted to the rollers. According to the method of motion transfer to the rollers, the roller pairs can be divided into drive ones, having a transfer mechanism and free ones, receiving movement due to the friction forces arising on the contact surface of the rollers [4, 20].

Free and driven rollers are distinguished by external forces applied to them, as a result of contact stresses distributed along the contact lines under the effect of these forces [20].

Let the layer of material touch the rollers in cross section B_1B_2. From Fig. 1 it follows that

$$R_1(1-\cos\varphi_{11})+R_2(1-\cos\varphi_{21})+\delta_1(1-\cos\gamma_1)+h_1-\delta_1=0,$$
$$R_2\sin\varphi_{21}=R_1\sin\varphi_{11})-\delta_1\sin\gamma_1,$$

As a result of the transformation of these expressions, we have

$$2R_1\sin^2\frac{\varphi_{11}}{2}+2R_2\sin^2\frac{\varphi_{21}}{2}+2\delta_1\sin^2\frac{\gamma_1}{2}+h_1-\delta_1=0, \quad (1)$$

$$R_2\sin\frac{\varphi_{21}}{2}\cos\frac{\varphi_{21}}{2}=R_1\sin\frac{\varphi_{11}}{2}\cos\frac{\varphi_{11}}{2}-\delta_1\sin\frac{\gamma_1}{2}\cos\frac{\gamma_1}{2} \quad (2)$$

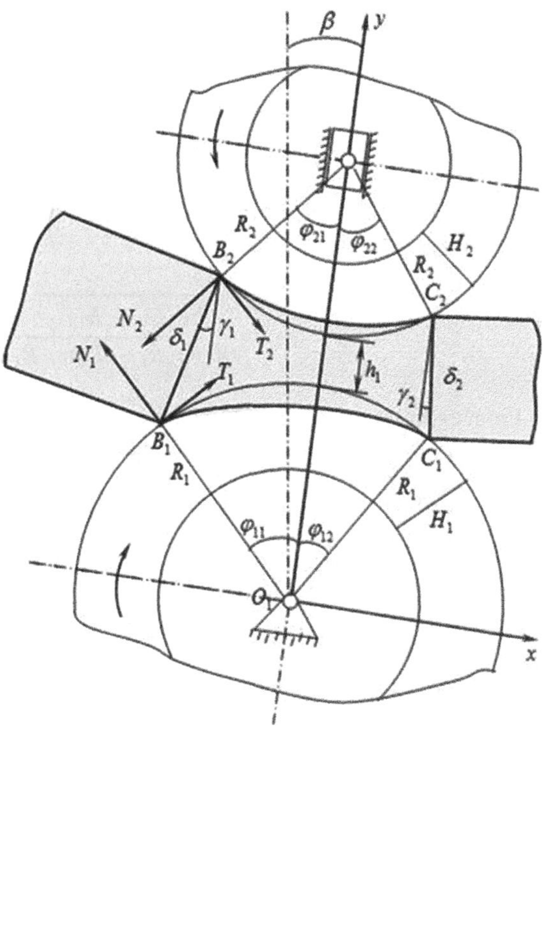

Fig. 1 Scheme of interaction of the layer of material with the roller pair in the second position

Assuming that $\sin\frac{\varphi_{11}}{2}\approx\frac{\varphi_{11}}{2}$, $\cos\frac{\varphi_{11}}{2}\approx 1$, $\sin\frac{\varphi_{21}}{2}\approx\frac{\varphi_{21}}{2}$, $\cos\frac{\varphi_{21}}{2}\approx 1$, $\sin\frac{\gamma_1}{2}\approx\frac{\gamma_1}{2}$, $\cos\frac{\gamma_1}{2}\approx 1$, we represent Eqs. (1) and (2) in a simplified form

$$R_1\varphi_{11}^2 + R_2\varphi_{21}^2 + \delta_1\gamma_1^2 + 2(h_1 - \delta_1) = 0, \tag{3}$$

$$\varphi_{21} = \frac{R_1}{R_2}\varphi_{11} - \frac{\delta_1}{R_2}\gamma_1 \tag{4}$$

After substitution φ_{21} of from Eq. (3) and transformations, Eq. (4) takes the form

$$R_1(R_1 + R_2)\varphi_{11}^2 - 2R_1\delta_1\gamma_1\varphi_{11} + \delta_1^2\gamma_1^2 + R_2\delta_1\gamma_1^2 + 2R_2(h_1 - \delta_1) = 0.$$

Solving this quadratic equation, we find.

$$\varphi_{11} = \frac{\delta_1\gamma_1}{R_1 + R_2} + \sqrt{\frac{2R_2(\delta_1 - h_1)}{R_1(R_1 + R_2)} - \frac{R_2\delta_1(\delta_1 + R_1 + R_2)\gamma_1^2}{R_1(R_1 + R_2)^2}}. \tag{5}$$

Calculations by formula (5) indicate that the value of the second term under the radical is small in comparison with the first term. On this basis, the second term can be ignored, thus simplifying the formula for determining the angle of contact φ_{11} :

$$\varphi_{11} = \sqrt{\frac{2R_2(\delta_1 - h_1)}{R_1(R_1 + R_2)}} + \frac{\delta_1\gamma_1}{R_1 + R_2} \tag{6}$$

Taking into account expression (6) from Eq. (4), we find expressions for determining the angle of contact φ_{21}:

$$\varphi_{21} = \sqrt{\frac{2R_1(\delta_1 - h_1)}{R_2(R_1 + R_2)}} + \frac{\delta_1\gamma_1}{R_1 + R_2} \tag{7}$$

From expressions (6) and (7) we find

$$\varphi_{11} + \varphi_{21} = \sqrt{\frac{2(R_1 + R_2)(\delta_1 - h_1)}{R_1 R_2}}. \tag{8}$$

From Eq. (8) it follows that the sum of the nip angles during a steady-state process does not depend on the feed inclination of the layer of material relative to the center line and on the inclination of the upper roller relative to the vertical line. It increases with an increase in the initial thickness of the layer of material and a decrease in the radii of the rollers.

The results of calculating the nip angle of the lower roller φ_{11} according to formula (6), made it possible to reveal the following:

- with the increase in R_2, δ_1 and γ_1 angle φ_{11} increases: the nature of the increase in φ_{11} with the increase in R_2 does not depend onR_1; angle φ_{11} increases faster at large values of δ_1; angle φ_{11} increases linearly with the increase in γ_1;
- with the increase in R_1, angle φ_{11} decreases, the nature of the decrease in φ_{11} with increasing R_1 does not depend on R_2.

The analysis of the results of calculating angle φ_{21} by formula (7) showed the following:

- with increasing R_2 and γ_1 angle α_1 increases, and angle α_2 decreases;
- with increasing R_1 angle α_1 decreases, and angle α_2 increases;
- the patterns of changes of α_1 and α_2 on parameters δ_1 and h_1 are the same.

Let us compose the force balance equations at the moment the layer of material touches the rollers:

$$\begin{cases} \sum X = -N_{1x} - N_{2x} + T_{1x} + T_{2x} = 0, \\ \sum Y = N_{1y} - N_{2y} + T_{1y} - T_{2y} = 0. \end{cases} \tag{9}$$

From the diagram of forces in Fig. 1, we find

$$\begin{aligned} &N_{1x'} = N_1 \sin \varphi_{11}, \quad T_{1x} = T_1 \cos \varphi_{11}, \quad N_{1y} = N_1 \cos \varphi_{11}, \\ &T_{1y} = T_1 \sin \varphi_{11}, \\ &N_{2x\prime} = N_2 \sin \alpha_{21}, \quad T_{2x} = T_2 \cos \alpha_{21}, \quad N_{2y} = N_2 \cos \alpha_{21}, \\ &T_{2y} = T_2 \sin \alpha_{21}. \end{aligned} \tag{10}$$

Taking these expressions into account, we rewrite system (9) in the following form

$$\begin{cases} N_1 \sin \varphi_{11} - T_1 \cos \varphi_{11} = -(N_2 \sin \varphi_{21} - T_2 \cos \varphi_{21}), \\ N_1 \cos \varphi_{11} + T_1 \sin \varphi_{11} = N_2 \cos \varphi_{21} + T_2 \sin \varphi_{21}. \end{cases} \tag{11}$$

We transform the system (11), expressing the friction forces T_1 and T_2 by normal forces N_1 and N_2 according to the Amonton-Coulomb law of friction:

$$\frac{\sin \varphi_{11} - f_1 \cos \varphi_{11}}{\cos \varphi_{11} + f_1 \sin \varphi_{11}} = -\frac{\sin \varphi_{21} - f_2 \cos \varphi_{21}}{\cos \varphi_{21} + f_2 \sin \varphi_{21}}$$

or

$$tg(\varphi_{11} + \varphi_{21}) = \frac{f_1 + f_2}{1 - f_1 f_2},$$

where f_1, f_2—are the coefficients of friction of the lower and upper rollers on the layer of material.

Bearing in mind that $f_1 = tg\upsilon_{11}$ and $f_2 = tg\upsilon_{21}$ (where υ_{11} and υ_{21}) are the angles of friction of the lower and upper rollers over the layer of material) we finally have:

$$\varphi_{11} + \varphi_{21} = \upsilon_{11} + \upsilon_{21}. \tag{12}$$

Thus, at the moment the layer of material touches the rollers, the sum of the angles of contact is equal to the sum of the friction angles, regardless of the inclination of the upper roll to the O_1y axis and of the inclination of the layer of material to the O_1x axis.

Taking into account Eq. (12), expressions (8) have the form:

$$\upsilon_{11} + \upsilon_{21} = \sqrt{\frac{2(R_1 + R_2)(\delta_1 - h_1)}{R_1 R_2}}. \tag{13}$$

Hence

$$\sqrt{2(\delta_1 - h_1)} = \sqrt{\frac{R_1 R_2}{R_1 + R_2}}(\upsilon_{11} + \upsilon_{21}). \tag{14}$$

Substituting this expression in formulas (6) and (7), we find

$$\begin{aligned} \varphi_{11} &= \frac{1}{R_1 + R_2}(R_2(\upsilon_{11} + \upsilon_{21}) + \delta_1\gamma_1), \\ \varphi_{21} &= \frac{1}{R_1 + R_2}(R_1(\upsilon_{11} + \upsilon_{21}) + \delta_1\gamma_1). \end{aligned} \tag{15}$$

The analysis of dependencies (15) showed that with an increase in the angles of friction of the lower and upper rollers over the layer of material and their sum, the angles of initial contact increase linearly.

The nip angles are estimated by the gripping condition, according to which the possibility of gripping the layer of material by pairs of rollers depends on the ratio of the pulling and pushing forces [20].

In a steady-state process, forces of normal pressure N_1, N_2 and friction forces T_1, T_2 (Fig. 1) act on the layer of material at points B_1 and B_2 on the side of the rollers.

Friction forces are directed in such a way that they tend to pull the layer of material into the gap between the rollers, and the forces of normal pressure prevent it. Obviously, the possibility of gripping depends on the ratio of the pulling and pushing forces. For the gripping, it is necessary that the components of the friction forces along the axis perpendicular to the line of centers T_{1x} and T_{2x} be greater than the components of the forces of normal pressures along the axis of the vertical line of centers N_{1x} and N_{2x} or, at least, be equal to them. Mathematically, this condition is formulated as:

$$N_{1x} + N_{2x} \leq T_{1x} + T_{2x}. \tag{16}$$

In addition to condition (16), we compose the force balance equation along the O_1y axis

$$N_{1y} - N_{2y} + T_{1y} - T_{2y} = 0. \tag{17}$$

By jointly solving inequality (16) and equality (17) by analogy with system (11), we obtain

$$\varphi_{11} + \varphi_{21} \leq \upsilon_{11} + \upsilon_{21}. \tag{18}$$

This inequality determines the condition of gripping in a steady-state process for the generalized two-roll module shown in Fig. 1 when both rollers are driven. Thus, in order to grip the layer of material by the rollers, the sum of the nip angles in a steady-state process must be less than the sum of the friction angles, or at least equal to it.

With inequality (18), and formulas (6) and (7), we determine

$$\begin{aligned} \varphi_{11} &\leq \frac{1}{R_1 + R_2}(R_2(\upsilon_1 + \upsilon_2) + \delta_1\gamma_1), \\ \varphi_{21} &\leq \frac{1}{R_1 + R_2}(R_1(\upsilon_1 + \upsilon_2) - \delta_1\gamma_1). \end{aligned} \tag{19}$$

Thus, in a steady-state process with free capture of a layer of material with two driven rollers, the angles of contact are estimated by inequalities (19).

If condition (18) is not met, then the gripping of the layer of material by the rollers does not occur. In this case, the layer of material is captured using pushing force U.

The angles of the final contact of the rollers φ_{12} and φ_{22} are determined similarly to the angles of the initial contact.

They are:

$$\begin{aligned} \varphi_{12} &= \frac{1}{R_1 + R_2}(R_2(\upsilon_{12} + \upsilon_{22}) + \delta_2\gamma_2), \\ \varphi_{22} &= \frac{1}{R_1 + R_2}(R_1(\upsilon_{12} + \upsilon_{22}) - \delta_2\gamma_2). \end{aligned} \tag{20}$$

where υ_{12}, υ_{22} are the angles of friction at the points C_1 and C_2, respectively,δ_2—is the thickness of the layer of material at the exit from the contact zone of the rollers, γ_2—is the angle of inclination of the layer of material relative to the line of centers at the exit from the contact zone.

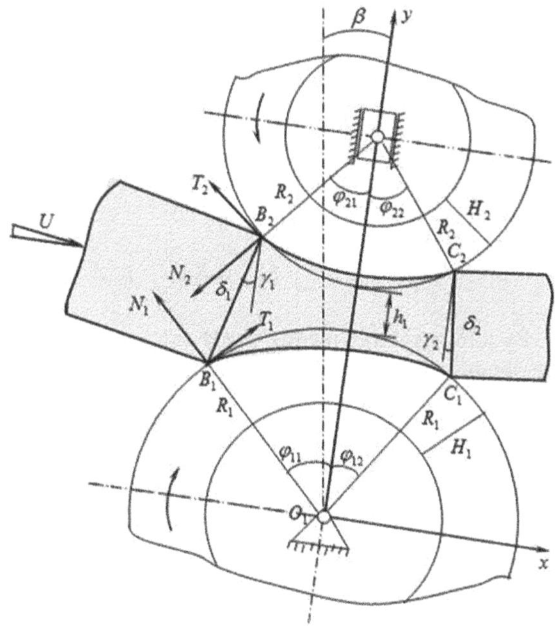

Fig. 2 Scheme of interaction of two-roll module with one driven roller

The analysis of geometric conditions of the interaction of the layer of material with the roller pairs made it possible to find the angles of contact of the two-roll module with two driven rollers. These formulas are general since they determine the angles of contact for all partial cases of interaction of the layer of material with two driven rollers.

Let, in the two-roll module considered, the upper roller be free (Fig. 2).

Likewise formulas (6) and (7), we find

$$\varphi_{11} = \sqrt{\frac{2R_2(\delta_1 - h_1)}{R_1(R_1 + R_2)}} + \frac{\delta_1\gamma_1}{R_1 + R_2}$$

$$\varphi_{21} = \sqrt{\frac{2R_1(\delta_1 - h_1)}{R_2(R_1 + R_2)}} - \frac{\delta_1\gamma_1}{R_1 + R_2} \tag{21}$$

In the study of two-roll modules with one driven roller, the critical angle of inclination of the layer of material relative to the center line is of great importance [4].

The critical angle of inclination of the layer of material in the considered two-roll module is $\gamma_{\kappa py} = \varphi_{21}$, that is

$$\gamma_{\kappa py} = \frac{\sqrt{2R_1(\delta_1 - h_1)}}{R_2(R_1 + R_2)} - \frac{\delta_1\gamma_1}{R_1 + R_2}. \tag{22}$$

In the critical inclination of the layer of material, the nip angle of the lower roller in a steady-state process has the following form:

$$\varphi_{11\kappa p} = \frac{\delta_1 + R_2}{\delta_1 + R_1 + R_2} \frac{\sqrt{2(\delta_1 - h_1)(R_1 + R_2)}}{R_1 R_2}. \tag{23}$$

Analysis of the graphs obtained on the basis of calculations of the critical angle of inclination according to formula (22), indicates that:

- an increase in the radius of the lower roller and the initial thickness of the layer of material leads to an increase in the critical angle of inclination;
- with an increase in the radius of the upper roller and the distance between the rollers, the critical angle of inclination decreases;
- the greater R_2 and the less R_1, the greater $\gamma_{1\text{кр}}$.

Let us compose the force balance equations for the layer of material at the contact with the rollers in the steady-state process according to Fig. 2:

$$\begin{cases} \sum X = -N_{1x} - N_{2x} + T_{1x} + T_{2x} = 0, \\ \sum Y = N_{1y} - N_{2y} + T_{1y} - T_{2y} = 0. \end{cases}$$

or taking into account expressions (10)

$$\begin{cases} N_1 \sin \varphi_{11} - T_1 \cos \varphi_{11} = -(N_2 \sin \varphi_{21} + T_2 \cos \varphi_{21}), \\ N_1 \cos \varphi_{11} + T_1 \sin \varphi_{11} = N_2 \cos \varphi_{21} - T_2 \sin \varphi_{21}. \end{cases} \tag{24}$$

The value of the friction force T_2 can be determined from equation [20]:

$$M_s = T_s r_s = N_2 f_s r_s = T_2 R_2.$$

where M_s—is the moment created by friction forces in the necks of a free roller; T_s—is the resultant force of friction in the necks; r_s—is the neck radius of the free roller; f_s—is the coefficient of friction in the necks.

We determine $T_2 = N_2 f_s \frac{r_s}{R_2}$. After substituting the values of T_2 and $T_1 = f_1 N_1$, into system (24) and jointly solving this equation, we obtain:

$$tg(\varphi_{11} + \varphi_{21}) = \frac{f_1 - f_s \frac{r_s}{R_2}}{1 + f_1 f_s \frac{r_s}{R_2}}.$$

We assume that $\frac{r_s}{R_2} f_s = \frac{r_s}{R_2} tg v_s = tg\left(v_s \frac{r_s}{R_2}\right)$, where v_s—is the angle of friction in the necks. Then we write:

$$\frac{f_1 - f_s \frac{r_s}{R_2}}{1 + f_1 f_s \frac{r_s}{R_2}} = \frac{tg v_1 tg\left(v_s \frac{r_s}{R_2}\right)}{1 - tg v_1 tg\left(v_s \frac{r_s}{R_2}\right)} = tg\left(v_1 - v_s \frac{r_s}{R_2}\right)$$

So,

$$\varphi_{11} + \varphi_{21} = v_1 - v_s \frac{r_s}{R_2}. \tag{25}$$

Taking into account Eq. (25), with (21) and (8), we have

$$\begin{aligned} \varphi_{11} &= \frac{1}{R_1 + R_2}\left(R_2\left(v_{11} - v_s\frac{r_s}{R_2}\right) + \delta_1\gamma_1\right), \\ \varphi_{21} &= \frac{1}{R_1 + R_2}\left(R_1\left(v_{11} - v_s\frac{r_s}{R_2}\right) - \delta_1\gamma_1\right), \end{aligned} \tag{26}$$

Likewise, we have

$$\begin{aligned} \varphi_{12} &= \frac{1}{R_1 + R_2}\left(R_2\left(v_{12} - v_s\frac{r_s}{R_2}\right) + m_1\delta_1\gamma_1\right), \\ \varphi_{22} &= \frac{1}{R_1 + R_2}\left(R_1\left(v_{12} - v_s\frac{r_s}{R_2}\right) - m_1\delta_1\gamma_1\right). \end{aligned} \tag{27}$$

At the moment of initial contact of the layer of material with the rollers, the free roller does not have a pulling effect on the layer of material. Moreover, the free roller to some extent counteracts the pulling of the layer due to the resistance to rotation in its supports. Let us formulate the free gripping condition in the form

$$N_{1x} + N_{2x} + T_{2x} \leq T_{1x}. \tag{28}$$

Solving inequalities (28) together with the second equation of system (24), we find the condition of gripping in a steady process of a two-roll module with one driven roller

$$\varphi_{11} + \varphi_{21} \leq v_{11} - v_s \frac{r_s}{R_2}. \tag{29}$$

Based on dependencies (27) and (28), we find expressions for evaluating the angles of contact in the steady-state process of a two-roll module with one driven roller: $\varphi_{11} + \varphi_{21} \leq v_{11}$

$$\begin{aligned} \varphi_{11} &\leq \frac{1}{R_1 + R_2}\left(R_2\left(v_{11} - v_s\frac{r_s}{R_2}\right) + \delta_1\gamma_1\right), \\ \varphi_{21} &\leq \frac{1}{R_1 + R_2}\left(R_1\left(v_{11} - v_s\frac{r_s}{R_2}\right) - \delta_1\gamma_1\right), \end{aligned} \tag{30}$$

Comparing conditions (19) with (30), we come to the conclusion that in a two-roll module with one driven roller, the gripping is more difficult to perform than in a two-roll module with two driven rollers since in most cases in two-roll modules the rollers are installed in rolling bearings, where the friction is small. As a result, the friction force T_2 can be ignored in comparison with other forces acting on the roller [20].

Then the gripping condition is

$$\varphi_{11} + \varphi_{21} \leq \upsilon_{11}. \tag{31}$$

Thus, if we assume that there is no resistance to rotation in the necks of a free roller, then the sum of the maximum nip angles in two-roll modules with one driven roller is two times less than with two driven rollers.

Let us compare the maximum thickness of the layer that can be gripped by the rollers of a two-roll module in the presence of a driven and free upper roller. We assume that the rollers of the two-roll module at the moment of gripping are located without a gap, that is $h_1 = 0$.

From expression (14), for a two-roll module with two driven rollers, for $h_1 = 0$, we have

$$\delta^n_{1\,\max} = \frac{R_1 R_2}{2(R_1 + R_2)}(\upsilon_{11} + \upsilon_{21})^2 \tag{32}$$

Likewise, considering expression (32), for a two-roll module for the upper free rollers we determine the following:

$$\delta^c_{1max} = \frac{R_1 R_2}{2(R_1 + R_2)}\upsilon^2_{11}$$

Determine the relation

$$\frac{\delta^n_{1\,\max}}{\delta^c_{1\,\max}} = \frac{(\upsilon_{11} + \upsilon_{21})^2}{\upsilon^2_{11}}$$

Hence, for $v_{11} = v_{21}$, we have:

$$\frac{\delta^n_{1\,\max}}{\delta^c_{1\,\max}} = 4.$$

Thus, the maximum thickness of a layer of material that can be gripped by the rollers of the two-roll module is almost four times less with a free upper roller than with a driven roller.

When the coefficient of friction in the necks of the free roller is zero for the free roller to rotate with a layer of material, the line, which is a continuation of the front end of the layer of material must pass through the axis of rotation of the free roller. For gripping at such an arrangement of the layer and the absence of friction in the necks, no additional external pushing forces are required [20].

Let the line, which is a continuation of the front end of the layer of material, pass through the axis of rotation of the free roller. In this case, the angle of inclination of the layer of material to the O_1x axis is critical $\gamma_{\kappa p}$, determined by formula (22). At the critical inclination of the layer of material, the angle of contact of the lower roller is determined by formula (23).

In the presence of friction in the necks of the free roller or at an arbitrary inclination of the layer of material relative to the O_1x axis, the conditions for rotation by the layer of the free upper roller is not fulfilled, and, therefore, an external pushing force is required for gripping.

Thus, the angles of contact in the steady-state process of a two-roll module with one driven roller were obtained, which were determined by formulas (27) and (28). They are evaluated by conditions (31). These formulas are general since they determine the angles of contact for all partial cases of interaction of the layer of material with roller pairs in a two-roll module with one driven roller.

3 Results

Analytical formulas were obtained for determining the angles of contact in the generalized two-roll module in the cases when both rollers are driven and when the upper roller is free. These formulas are general since they determine the angles of contact for all partial cases of interaction of a layer of material with roller pairs in a two-roll module of technological machines.

4 Conclusions

The angles of contact and conditions for their assessment in the steady-state process of the generalized two-roll module are determined in the study, which takes into account all possible asymmetries of the interaction of the layer of material with the roller pairs of technological machines.

Based on the analysis of calculations using the formulas obtained for the angles of contact, the following was revealed:

- the sum of the nip angles in a steady-state process does not depend on the feed inclination of the layer of material relative to the center line and on the inclination of the upper roller relative to the vertical line;
- the sum of the angles of contact is equal to the sum of the friction angles, regardless of the inclination of the upper roller to the ordinate axis and the inclination of the layer of material to the abscissa axis;
- at an increase in the angles of friction of the lower and upper rollers over the layer of material and their sum, the angles of initial contact increase linearly;

- in a two-roll module with one driven roller, gripping is more difficult to perform than in a two-roll module with two driven rollers;
- the maximum thickness of the layer of material that can be gripped by the rollers of the two-roll module is almost four times less with a free upper roller than with a driven roller.

References

1. Alexa, V., Ratiu, S., Kiss, I.: IOP Conf. Series: Materials Science and Engineering, p. 106012019 (2016)
2. Alexa, V., Ratiu, S.A., Kiss, I., Ciota, G.: J IOP Conf. Series: Materials Science and Engineering. 200 012038 (2017)
3. Khurramov, S.R.: IOP Conf. Series: Earth and Environmental Science, vol. 614, p. 012096. (2020)
4. Khurramov, Sh.R.: Journal of Physics: conf. Series **1546**, 012132 (2020)
5. Djuraev, A., Kuliev, T.M.: Int. J. Adv. Sci. Technol. **29**(8), 453–460 (2020)
6. Sh.R.Khurramov, F.Z.Kurbanova. IOP Conf. Series: Earth and Environmental Science **614** 012096 (2020)
7. Bahadirov, G.A., Sultanov, T.Z., Abdukarimov, A.: IOP Conf. Series: Earth and Environmental Science. vol. 614, p. 012102. (2020)
8. Djuraev, A., Kuliev, T.M.: Int. J. Adv. Sci. Technol. **29**(8), 444–452 (2020)
9. Khurramov, S., Bahadirov, G., Buriev, E., Abdukhalikova, D.: E3S Web of Conferences, vol. 264 p. 01019. (2021)
10. Bahadirov, G.A., Sultanov, T.Z., Abdukarimov, A.: IOP Conf. Ser.: Earth Environ. Sci. 614012101 (2020)
11. Khurramov, S.R.: J. Phys.: Conf. Ser. **1889**, 042036 (2021)
12. Xiawei, F., Xiochen, W., Quan, Y., Juquan, S.: Adv. Mech. Eng. 11–8 (2019)
13. Khurramov, S.R.: J. Phys.: Conf. Ser. **1901**, 012115 (2021)
14. Alexa, V., Nussabon, A.J.: Ann. Fac. Eng. Hunedoara **4**, 7–10 (2006)
15. Khurramov, S.R., Abdukarimov, A., Khalturaev, F.S., Kurbanova, F.Z.: IOP Conf. Series: Materials Science and Engineering, vol. **916**, p. 012051. (2020)
16. Khurramov, S.R.: J. Phys.: Conf. Ser. **1789**, 012004 (2021)
17. Farhatnia, F., Salimi, M.A.-J.: J. Eng., Sci. Technol. **3**(4), 147–163 (2011)
18. Khurramov, Sh. R., Bahadirov, G.A.: J. Phys.: Conf. Ser. **1889**, 042029 (2021)
19. Gudur, P., Saiunkhe, M.A., Dixit, U.S.: J. Mech. Sci. **50**, 315–327 (2008)
20. Kurbanova, F.Z., Buriev, E.S., Rasulev, A.: J. Phys.: Conf. Ser. **1889**, 042032 (2021)
21. Khalturaev, F.S., Musirov, M.U.: J. Phys.: Conf. Ser. **1889**, 042024 (2021)

Numerical Study of the Lubricant Viscosity Grade Influence on Thrust Bearing Operation

Nikolay Sokolov, Mullagali Khadiev, Pavel Fedotov, and Eugeny Fedotov

Abstract The article presents studies of the influence of the viscosity grade of the supplied oil ISO VG32 and VG46 in a wide range of rotor speeds and operating clearances on the local and integral characteristics of a thrust plain bearing with fixed pads of the compressor. The studies were carried out using the Sm2Px3Txτ calculation program based on the results of numerical experiments of the bearing. The program is built by numerical implementation of a non-stationary periodic thermoelastohydrodynamic (PTEHD) mathematical model of the thrust bearing operation. The research results indicate a significant influence of the oil viscosity grade on the main characteristics and temperature conditions of the thrust bearing. When changing from ISO VG46 to the lighter oil VG32, there is a noticeable reduction in bearing pad temperatures and power loss. However, the level of this change is determined by the specified operating clearance between the rotating collar and the bearing pads. The influence of oil viscosity grade and the profile of the active surface on the temperature regime of the pad is analyzed. The value and location of the maximum temperature of the thrust bearing pad is determined, as well as the possibility of applying the standard point 75/75 from API-670 in practice.

Keywords Thrust bearing · Mathematical model · Lubricating and boundary films · Fixed pad · Numerical method · Boundary value problem · Viscosity grade · Circumferential velocity · Clearance height

N. Sokolov (✉) · M. Khadiev
Kazan National Research Technological University, 10 Popov, Kazan 420029, Russia
e-mail: sokol-88@list.ru

P. Fedotov
Kazan Federal University, 35 Kremlevskaya, Kazan 420008, Russia

E. Fedotov
AST Volga Region LLC, 50 Petersburgskaya, Kazan 420107, Russia

A. G. Kravets et al. (eds.), *Cyber-Physical Systems Engineering and Control*,
Studies in Systems, Decision and Control 477,
https://doi.org/10.1007/978-3-031-33159-6_16

1 Introduction

The hydrodynamic fixed-geometry thrust bearing of a centrifugal and screw compressor is a critical element of the casing design, the quality of which determines the reliability of the compressor as a whole. The temperature regime of the bearing is largely determined by the operating parameters (external factors affecting it), which include the rotor speed, temperature and viscosity of the supplied lubricant, lubricant consumption, axial load from the pressure drop of the compressible gas, as well as the nature of the change in this load over time. During the operation of a thrust bearing, diagnostic signals are needed to monitor the state of its operation. The generally accepted criteria for performance under high heat conditions, and therefore the limiting parameters for plain bearings, are the minimum thickness and maximum temperature of the lubricating film. The main diagnostic signal, despite the large temporal inertia, remains the temperature of the active surface of the pad [1]. For example, a chromel–alumel or chromel-kopel thermocouple in the solid of the pad near the babbitt layer of the bearing. In some cases, due to the difficulty of removing the airbag signal cable from the housing and due to frequent breaks during assembly and disassembly, the temperature of the lubricant at the drain from the bearing housing is used, despite incomplete information [2].

The limiting temperatures of the lubricating film of a plain bearing at increased circumferential velocities and loads can have a significant effect on the mechanical properties of the anti-friction coating of the pads. In the case of using a babbit coating, its fatigue strength is significantly reduced, which leads over time, under the influence of a periodic external load, to the appearance of cracks on the active surface of the pad and further chipping of the babbit [3]. Bearings with babbitt coating B-83 or Tegostar 738 (contains 81.3% tin Sn), which are most often used in compressors, are recommended for use at lubrication temperatures not exceeding 115 °C [3]. With an increase in the tin content, the allowable operating temperature of the material decreases. It should also be taken into account that increased temperature operating conditions of bearings lead to aging and the appearance of coke-like products in the oil over time [4]. In accordance with this, according to [5], the limiting temperature for mineral oils most used in compressor engineering is a temperature of 120–140 °C. The limiting temperature for B-83 babbitt is 110–115 °C [5]. Thus, for the journal and thrust bearings of the compressor, the allowable operating temperature can be considered to be $[t_{max}] = 110$ °C [3].

Lubrication is supplied to the thrust bearing to ensure the formation of a hydrodynamic film and the removal of heat generated during operation due to the viscous shear of the lubricant layers. At high shear rates, this heat dissipation (or power loss) can be significant in terms of overall compressor power input and energy efficiency. In connection with the study of the general temperature regime of operation of the thrust bearing, of interest are works that consider the change in the temperature field of the bearing when the operating parameters change. Such works include, for example, experimental studies [6–8], which studied the effect of the viscosity grade of the supplied oil on the temperature distribution of tilting pads and the rotating element (collar or trunnion of the rotor), on the change in oil inlet and outlet temperatures, flow rates and eccentricity depending on the applied load and circumferential velocity. The authors themselves confirm that the obtained isotherms of the active surface of the pads in a wide range of speeds and loads were obtained by extrapolation from the readings of thermocouples and, therefore, have inaccuracies.

To avoid such inaccuracies when analyzing the operation of the hydrodynamic fixed-geometry thrust bearing the Sm2Px3Txτ program [9] allows. The program is based on the numerical implementation of a non-stationary periodic thermoelasto-hydrodynamic (PTEHD) model of operation of a thrust bearing. This formulation allows us to consider jointly occurring hydrodynamic and thermal processes in the lubricant film and solid elements of the bearing. The advantage of the PTEHD mathematical model is the description of a more complete thermal picture of the bearing and, consequently, obtaining more reliable output data [10]: local characteristics (maximum temperature and minimum thickness of the bearing lubricating film) and integral characteristics (load-bearing capacity, frictional power losses, lubrication costs through the inlet and outlet sections of the lubricating film, heat flows through the sections of structural elements and the lubricant film, etc.), as well as the distribution of pressure and temperatures in the lubricating and boundary films and solids, including depending on time.

A feature of the PTEHD theory when setting periodic thermal boundary conditions is the direct calculation of the lubricant temperature distribution in the inlet section of the lubricating film, taking into account the mixing of two lubricant flows: a heated flow from the previous pad, which moves under the action of a rotating collar (sticking condition), and a cold lubrication from the inter-pad groove (IPG). Such a setting also makes it possible to calculate the temperatures of the pad and the rotating collar, taking into account heat exchange with the external environment. The latter is especially important, since the collar acts as a heat accumulator, participating in the redistribution of heat. When passing through the areas of the lubricating and boundary films in the process of gradual heating of the lubricant in the direction of rotation, the heat flow changes its direction from the collar to the lubricating film and back [11]. The temperature of the rotating collar in this case assumes a certain averaged constant value.

This article presents a description of some differential equations of the PTEHD mathematical model, the basis of numerical implementation, and describes the results of numerical experiments of the hydrodynamic fixed-geometry thrust bearing of a centrifugal or screw compressor: the effect of oil supply with different viscosity grade (ISO VG32 and VG46) on the general temperature state and bearing power loss at different circumferential velocities and height of the working clearance.

2 Description of the Mathematical Model

The developed periodic thermoelastohydrodynamic (PTEHD) mathematical model is based on the fundamental laws of conservation of mass, momentum and energy, which ensures the adequacy of the formulation and accuracy of calculations. Two profiles of the active surface of pads are considered as the most used in centrifugal and screw compressors: parallel taper land (Fig. 1a) and taper land (Fig. 1b) with a flat land. The lubricating area of L_2 the bearing is divided into a converging bearing lubricating film of the pad and an expanding boundary film of the IPG. The boundary film between the pads is determined by a conditional boundary, which allows geometrically smoothly changing its thickness. In order to get acquainted with the description of the mathematical model in the PTEHD formulation in relation to the operation of a thrust bearing with fixed pads and some physics of the hydrodynamic process, one can refer to the articles [12, 13].

The PTEHD mathematical model is presented in dimensional and dimensionless forms through relative (i.e., dimensionless) quantities related to characteristic

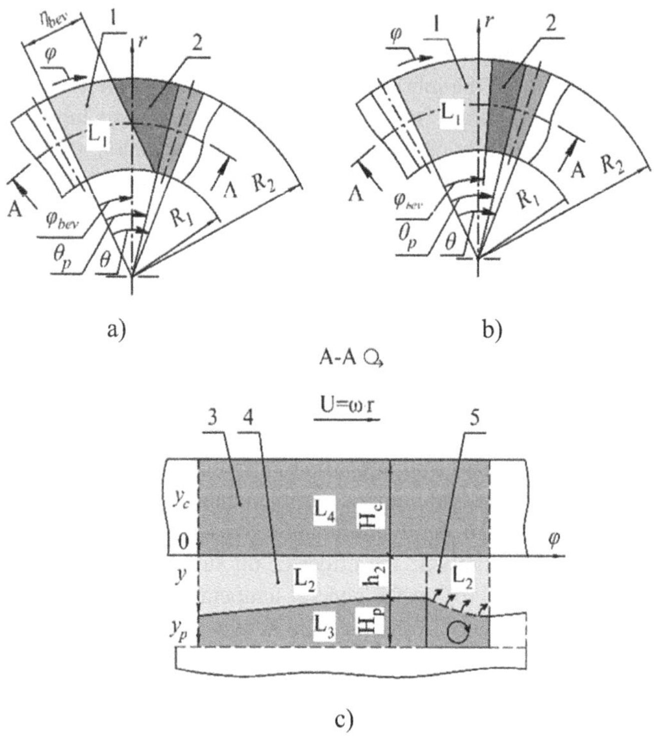

Fig. 1 Calculation scheme of the thrust bearing: **a** pad profile with a parallel taper land; **b** pad profile with a taper land; **c** section along A–A along the middle radius: 1—bevel part, 2—flat part of the pad; 3—rotating collar; 4—lubricating film; 5—boundary film

dimensions (the «-» sign above the quantity). The dimensionless view is necessary to solve the equations by numerical methods and to simplify the parametric analysis of the thrust bearing. The origin of the coordinates of the lubricating and boundary films is located on the surface of the collar. The main governing equations with the corresponding boundary conditions are:

1. the generalized Reynolds equation describing the two-dimensional distribution of pressure in the lubricating film of the $\overline{L}_1(-1 \le \overline{r} \le 1,\ 0 \le \overline{\varphi} \le \overline{\theta}_p, 0 \le \overline{y} \le 1)$ area. The equation is derived with a minimum of restrictive assumptions, in which the density and viscosity of the lubricant are functions of all three coordinates (the form of the equation is close to the derivation by the author Dowson D. for the Cartesian coordinate system [14]). In dimensionless non-stationary form, the equation takes the following form

$$
\begin{aligned}
&-\lambda^2 \frac{\partial}{\partial \bar{r}}\left[(\sigma \bar{r}+1)\bar{h}^3 \bar{f}_0 \frac{\partial \bar{p}}{\partial \bar{r}}\right] - \frac{\partial}{\partial \bar{\varphi}}\left[\frac{\bar{h}^3}{(\sigma \bar{r}+1)} \bar{f}_0 \frac{\partial \bar{p}}{\partial \bar{\varphi}}\right] \\
&= -\mathrm{Re}\psi\sigma\lambda^2 \frac{\partial\left(\bar{h}^3 \bar{f}_1\right)t}{\partial \bar{r}} + \bar{\omega}(\sigma \bar{r}+1)\frac{\partial\left(\bar{h}\bar{f}_2\right)}{\partial \bar{\varphi}} + Sh(\sigma \bar{r}+1)\bar{A},
\end{aligned}
\tag{1}
$$

where $\overline{r}, \overline{\varphi}, \overline{y}$ are dimensionless coordinates; $\overline{p} = ph_{20}^2/\left(\mu_0\omega_* R_{av}^2\theta\right)$ is the local dimensionless pressure; $\overline{A} = \frac{\partial}{\partial \overline{\tau}}\left(\overline{h}\int\limits_0^1 \overline{\rho} d\overline{y}\right) - \overline{\rho}_{\overline{y}=1}\frac{\partial \overline{h}}{\partial \overline{\tau}}$ is a non-stationary multiplier; $\overline{f}_0, \overline{f}_1, \overline{f}_2$ is functions that take into account the variability of lubricant viscosity over the film thickness; λ, σ is relative length and width of the cushion; $\psi = h_{20}/(R_{av}\theta)$ is relative thickness; θ is the angular length of the cushion with the IPG; $\overline{\omega}, \omega_*$ is current and characteristic (usually maximum) angular velocities of the collar; R_{av} is the average radius of the pad; h_{20} is the characteristic thickness of the bearing film; μ_0 is viscosity at the temperature t_0 of the lubricant supply to the IPG; $\overline{\tau} = \tau/\tau_*$ is dimensionless time;
2. the internal energy balance equation describing the three-dimensional temperature distribution in the lubricating and boundary films of the $\overline{L}_2(-1 \le \overline{r} \le 1,\ 0 \le \overline{\varphi} \le \overline{\theta}_p, \overline{\theta}_p \le \overline{\varphi} \le 1, 0 \le \overline{y} \le 1)$ area. In a divergent dimensional nonstationary form, the energy equation takes the following form

$$
\begin{aligned}
&c_p\left(\rho\frac{\partial t}{\partial \tau} + t\frac{\partial \rho}{\partial \tau}\right) + \frac{1}{r}\frac{\partial}{\partial r}(c_p\rho r V_r t) + \frac{\partial}{\partial \varphi}\left(\frac{c_p\rho}{r}V_\varphi t - \frac{\lambda_{oil}}{r^2}\frac{\partial t}{\partial \varphi}\right) + \\
&+\frac{\partial}{\partial y}\left(c_p\rho V_y t - \lambda_{oil}\frac{\partial t}{\partial y}\right) = \mu\left[\left(\frac{\partial V_\varphi}{\partial y}\right)^2 + \left(\frac{\partial V_r}{\partial y}\right)^2\right],
\end{aligned}
\tag{2}
$$

where t is the local temperature, c_p, λ_{oil} is the isobaric heat capacity and thermal conductivity of the lubricant, and ρ is the local density of the lubricant. Equation (2) was transformed into a dimensionless form at the stage of numerical implementation using a dimensionless temperature $\overline{t} = c_{po}\rho_0 h_{20}^2(t - t_0)/(\mu_0\omega_* R_{av}^2\theta)$ while maintaining the divergent form;

3 the three-dimensional distribution of temperatures in the areas of the pad $\overline{L}_3(-1 \leq \overline{r} \leq 1, 0 \leq \overline{\varphi} \leq \overline{\theta}_p, 0 \leq \overline{y}_p \leq 1)$ and the rotating collar $\overline{L}_4(-1 \leq \overline{r} \leq 1, 0 \leq \overline{\varphi} \leq 1, 0 \leq \overline{y}_c \leq 1)$ is described by its own heat transfer equations with the corresponding boundary conditions: at the outer boundaries, heat transfer is taken into account by the Newton-Richmann boundary conditions; between the lubricating film and the pad, as well as between the lubricating and boundary films and the collar, the conditions for the continuity of temperatures and heat fluxes (conjugation condition) are set.

The lubrication rates $\overline{V}_r$ and $\overline{V}_\varphi$ are derived from the truncated Navier–Stokes equations after estimating dimensionless quantities using the N.A. Slezkin method [15] and taking into account the condition that the pressure gradient along the clearance height $\overline{h}$ is equal to zero. In contrast to previous studies [13], the velocity $\overline{V}_y$ was obtained by solving the truncated Navier–Stokes equation along the axis y, which in a dimensionless form takes the following form

$$\mathrm{Re}\psi^3\overline{\rho}Sh\frac{\partial\overline{V}_y}{\partial\overline{\tau}} = 2\psi^2\frac{\partial}{\partial\overline{y}}\left(\overline{\mu}\frac{\partial\overline{V}_y}{\partial\overline{y}}\right) - \\ -\frac{2\psi^2}{3}\frac{\partial}{\partial\overline{y}}\left[\overline{\mu}\left(\frac{\lambda}{(\sigma\overline{r}+1)}\frac{\partial}{\partial\overline{r}}[(\sigma\overline{r}+1)\overline{V}_r] + \frac{1}{(\sigma\overline{r}+1)}\frac{\partial\overline{V}_\varphi}{\partial\overline{\varphi}} + \frac{\partial\overline{V}_y}{\partial\overline{y}}\right)\right]. \tag{3}$$

The boundary conditions for Eq. (3) in the area of the lubricating film are assumed to be equal to zero velocity $\overline{V}_y$ on the surfaces of the pad and collar (impermeability condition). In the boundary film of the inter-pad groove at $\overline{\theta}_p \leq \overline{\varphi} \leq 1, \overline{y} = 1$, the velocity is calculated through the outer normal $\vec{\nu}$ to the conditional boundary film boundary. In dimensionless form, the condition takes the following form

$$\overline{V}_y = \psi\left[\lambda\frac{\partial\overline{h}}{\partial\overline{r}}\overline{V}_r + \frac{1}{\sigma\overline{r}+1}\frac{\partial\overline{h}}{\partial\overline{\varphi}}\overline{V}_\varphi\right]. \tag{4}$$

When modeling the operation of a hydrodynamic bearing, many authors, due to mathematical complexity, usually use the calculation of the temperature at the inlet to the lubricating film at $\overline{\varphi} = 0$ by two methods: using the heat balance equation when setting the temperature difference at the inlet and outlet of the bearing [3], as well as based on the mixing model in the inter-pad groove in accordance with the energy balance of three flows and the introduction of the hot oil entrainment coefficient λ, which varies depending on the bearing design (method of supplying lubricant to the film) and operating parameters [8, 16, 17]. In the PTEHD mathematical model, the boundary condition for the energy Eq. (2) at the inlet to the lubricant film is expressed in a periodic form, which implies the equality of temperatures and heat fluxes of the lubricant (due to convection and thermal conductivity). In dimensional form, in the absence of skew of the pads and the beating of the collar, the condition takes the following form

$$t|_{\varphi=0} = t|_{\varphi=\theta}, \left(\frac{c_p\rho}{r}V_\varphi t - \frac{\lambda_{oil}}{r^2}\frac{\partial t}{\partial\varphi}\right)\bigg|_{\varphi=0} = \left(\frac{c_p\rho}{r}V_\varphi t - \frac{\lambda_{oil}}{r^2}\frac{\partial t}{\partial\varphi}\right)\bigg|_{\varphi=\theta} \tag{5}$$

The differential equations of the mathematical model in the PTEHD formulation and their boundary conditions are interconnected through such physical properties of the working lubricant as viscosity, density, heat capacity and thermal conductivity, as well as through the shape of the clearance, which includes the geometric profile of the active surface of the pad, and some regime parameters.

3 Construction of a Grid Scheme of the Discontinuous Galerkin Method

For numerical solution of Eq. (1), a scheme of the summatorial identities method was constructed, and for Eq. (2), a scheme of the Galerkin discontinuous method with rectangular elements. The heat transfer equations of the pad and collar are solved by finite element methods (FEM). In the constructed scheme for Eq. (2), piecewise constant within the computational domain Ω_h and piecewise linear near the boundary Γ_y of the space of approximating functions were used. The choice of this type of approximating functions allows us to significantly reduce the number of resources required for the calculation. The way of constructing schemes of this type is given in [18]. The mesh scheme in the operator form at a fixed ρ form has the following form

$$B\left(\frac{\partial(\rho_h u_h)}{\partial\tau}\right) + (A_v + A_q + A_\gamma)u_h = F + F_\gamma, \tag{6}$$

where ρ_h, u_h are the grid approximations of the density and temperature functions, respectively. The operators in Eq. (6) are defined by the following forms

$$\begin{aligned} A_v u_h \cdot w_h = \sum_{K\in\Im_h}\int_K (-u_h\overline{V}\cdot\overline{\nabla}w_h)dx + \\ +\sum_{\gamma\backslash\Gamma_y}\int_\gamma \left[u_{h,+p}(\overline{V}\cdot p)^- - u_{h,-p}(\overline{V}\cdot p)^+\right](w_{h,+p} - w_{h,-p})dx, \end{aligned} \tag{7}$$

$$A_q u_h \cdot w_h = \sum_{K\in\Im_h}\int_K q_h\cdot\overline{\nabla}w_h dx + \sum_{\gamma\backslash\Gamma_y}\int_\gamma (w_{h,+p} - w_{h,-p})q_{h,+p}\cdot p dx \tag{8}$$

$$A_\gamma u_h \cdot w_h = \sum_{\gamma \in \Gamma_y} \int_\gamma \omega_\alpha u_h w_h dx, \quad B u_h \cdot w_h = \sum_{K \in \Im_h} \int_K b u w_h dx, \tag{9}$$

$$F \cdot w_h = \sum_{K \in \Im_h} \int_K f w_h dx, \quad F_\gamma \cdot w_h = \sum_{\gamma \in \Gamma_y} \int_\gamma \lambda w_h dx. \tag{10}$$

The operators in Eq. (6) are defined for any $\forall w_h$ from the space of approximating functions, $w = (w_{h,r}, w_{h,\varphi}, w_{h,y})$. In equalities (7)-(10) $\Im_h$ is a set of partition elements of the area $\overline{\Omega}$; $K \in \Im_h$ is partition element; p is a unit normal to the boundaries of the elements of the partition of the area, oriented so that $e \cdot p > 0$, $e = (1, 1, 1)/\sqrt{3}$; q_h is grid approximation of the dissipative part of the equation; $(w)^\pm$ is the positive or negative part of the w function. It should be noted that the spaces of approximating functions generally contain discontinuous functions. The symbols $w_{\pm p}$ denote the limiting values of the functions of the partition elements adjacent to the boundary from the side $\pm p$.

To solve problem (6), an iterative method is constructed

$$B\left(\frac{\rho^k u_h^{k+1} - \breve{\rho}\breve{u}_h}{h_\tau}\right) + (A_v + A_q + A_g)u_h^{k+1} = F + F_g, \quad k = 0, 1, \ldots, \tag{11}$$

where the symbol $\breve{w}$ denotes the value of the function w from the previous time layer, and as u_h^0 is taken the initial value for the energy Eq. (2) in the lubricating film. To solve the resulting system of Eqs. (1)–(3), etc., the stabilized method of biconjugate gradients with incomplete LU decomposition is used as a preconditioner [13].

4 Results of Numerical Experiments

Before carrying out numerical calculations, the following values were taken as initial data [13]: inner and outer diameters are $D_1 = 70$ mm and $D_2 = 115$ mm; number of pads is $Z = 8$; the angular length of the bevel and pad are $\theta_{bev} = 29.1°$ and $\theta_p = 38.8°$; bevel width and depth are $h_{bev} = 20$ mm and $\delta_{bev} = 0.05$ mm. The thickness of the rotating collar is $H_c = 25$ mm, and the thickness of the pad is $H_p = 5$ mm. The lubricant supply temperature is $t_0 = 40$ °C. At the conditional boundary of the boundary film, the temperature of the lubricant in the IPG is set, i.e. $t = t_0$. The sizes of the approximating grids are $N_r = 51$, $N_\varphi = 71$, $N_y = 51$, $N_{yp} = 9$, $N_{yc} = 9$.

As a lubricant in the experiments, the characteristics of turbine oils with different viscosity grades were used: oil Tp-22S according to the standard TU 38.101821–83 or Tp-22B according to the standard TU 38.401–58-48–92 (ISO VG32), as well as Tp-30 according to standard GOST 9972–74 (ISO VG46). Some of their previously measured physical characteristics are listed in Table 1.

Table 1 Physical characteristics

Viscosity grade according to ISO 3448	Viscosity at 40 °C, mm^2/s (cSt)	Viscosity at 100 °C, mm^2/s (cSt)	Density at 40 °C, kg/m^3	Specific heat capacity at 40 °C,	Viscosity grade according to ISO 3448
ISO VG32	36	4.4	882.74	1923	0.128
ISO VG46	48.9	5.94	886.2	1906	0.13

The calculations are carried out in dimensional form to facilitate the description of general patterns. The results of numerical experiments are presented as a dependence of the maximum temperature t_{max} of the lubricating film (Fig. 2a) on the circumferential velocity at different clearance heights (equivalent to the applied axial load) for oils of two viscosity grades: ISO VG32 and VG46. The circumferential velocity is determined on the average radius of the pads $R_{av} = 92.5$ mm. A parallel taper land was chosen as the pad profile (Fig. 1a); for a taper land (Fig. 1b), the results are similar. Thrust bearing power losses are determined by integrating the dissipative energy dissipation function D over the volume of the lubricating and boundary films, i.e. the right side of the energy Eq. (2).

As can be seen from Fig. 2a, the temperature t_{max} increases significantly with an increase in circumferential velocity. This is due to the release and accumulation of heat due to the irreversible dissipation of mechanical energy during the viscous shear of the lubricant layers as it moves in the direction of collar rotation, which leads to an increase in the temperature level along the lubricant flow [19]. At the same lubricant supply temperature $t_0 = 40$ °C, oil enters the working area of the bearing with different viscosities, characterized by a viscosity grade [6–8, 20]. In this case, the viscosity of the lubricant is a value on which the total heat release due

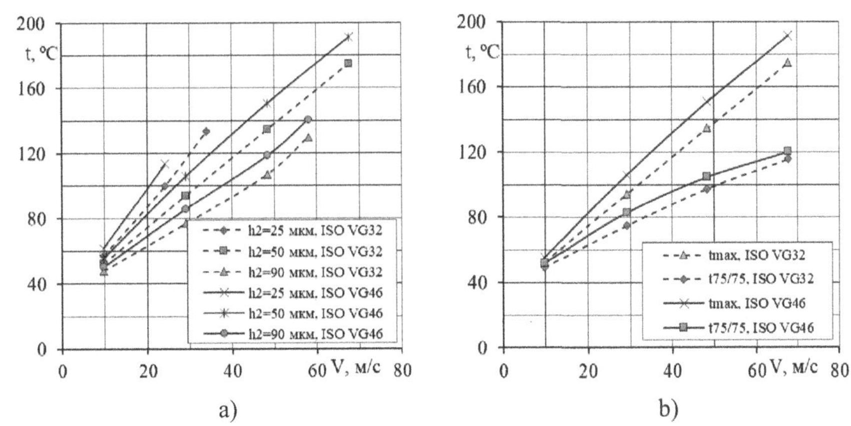

Fig. 2 Dependence of temperatures on the average circumferential velocity of rotation, $t_0 = 40$ °C: **a** t_{max} at different clearances h_2; **b** t75/75 at $h_2 = 50$ μm

to shear in the lubricating and boundary films of the bearing depends [3, 11, 17, 19]. The resulting difference between the temperature curves of oils of viscosity grades ISO VG32 and VG46 clearly characterizes this pattern. From Fig. 2a, you can also see that the temperature curve of the lighter ISO VG32 oil looks flatter due to less heat generation. However, the very difference between the temperature curves of oils of two viscosity grades, the influence of circumferential velocity and working clearance has little effect, similar to studies [6, 7]. The difference between the curves with an increase in the average circumferential velocity gradually becomes constant and reaches 13° with a clearance of 25 μm, 17° with a clearance of 50 μm, and 11° with a clearance of 90 μm. With a decrease in the height of the clearance h_2 between the pads and the collar (see Fig. 1), the maximum temperature increases. For example, for a thicker ISO VG46 oil with a minimum allowable clearance of 25 μm [3, 17], the temperature t_{max} already reaches the limit value of 113 °C at a velocity of 24.22 m/s (corresponding to a rotational speed of 5000 rpm). However, for ISO VG32 oil, due to reduced heat generation, the bearing operation range is more extended: with a clearance of 25 μm, the temperature reaches 133 °C at a velocity of 33.9 m/s (corresponding to a rotational speed of 7000 rpm). Thus, the release of heat in the lubricating and boundary films of the thrust bearing depends on a combination of three quantities: the rotational speed of the collar (i.e., circumferential velocity), the operating clearance, temperature, and the viscosity of the supplied oil itself.

In the practice of diagnosing the operation of a thrust bearing, it is recommended to install a temperature sensor or thermocouple at the 75/75 standard point proposed by Elwell [19, 21] and prescribed in API 670 (4th edition): the sensor should be located at a distance of 75% of the width of the pad radially from the inner diameter (i.e. $r = 51.875$ mm) and 75% of the pad length from the leading edge (i.e. $\varphi = 0.508$ rad.). As can be seen from Fig. 2b, with this method, the sensor does not actually measure the maximum temperature: with an increase in the average circumferential velocity, the difference between the temperature curves tmax and t75/75 increases. With a clearance of 50 μm for ISO VG32 oil, the difference reaches 59°, for ISO VG46 oil −71°. Naturally, this is due to the location of the maximum temperature tmax of the lubricating film near the active surface of the thrust bearing pad. To understand this, it is necessary to consider the influence of the determining factors on the temperature field (isotherms) of the pad.

It is also of interest to consider the change in power losses depending on the height of the clearance for different viscosity grades. As can be seen from Fig. 3a, with an increase in the circumferential velocity and a decrease in the clearance h_2, the bearing power losses increase significantly, which is associated with an increase in the velocity gradients V_r and V_φ in the height of the clearance between the pads and the collar. For the thicker ISO VG46 oil, power losses are increased, which can be explained by an increased viscosity index μ of the incoming oil and significant shear losses. For ISO VG46 oil, at different clearances and at a certain combination of circumferential velocities, the losses can coincide (point A). It should be noted that the paper considers only hydrodynamic power losses in the lubricating and boundary films of the bearing without considering the losses due to turbulence in the IPG and

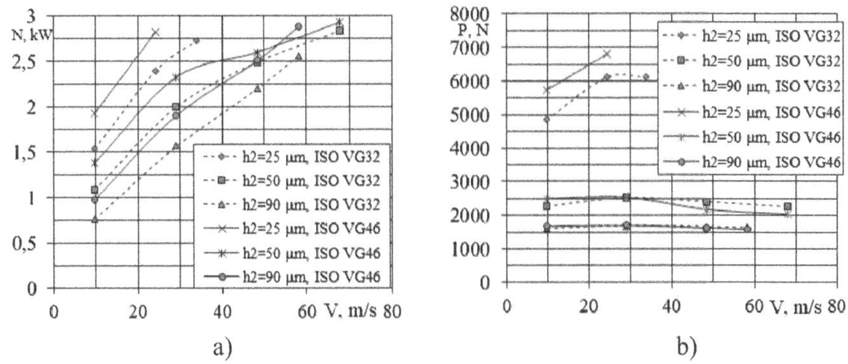

Fig. 3 Dependence of power losses **a** and bearing capacity **b** of the thrust bearing on the average circumferential rotation velocity at different clearances h_2 and viscosity grade, $t_0 = 40$ °C

the parasitic friction of the free surfaces of the rotating collar against the moving lubricant, the contribution of which to the total losses can be significant [3, 6, 7].

The change in the bearing capacity of the thrust bearing is also considered: an increase in circumferential velocity increases the load capacity only at small clearances (Fig. 3b). For ISO VG32 oil, a further increase, on the contrary, leads to some reduction in losses. With an increase in the clearance h_2, the influence of circumferential velocity and oil viscosity grade (i.e. the viscosity of the oil being surrounded at a specified temperature) becomes minimal. Apparently, this is due to the mutual influence of several mechanisms. For example, with increased clearances due to reduced hydraulic resistance, the lubricant flow moves between the pads and the collar at reduced velocities and viscosity. However, the inevitable release of heat during shear losses, in turn, leads to a decrease in viscosity (dilution) of the working volume of the oil and a decrease in heat generation. The shape of the bearing capacity curve depends on the combination of these inverse mechanisms. With a clearance of 50 μm at a velocity of more than 28 m/s, there is some difference between the two viscosity grades up to 10%. With a clearance of 90 μm, the bearing capacity of a thrust bearing in the two viscosity grades ISO VG32 and VG46 is almost the same.

The following articles will present the results of changing the main characteristics of the thrust bearing under dynamic loading, i.e. during axial movement of the rotating collar, as well as in the modes of starting and stopping (running out) of the rotor.

5 Conclusions

Based on the numerical studies carried out using the Sm2Px3Txτ calculation program, the following conclusions can be drawn:

1. The maximum temperature t_{max} of the lubricating film increases significantly with an increase in the average circumferential rotation velocity due to the release and accumulation of heat during the dissipation of the lubricant. The release of heat in the lubricating and boundary films of the thrust bearing depends on the velocity of the collar (i.e. circumferential velocity), the size of the working clearance, temperature and the viscosity of the oil supplied directly.
2. The temperature regime of the thrust bearing is significantly affected by the viscosity grade of the supplied oil (oils of viscosity grades ISO VG32 and VG46 are considered). At the same time, the influence of the velocity itself and the height h_2 of the clearance becomes insignificant on the difference between the temperature curves of the two grades of oil with an increase in circumferential velocity.
3. The use of a thicker ISO VG46 oil leads to higher thrust bearing power losses than ISO VG32 oil: at a circumferential velocity of 29 m/s, the difference reaches 16%. With a further increase in velocity at some clearances, the difference can be significantly reduced.
4. The bearing capacity of the thrust bearing depends on the circumferential velocity only at minimum clearances (up to 20–25 μm). With an increase in the clearance between the pads and the collar, the bearing capacity due to internal heat exchange and hydrodynamic processes becomes practically independent of the collar velocity and viscosity grade.
5. In general, ISO VG46 viscosity grade oil is recommended for use at low rotor speeds of a centrifugal or screw compressor, when power losses and overall low temperature operation are not so great. With increasing circumferential velocities and energy consumption, an ISO VG32 viscosity grade oil should be used. However, in this case, it is necessary to more accurately calculate the minimum thickness of the lubricating film for a given axial load due to the reduced load capacity (at least 20–25 μm).

References

1. Harnoy, A.: Bearing design in machinery. Marcel Dekker, New York (2003)
2. Capitao, J.W., Gregory, R.S., Whitford, R. P. Effects of high-operating speeds on tilting pad thrust bearing performance. Journal of lubrication, Tech. Jan. 98(1), pp. 73–79 (1976). https://doi.org/10.1115/1.3452779
3. Khadiev, M.B., Khamidullin, I.V. Compressors in technological processes. Calculation of plain bearings of centrifugal and screw compressors. Kazan: KNITU, Russia. 260 p. (2021)
4. Klamann, D. Lubricants and related products: Synthesis, properties, applications, international standards. Verlag Chemie. 489 p. (1984)

5. Wang, Q.J., Chung, Y.-W. Encyclopedia of tribology. Springer Science + Business Media New York (2013). https://doi.org/10.1007/978-0-387-92897-5
6. Glavatskih, S., Scan, M. Influence of oil viscosity grade on thrust pad bearing operation. Proceedings of the Institution of Mechanical Engineers, Part J: Journal of Engineering Tribology, Vol. 5, No. 218, 1994–1996, pp. 401–412 (2004). https://doi.org/10.1243/1350650042128085
7. Brockwell, K., Dmochowski, W., Scan, M.: An investigation of the steady-state performance of a pivoted shoe journal bearing with ISO VG 32 and VG 68 oils. Tribol. Trans. **47**, 480–488 (2004). https://doi.org/10.1080/05698190490493382
8. Glavatskih, S., Fillon, M., Larsson, R.: The Significance of oil thermal properties on the performance of a tilting-pad thrust bearing. J. Tribol. **124**(2), 377–385 (2002). https://doi.org/10.1115/1.1405129
9. Fedotov, P.E., Fedotov, E.M., Sokolov, N.V., and Khadiev, M.B. Sm2Px3Txτ – Dynamically loaded thrust plain bearing when formulating a direct problem. Certificate of the state registration of a computer program, No. 2020615227, Russia (2020)
10. Khadiev, M.B. Hydrodynamic, thermal and deformation characteristics of lubricating films of support and sealing units of turbomachines. Dissertation for the Doctor of technical sciences, Kazan State Technological University, Russia. 410 p. (2002)
11. Hori, Y. Hydrodynamic lubrication. Springer-Verlag, 231 p. (2006)
12. Sokolov, N.V., Khadiev, M.B., Maksimov, T.V., Fedotov, E.M., and Fedotov, P.E. Mathematical modeling of dynamic processes of lubricating layers thrust bearing turbochargers. Journal of Physics: Conf. Ser., Vol. 1158, No. 04219 (2019). https://doi.org/10.1088/1742-6596/1158/4/042019
13. Sokolov, N.V., Khadiev, M.B., Fedotov, P.E., and Fedotov, E.M. Three-dimensional periodic thermoelastohydrodynamic (PTEHD) modeling of hydrodynamic processes of a thrust Bearing. International scientific and technical engine conference (EC 2021), IEEE Xplore (2021)
14. Dowson, D.: A generalized Reynolds equation for fluid-film lubrication. Int. J. Mech. Sci. **4**(2), 159–170 (1962). https://doi.org/10.1016/S0020-7403(62)80038-1
15. Slezkin, N.A. Dynamics of a viscous incompressible fluid. State publishing house of technical and theoretical literature, Moscow, Russia. 520 p. (1955)
16. Heshmat, H., Pinkus, O.: Mixing inlet temperature in hydrodynamic bearings. J. Tribol. **108**, 231–244 (1986). https://doi.org/10.1115/1.3261168
17. He, M., Byrne, J.M., Cloud, C., and Vazquez, J. Steady state performance predictions of directly lubricated fluid film journal bearings. Proceedings of the 41st Turbomachinery Symposium. Texas A&M University. Turbomachinery Laboratories. Huston, Texas (2012). https://doi.org/10.21423/R1PM0P
18. Fedotov, E.M.: Limit Galerkin-Petrov schemes for the nonlinear convection-diffusion equation. Springer Link, Differential equations **46**(7), 1042–1052 (2010). https://doi.org/10.1134/S0012266110070116
19. He, M., Byrne, J.M. Fundamentals of fluid film thrust bearing operation and modeling. Asia Turbomachinery and Pump Symposium, pp. 1–26. (2018)
20. Mang, T. Encyclopedia of lubricants and lubrication. Springer-Verlag, Berlin Heidelberg (2014). https://doi.org/10.1007/978-3-642-22647-2
21. Elwell R.C. Thrust bearing temperature. Machine design, pp. 79–81 and 91–94 (1971)

Cyber-Physical Systems Computing and Control

The Method for Increasing the Software Efficiency for Computing Systems with a Hierarchical Memory Structure

Vitaly Egunov and Alla G. Kravets

Abstract This chapter considers the problem of increasing software efficiency in terms of reducing the costs of their development and operation in the process of solving production and research tasks. The existing approaches to solving this problem are analyzed on the example of parameterized algorithms for implementing matrix–vector multiplication and matrix–matrix multiplication operations of the BLAS interface. To achieve this goal of increasing software efficiency, a new method is proposed, designed to improve data caching algorithms in software development for computing systems with a hierarchical memory structure. In the context of the proposed method, an analytical approach to evaluating the software's effectiveness in terms of using a memory subsystem with a hierarchical structure is implemented. The two-side Householder transformation for transforming the general-form matrix to the Hessenberg form was optimized using the developed method. The obtained optimized algorithms are presented. The use of these algorithms can significantly reduce the software execution time, which is confirmed by computational experiments.

Keywords Software efficiency · Memory hierarchy · Cache memory · Reflection transformation · Householder transformation · Cache miss · Cache hit · Efficiency evaluation

1 Introduction

Improving software efficiency is an urgent task and involves many different aspects. This is both reducing the software development time and the execution time [1, 2], increasing energy efficiency [3, 4] and using techniques and patterns of software design to increase its efficiency [5]. Moreover, portability should ensure high performance, which is impossible without standardization of the interfaces of the basic

V. Egunov (✉) · A. G. Kravets
Volgograd State Technical University, 28, Lenina Ave., Volgograd 400005, Russia
e-mail: vegunov@mail.ru

A. G. Kravets et al. (eds.), *Cyber-Physical Systems Engineering and Control*,
Studies in Systems, Decision and Control 477,
https://doi.org/10.1007/978-3-031-33159-6_17

primitives [6]. One of the first such standards was Basic Linear Algebra Subroutines (BLAS), which became the interface standard for libraries implementing linear algebra operations [7, 8]. At the same time, every time a new computing architecture appears, a high-performance implementation of BLAS for it appears. As an example, hardware vendor libraries such as Intel's MKL [9], AMD's ACML [10], and IBM's ESSL [11] can be cited. There are many alternative Open Source solutions, for example, GotoBLAS [12], OpenBLAS [13], and ATLAS [14]. However, all these solutions require significant development efforts to achieve the required performance for different supported architectures. Accordingly, the effectiveness of the resulting software largely depends on the qualifications of the programmer. At the same time, the abundance of modern computing architectures and the increasing problems' complexity often hurt the development process.

Parameterized algorithms are proposed for the effective implementation of basic BLAS primitives on various architectures. Meanwhile, much attention is paid to matrix–vector multiplication (MVM) and matrix–matrix multiplication (MMM), which are the basis for the implementation of most BLAS procedures of the second and third levels [15]. Developers need to configure some parameters to improve performance on the target architecture.

As mentioned above, the "manual" tuning of parameterized algorithms requires expert knowledge in this field and considerable effort and time from the developer. Several auto-tuning solutions have been proposed, for example, ATLAS and PHiPAC [16]. However, ATLAS uses an internal kernel optimized by a developer expert and requires significant manual coding. Thus, despite the possibility of "auto-tuning", the main optimization steps are hidden in the kernel, the development of which still requires considerable effort from an expert developer. Many other implementations, such as GotoBLAS, are also based on an internal high-performance kernel developed by an expert developer, implementation aspects are hidden in this kernel.

A new implementation of BLAS–BLIS (BLAS-like Library Instantiation Software) was presented [17]. It is claimed that BLIS provides an opportunity to rapidly unfold high-performance implementations of BLAS on target architectures, and design and analyze the effectiveness of these implementations on future promising architectures. The BLIS approach is similar to GotoBLAS, i.e. there is an efficient kernel implementing MMM, and BLAS level 3 functions are implemented on this kernel. Thus, it is necessary to develop a kernel and configure the optimization parameters of the cycle around this kernel.

All these approaches are united by the presence of an internal high-performance kernel, hidden from the developer. This makes it difficult to use these libraries effectively, possibly except for BLIS. Kernel development is still difficult, heuristic approaches are widely used in the development process.

The presence of a black-box kernel complicates the analytical analysis of the obtained solutions. Different models of computing systems can be used to tune the algorithm. As an example of a model used to tune algorithm parameters, we can cite the model of the target architectural platform described in [6], which allows calculating parameter values instead of iterating over their values and testing the resulting implementations on target architectures. At the same time, the parameters selected using the architecture model can give a good result, however, requires a highly qualified developer who must be a high-performance software development expert, and knows existing and promising computing architectures.

As an alternative approach, a set of formal procedures and methods of software design based on a mathematical model of a computing system can be proposed. This approach can preserve the advantages of the described methods, in particular, using the setting of the key algorithm parameters by the algorithm that does not provide for their enumeration. It also eliminates the concept of the black-box kernel and reduces the requirements for the developer's qualifications.

2 The Method for Increasing the Software Efficiency

Software efficiency refers to several indicators. From energy efficiency and minimizing software execution time to optimal use of system computing resources and software portability. To increase efficiency, parallelization and vectorization of calculations are used, data caching strategies are improved in systems with a hierarchical memory structure, load balancing in heterogeneous computing systems, and the use of modern efficient coding techniques.

This study proposes a new software development method designed to improve data caching algorithms for computing systems with a hierarchical memory structure. The application of the method includes several stages. In the first stage, it is necessary to estimate the software execution time:

- Estimation of the number of memory accesses (MA) and arithmetic-logical operations;
- MA are divided into groups with an equal value of the cache miss probability;
- Estimation of the number of MA with an equal value of the probability of a cache miss;
- Estimation of the software execution time.

In the second stage, an attempt is made to adjust the algorithm for reducing the software execution time:

- Search for algorithm sections that hurt the software execution time due to an inefficient data caching algorithm;
- An attempt to correct the original algorithm, which leads to a new division of MA into groups;
- Re-evaluation the software execution time.

This method can be performed several times. The decision on re-application can be made based on the analysis of the received groups of operations, the ratio of the initial and received software execution time, as well as the potential for improving the partitioning of MA into groups.

The software execution time is proposed to be estimated as follows:

$$T = \sum_{s=1}^{N_s} \left(N_s^1 * t_{op} + \sum_{g=1}^{NG_s} N_{s,g}^2 \left(\alpha_{s,g} * t_{mem} + \left(1 - \alpha_{s,g} \right) * t_{cache} \right) \right), \quad (1)$$

where

N_s number of conversion steps;
s conversion step number;
N_s^1 number of arithmetic-logical operations required at the s-th step to perform data transformations;
t_{op} arithmetic-logical operations execution time;
NG_s number of MA groups performed within the s-th conversion step;
g number of the group of MA performed as part of the s-th conversion step;
$N_{s,g}^2$ number of MA in the operation group number g;
$\alpha_{s,g}$ the probability of a cache miss when performing a MA in the group of operations g when performing the s-th conversion step;
t_{mem} accessing the main memory time on the target computing system;
t_{cache} accessing the cache memory tine on the target computing system.

An important aspect of the described software development method is to estimate the probability of a cache miss for operations from different groups. Each cache line causes a Compulsory Miss on the first access. Thus, the number of Compulsory Misses is equal to the number of cache lines that were used at least once. It is possible to avoid such mistakes only if the necessary memory block is pre-selected. In this model, data pre-selection is not taken into account, however, in general, it can have a positive impact on the cache hit rate and the efficiency of cache memory.

Thus, "cold" misses are generated in the following cases:

- Out-of-cycle data access, without subsequent repeated requests;
- The first pass through the data while performing cyclic access.

With cyclic access, the situation will change. Suppose, for example, some array is processed in the loop, then "cold" misses will be generated within the first pass. On the second and subsequent passes, the probability of a cache miss will depend on whether the array is placed in cache memory entirely. If so, by the time of the second pass, all elements of the array will be buffered and no misses will be generated. If not, by the time of the second pass, the first buffered elements will be pushed out of the cache and they will need to be reloaded, a cache miss will be generated. In this regard, while forming groups of memory access operations in the software development method described above, as well as when estimating the probabilities of cache misses in the resulting groups, it is necessary to answer several questions. The first question relates to "cold" misses, and the second and third—are to the processing of an array of size N_B.

- What is the probability of Compulsory Misses?
- What is the maximum size of N_B, with which only "cold" misses are generated and there is no displacement of previously buffered elements (Capacity Miss, Conflict Miss)?

– What is the probability P_{CH} of cache hits when processing an array of a size N_B?

Let's first consider the situation when the data is located in memory continuously. In this case, a cache miss is generated when accessing the first element in the next cache line, access to the remaining elements will be carried out from cache memory. The probability of a cache hit in this case can be estimated as follows.

$$P_{CH} = \frac{n-1}{n}, n = \frac{L}{B}, \tag{2}$$

where L is the cache line length in bytes, and B is the size of the used variable type.

For example, if the cache line length is $L = 64$, using the double type *(size of(double)* $= 8$*)*, we get $n = 64/8 = 8$, then the probability of a cache hit is $P_{CH} = 7/8$. When using the float type *(size of(float)* $= 4$*)*, we get $n = 64/4 = 16$, then $P_{CH} = 15/16$.

When changing the access scheme, for example, reading values, not one after the other, but with some intervals, the probability of cache hits can be estimated as

$$P_{CH} = \frac{k-1}{n}, \tag{3}$$

where k is the number of used elements in the cache line.

Expressions (2) and (3) answer the first question formulated above. Moreover, these expressions can also be used in the case of data that are not located continuously. The probability of "cold" misses will depend on the cache line length, the data type size, and the intervals between the data being read, which determine the number of elements used in the cache line.

Evaluation using (2) and (3) can already become extremely useful, despite their obvious insufficiency. It can identify inefficient data access schemes that reduce the efficiency of data use and help to adjust the data caching algorithm. In addition, in the case of out-of-cycle data access, this estimation can be considered complete and sufficient.

In the case of array processing, it is necessary to answer the next two questions, which are interrelated. The analysis in this case should take into account several factors:

- Data distribution scheme;
- Cache memory capacity;
- Cache memory type.

In general, the data is not located in memory continuously. As an example, we can give the processing of a submatrix of a rectangular matrix N*M. In this case, used and unused data blocks will alternate in memory. This case can no longer be analyzed without considering the features of the computing system, based only on the data distribution scheme and the cache line length. The cache memory can be built according to one of the following principles:

- Direct mapped cache;
- Fully associative cache mapping;
- Set associative cache mapping.

Direct mapped cache maps the main memory block to only one cache memory location. Fully associative cache maps the main memory block to any cache location. Set associative cache mapping is a compromise solution. This cache maps the main memory block to a subset of cache locations. Direct mapping is used to search for blocks, and fully associative search is used to search within a set.

Therefore, for example, for a fully associative cache mapping, the maximum size of N_B, at which only "cold" misses are generated and there is no displacement of cache lines, coincides with the cache memory size. For a set associative cache mapping and a direct mapped cache, the situation will be different. Let there be a set associative cache with the following characteristics: S is the cache size, W is the associativity degree, N is the number of sets, and C is the cache line size. Then the contents of the main memory cells with addresses differing by N * C bytes will be loaded into one set. Thus, cache misses can start to be generated before the cache is full.

Modern computing systems have several levels of cache memory. As the level increases closer to the Central Processing Unit (CPU), the cache size decreases, and the data access rate increases. As a result, optimization must be performed in such a way that the most used data remains in the fastest cache for as long as possible.

It is also worth noting that estimating the probability of cache misses only using expressions (2) and (3) does not allow optimization for a specific cache level, for example, L1. These expressions allow us to choose the optimal strategy for accessing data. This approach can be called a "high-level" optimization of the data caching algorithm. In this chapter, this approach was used, and it will be shown that it gives significant results even without considering the features of the various levels of cache memory.

3 Results and Discussions

The method described above was applied to the problem of reducing a general-form matrix to a Hessenberg form in the process of solving the eigenvalue problem [18]. The Householder method, also called the Householder reflection transformation [19, 20], was used as a basic method:

$$A = QRQ^T, \tag{4}$$

where A is the original matrix, R is the Hessenberg matrix, Q is the orthogonal matrix, which is the product of the Householder reflection matrices, and each matrix is uniquely determined by the reflection vector.

To transform the n*n matrix to the upper Hessenberg form, it is necessary to perform (n-2) steps, the execution of the k–th step can be represented as follows.

In Table 1, the following designations were used: **a**—columns of the original matrix, **b**—its rows. In the "Processing sequence" column, the arrows show the sequence of transformation of rows (red arrows) and columns (blue arrows) of the original matrix within one step of the reduction to the upper Hessenberg form (Table 2).

Allocation groups of MA with an equal probability of cache miss and estimate the number of operations in each group:

- The probability of a cache miss during processing by elements of the original matrix when calculating the modulus of the leading column of the original matrix and during the formation of the elements of the reflection vector α_V; the matrix size is sufficient to not fit entirely into the cache, we take this parameter equal to 1; the number of operations is (n – k – 1);
- The probability of a cache miss when calculating the elements of the reflection vector α_{VR}; with a reasonable matrix size, when the reflection vector is completely placed in cache memory, this parameter can be taken equal to 0; the number of operations is (n – k – 1);
- The probability of a cache miss when multiplying the original matrix by the reflection matrix on the right α_C; the matrix does not fit entirely into the cache memory, and since access is carried out along the columns of the original matrix, we assume the probability of a cache miss to be $\alpha_C = 1$, although this indicator may be slightly smaller, especially when performing steps with a large number and taking into account parameter α_V; the number of operations is 2 * (n – k – 1) * (n – k – 1);

Table 1 Execution of the k-th step of the Householder transformation

Operations of the k-th transformation step	Processing sequence
$s_k = -sign(a_{k+1,k})\left(\sum_{i=k+1}^{n} a_{ik}^2\right)^{\frac{1}{2}}$, $\varphi_k = \left(s_k^2 - s_k a_{k+1,k}\right)^{-1}$ $\boldsymbol{u}_k^T = (0, \ldots, 0, a_{k+1,k} - s_k, a_{k+2,k}, \ldots, a_{nk})$ For j = k to n $\lambda_j = \varphi_k \boldsymbol{u}_k^T \boldsymbol{a}_j$ $\boldsymbol{a}_j = \boldsymbol{a}_j - \lambda_j \boldsymbol{u}_k$ For j = k to n $\lambda_j = \varphi_k \boldsymbol{b}_j \boldsymbol{u}_k$ $\boldsymbol{b}_j = \boldsymbol{b}_j - \lambda_j \boldsymbol{u}_k^T$	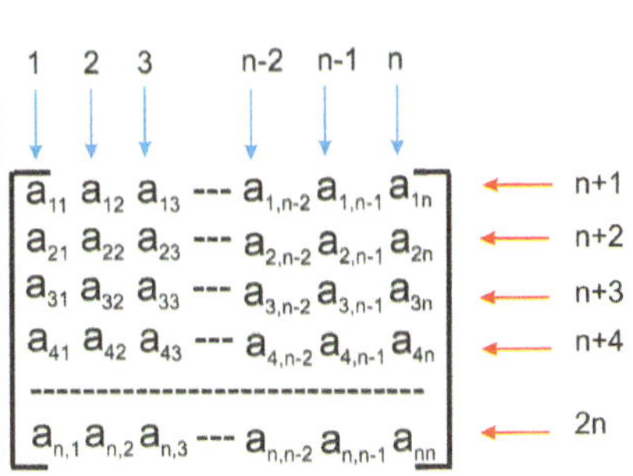

Table 2 ROH and SPH

	Operations of the k-th transformation step	Processing sequence
ROH	$\boldsymbol{sc} = \varphi_k \boldsymbol{u}_k^T \boldsymbol{A}$ For r = k + 1 to n For c = k to n $a_{r,c} = a_{r,c} - sc_c * u_r$ For j = k to n $\lambda_j = \varphi_k \boldsymbol{b}_j \boldsymbol{u}_k$ $\boldsymbol{b}_j = \boldsymbol{b}_j - \lambda_j \boldsymbol{u}_k^T$	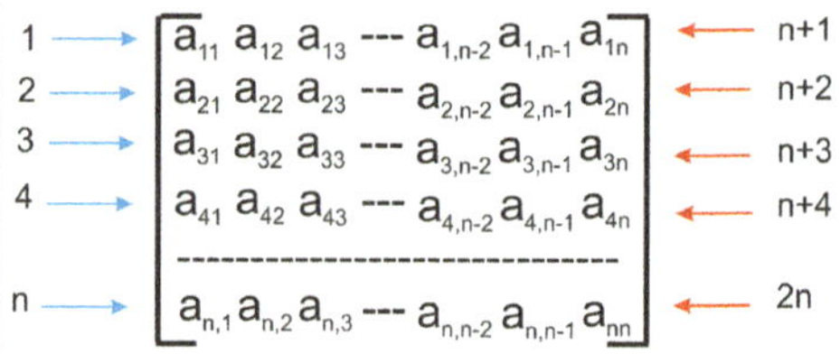
SPH	$\boldsymbol{v} = \varphi_k \boldsymbol{u}_k^T \boldsymbol{A}$ $\boldsymbol{D} = diag(v_0, v_1, \ldots, v_n)$ $\boldsymbol{U}[:\, j] = [\boldsymbol{u}_k]$ $\boldsymbol{W} = \boldsymbol{UD}$ $\mathbf{b_j} = \mathbf{b_j} - \mathbf{w_j}$ $\lambda'_\mathbf{j} = \psi_k \mathbf{b_j u_k}$ $\mathbf{b_j} = \mathbf{b_j} - \lambda'_\mathbf{j} \mathbf{u_k^T}$ $\quad j = \overline{k, n}$	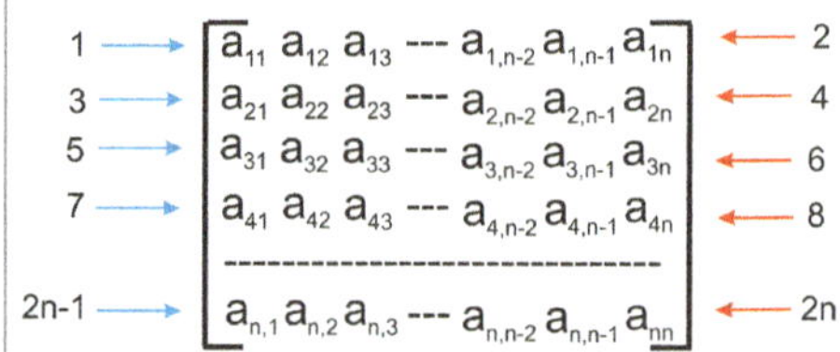

- The probability of a cache miss when multiplying the original matrix by the reflection matrix on the left α_R; the matrix does not fit entirely into the cache memory, and since access is carried out along the rows of the original matrix, we assume the probability of a cache miss to be $\alpha_R; = 1/\mathrm{m}$, although this indicator may also be slightly smaller; the number of operations is 2 * (n – k – 1) * (n – k – 1);
- The probability of a cache miss when accessing the elements of the reflection vector in the process of multiplication by reflection matrices α_{VEC}; the probability of a cache miss is either 0 if the vector is completely placed in the cache (the elements of the vector were buffered during its formation), or more, up to 1 / m with a large matrix dimension; assume the probability is $\alpha_{VEC}; = 0$; the number of operations is 2 * (n – k – 1) * (n – k – 1).

We will not estimate the number of arithmetic-logical operations in this case, because only permutation optimization will be used.

After analyzing the formed groups of MA, we can conclude that the parameters α_C and α_R, have the greatest impact on the efficiency of working with memory since the number of operations in these groups is proportional to the square of the matrix dimension; In addition, the group of operations α_R is well optimized, because the probability of a cache miss is 1/m, therefore for to increase the efficiency of the implementation of this algorithm, it is necessary to optimize the group of operations α_C.

Two iterations of the proposed software development method were performed, resulting in two new optimized schemes for performing the transformation: Row-Oriented Householder (ROH) and Single-Pass Householder (SPH).

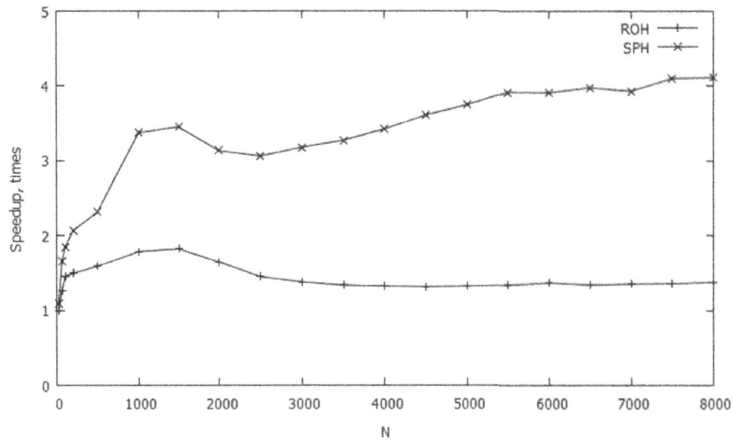

Fig. 1 The speedup ROH and SPH relative to the classical householder transformation

As mentioned above, to optimize the classical Householder transformation scheme, it is necessary to optimize the group of operations α_C processing columns of the original matrix, where a large number of cache misses are generated. The ROH and SPH transformations are much more efficient than the basic Householder transformation since:

- Within ROH and SPH, data is not processed along columns, only along rows, which reduces the α_C indicator to a value of 1/m;
- In the fastest SPH algorithm, this indicator is reduced by half, since the conversion is performed within one pass along the matrix versus two in the classical scheme and ROH.

Computational experiments have shown the acceleration of the obtained algorithm relative to the classical Householder transformation up to 2 times for ROH and up to 4 times for SPH (Fig. 1).

4 Conclusions

This chapter presents a new software development method designed to improve data caching algorithms for computing systems with a hierarchical memory structure. In the context of the proposed method, an analytical approach is implemented to estimate the effectiveness of the software from the point of view of using the memory subsystem. The proposed method is applied to the two-side Householder transformation for reducing the general-form matrix to the Hessenberg form. Computational experiments have shown the effectiveness of this method. New effective implementations of the basic transformation were obtained.

At the same time, it is worth noting that only the reduction of the number of "cold" cache misses was used in the work. In this regard, the result of further work in this direction will attempt to formalize the estimation of the probability of cache misses for cyclic data access, using the results obtained to optimize computationally complex transformations.

References

1. Tomov, S., Nath, R., Dongarra, J.: Accelerating the reduction to upper Hessenberg, tridiagonal, and bidiagonal forms through hybrid GPU-based computing. Parallel Comput. **36**, 645–654 (2010)
2. Buttari, A., Langou, J., Kurzak, J., Dongarra, J.: Parallel tiled QR factorization for multicore architectures. Concurr. Comput. Pract. Exp. **20**, 1573–1590 (2008)
3. Kravets, A.G., Egunov, V.: The software cache optimization-based method for decreasing energy consumption of computational clusters. Energies **15**, 7509 (2022). https://doi.org/10.3390/en15207509
4. Viet, N.T., Kravets, A.G.: The new method for analyzing technology trends of smart energy asset performance management. Energies **15**(18), 6613 (2022). https://doi.org/10.3390/en15186613
5. Pandey, N., Verma, O.P., Kumar, A.: A framework for usage pattern–based power optimization and battery lifetime prediction in smartphones. Pers. Ubiquit. Comput. **26**, 821–836 (2022). https://doi.org/10.1007/s00779-019-01213-4
6. Low, T.M., Igual, F.D., Smith, T.M., Quintana-Orti, E.S.: Analytical modeling is enough for high-performance BLIS. ACM Trans. Math. Softw. **43**, 1–18 (2016).https://doi.org/10.1145/2925987
7. Lawson, C.L., Hanson, R.J., Kincaid, D.R., Krogh, F.T.: Basic linear algebra subprograms for fortran usage. ACM Trans. Math. Softw. **5**(3), 308–323 (1979). https://doi.org/10.1145/355841.355847
8. Dongarra, J.J., Du Croz, J., Hammarling, S., Duff, I.S.: A set of level 3 basic linear algebra subprograms. ACM Trans. Math. Softw. **16**(1), 1–17 (1990). https://doi.org/10.1145/77626.79170
9. Intel. Math Kernel Library. https://software.intel.com/en-us/intel-mkl (2015)
10. AMD. AMD Core Math Library. http://developer.amd.com/tools-and-sdks/cpu-development/amd-core-amth-library-acml/ (2015)
11. IBM. Engineering and Scientific Subroutine Library. http://www-03.ibm.com/systems/power/software/essl/ (2015)
12. Goto, K., van de Geijn, R.: High performance implementation of the level-3 BLAS. ACM Trans. Math. Softw. **35**(1), 1–14. https://doi.org/10.1145/1377603.1377607
13. Xianyi, Z., Qian, W., Yunquan, Z.: Model-driven level 3 BLAS performance optimization on Loongson 3A processor. In: 2012 IEEE 18th International Conference on Parallel and Distributed Systems, Singapore, Boston, Massachusetts, USA, IEEE Xplore Digital Library, pp. 684–691 (2012). https://doi.org/10.1109/ICPADS.2012.97
14. Clint Whaley, R., Dongarra, J.J.: Automatically tuned linear algebra software. In: Proceedings of SC'98
15. Frison, G.: Algorithms and methods for high-performance model predictive control. Ph.D. thesis (2016). https://backend.orbit.dtu.dk/ws/portalfiles/portal/124371046/phd402_Frison_G.pdf (дата обращения: 5.09.2020)
16. Bilmes, J., Asanovic, K., Chin, C.-W., Demmel, J.: Optimizing matrix multiply using phipac: a portable, high-performance, ANSI C coding methodology. In: Proceedings of the 11th International Conference on Supercomputing (ICS '97). ACM, New York, NY, USA, pp. 340–347. https://doi.org/10.1145/263580.263662

17. Van Zee, F.G., Smith, T., Marker, B., Low, T.M., van de Geijn, R.A., Igual, F.D., Smelyanskiy, M., Zhang, X., Kistler, M., Austel, V., Gunnels, J., Killough, L.: The BLIS framework: experiments in portability. ACM Trans. Math. Soft. (2014). In review
18. Ortega, J.M.: Introduction to Parallel and Vector Solution of Linear Systems, p. 305. Plenum Press, New York, NY, USA (1988). ISBN 0-306-42862-8
19. Andreev, A.E., Egunov, V.A.: Solving of eigenvalue and singular value problems via modified householder transformations on shared memory parallel computing systems. In: Supercomputing: RuSCDays, Proceedings of the 5th Russian Supercomputing Days, Moscow, Russia, 23–24 September 2019, vol. 1129, pp. 131–151. Springer, Cham, Switzerland (2019)
20. Egunov, V.A., Andreev, A.E.: Implementation of QR and LQ decompositions on shared memory parallel computing systems. In: 2016 2nd International Conference on Industrial Engineering, Applications and Manufacturing (ICIEAM) (Chelyabinsk, Russia, 19–20 May 2016). [Publisher: IEEE], p. 5 (2016). https://doi.org/10.1109/ICIEAM.2016.7911607

Extreme Regulator in the Control Loop of a Non-stationary Object with Discrete Time

V. B. Gusev

Abstract The problem of finding the extremum for a nonstationary objective function is considered when its values, depending on the components of the control vector, are given only in a discrete set of points. To find a solution, a discrete gradient method of unconditional optimization is proposed. The conditions for its application are formulated. An estimate of the lower limit of the solution error due to the magnitude of the step along the argument, the rate of change of the objective function, as well as the estimation of its second derivatives is obtained. The application of the method is demonstrated on a numerical model of an extreme controller designed to control a nonstationary object with a nonlinear objective function.

Keywords Non-stationary system · Extreme regulator · Discrete set of time points · Unconditional optimization · Lower limit of the solution error

1 Introduction

At managing an object in conditions of uncertainty, when there is no formalized description of the laws of its behavior, the most appropriate method is extreme regulation, consisting in establishing and maintaining such a mode of operation in which an extreme (minimum or maximum) value of the criterion characterizing the quality of the functioning of the object is achieved. The task of building control mechanisms for models of autonomous systems has much in common with the problem of an extreme regulator that automatically searches for and maintains the extreme value of the regulated value. A limitation of the applicability of an extreme regulator is that in the case of a limited amount of information about an object, its inertial properties may not be taken into account. In addition, when it runs, it is impossible to control the long-term effects of management.

V. B. Gusev (✉)
V.A. Trapeznikov Institute of Control Sciences of the Russian Academy of Sciences, Moscow, Russia
e-mail: gusvbr@mail.ru

A. G. Kravets et al. (eds.), *Cyber-Physical Systems Engineering and Control*, Studies in Systems, Decision and Control 477,
https://doi.org/10.1007/978-3-031-33159-6_18

Extreme regulation formed an independent direction within the framework of the theory of nonlinear automatic control systems [1], and extreme regulators were widely used in the 60s of the last century. For example, they were used in relay extreme systems [2], pulsed self-adjusting and extreme automatic control systems [3]. Extreme regulators were also used in setting up resonant circuits and automatic measuring devices, in finding the optimal parameters of customizable models, in controlling chemical reactors, heating installations, flotation and crushing processes [4].

In the work [5] to control the supply of air to the furnace, a method of extreme control of the next type is proposed

$$\Delta x(k+1) = h \cdot \text{sign}\,[\Delta y(k+1) \cdot \Delta x(k)], \tag{1}$$

where x is the control parameter, y is the measured response parameter to be optimized, k is the point in time.

In the work [6] for modeling the metabolic process, a heuristic algorithm of extreme regulation is proposed, using a random search for the best response at each step of iterations.

Let the model of a non-stationary autonomous system apply the optimization criterion at each point of the trajectory of the view

$$F(y, u) \rightarrow \underset{u}{\text{extr}} \tag{2}$$

where $F(y, u)$ is the objective function, smooth by u, $y(t, p, x, u)$ is the system state vector, $u(t)$ is the control vector, p is the parameter vector, $x(t)$ is the vector of the state of the external environment. Optimal control can then be determined by means of the extremum necessary condition

$$dF(y(t, p, x(t), u(t)), u(t))/du = 0. \tag{3}$$

The methods of the proportional regulator [7] are not suitable for finding a solution, since they assume zero control at the points of the optimum. If only the values of the objective function are known at the appropriate moments, we will use the following entry:

$$f(t, u) = F(y(t, p, x(t), u), u). \tag{4}$$

Estimation of the derivative $df(t, u)/du$ in the case where measurements are made at discrete moments in time, it can only be obtained approximately, for example, using the spline representation of the function $f(t, u)$, which is usually not unambiguous. The consequence of such assumptions is to expect an increase in the error of the solution compared to methods using accurate estimates of derivatives of the objective function. Moreover, the error of the solution over time will not decrease to zero, that is, it is irreparable. Obviously, its value should increase with the increase in the sampling interval and the rate of change of the target indicator.

It is necessary to have an algorithm of extreme regulation with discrete time, providing an allowable (controlled) value of the control error, as well as to determine the conditions that must be met by the dynamic characteristics of the controlled process.

Similar algorithms have been investigated for the case of continuous time when the objective function $f(t, u)$ has continuous derivatives on the arguments and convex by u, where t is a scalar, u is a vector. The task was to find the unconditional extremum of the objective function at each moment t.

$$f(t, u) \underset{u}{\rightarrow} extr. \tag{5}$$

The necessary condition for the optimum of the objective function in this case is:

$$\nabla u f(t, u) \underset{\mathbf{u}}{\rightarrow} 0. \tag{6}$$

The convergence of gradient algorithms of unconditional optimization for non-stationary problems was considered in the work [8]. It assumed an accurate knowledge of the gradient of the objective function $\nabla u f(tk, u(tk))$, which made it possible to prove the convergence of the iterative process for discrete time

$$u(tk+1) = u(tk) - \gamma k \nabla u f(tk, u(tk)), \tag{7}$$

where convergence conditions contain the requirement:

$$\|u(tk+1) - u(tk)\| \leq a, a > 0. \tag{8}$$

The work [9] develops the results [8] by weakening the condition of convergence of the iterative process to the following:

$$\|\nabla u f(tk+1, u) - \nabla u f(tk, u)\| \leq a, a > 0. \tag{9}$$

A special case of an extreme regulator with a nonlinear objective function is described in [7] for a specific linear dynamical system described by an ordinary first-order differential equation. The magnitude of the step here was chosen to be constant, and its sign was inverse to the sign of the derivative. This publication also suggested that it was possible to accurately calculate the values of the derivatives of the objective function. The algorithm below does not require this.

2 Problem Statement

Next, we will assume that time takes a discrete series of values $t1, t2, ..ti, ..$ The task is to determine the value of the control function $u(tk+1)$, using values $u(tj)$,$f(tj, u(tj))$, $j \leq k$.

The aim is to develop a method for approximating to the extremum a non-stationary objective function given at a discrete time that provides an allowable (controllable) error value. It is also necessary to determine the conditions that must be met by time intervals, the target indicator used, the parameters of the influence of the external environment on it. Such a method can be used to construct an extreme regulator of a system available for observation at discrete moments in time.

As a demonstration of the method of approximate search of the extremum for a non-stationary system with discrete time, a numerical implementation of the extreme regulator method for a simplified model of vehicle movement in a non-uniform environment is used. It is represented by finite difference relations in discrete time. However, when searching for a solution, in accordance with the formulation of the problem, only the values of the objective function and control at the previous and current moments of time are used. An example of the application of this type of method for a model with economic interpretation is given in [10].

3 Calculation Method

Let the function $f(t, u)$ has continuous partial derivatives of the 1st order by u, as well as on a discrete set of values tk in increments Δt has at least one point of stationarity $\tilde{u}$: $\nabla u f(t, \tilde{u}) = 0$. Denote: . $u_i^k = ui(tk)$. i-th vector component u^k, $i = 1, 2, .., n$; $fk = f(tk, u^k)$. Finite-difference approximations of first-order partial derivatives for data on two points are of the form $\partial f/\partial u_i^k \cong (f_k - f_{k-1})/\left(u_i^k - u_i^{k-1}\right) = \Delta f_k/\Delta u_i^k$ or in vector form $\nabla u f k \cong \overline{\nabla} u f k$, $i = 1, 2, .., n$. Vector with components $\Delta u_i^k/\Delta f_k$, $i = 1, 2, ..n$ we will denote $\overline{\nabla} f\, u^k$. Using a linear segment of the Taylor–Lagrange series, where the residual term is given at an intermediate point $\tilde{u}, \tilde{u}_i \in \left[u_i^k, u_i^{k+1}\right]$, $i = 1, 2, .., n$, $t_k \leq \tilde{t} \leq t_{k+1}$, obtain an exact expression for the finite difference estimation of the derivative

$$\begin{aligned}&(f(t_{(k+1)}, u^{k+1}) - f(t_k, u^k))/(u_i^{k+1} - u_i^k) - \partial f(\tilde{t}, \tilde{u})/\partial u_i \\ &= \partial f(\tilde{t}, \tilde{u}_i)/\partial t \cdot (t_{i+1} - t_i)/(u_{i+1}^k - u_i^k)\end{aligned} \tag{10}$$

According to the necessary condition of optimality, we will put $\nabla u f\left(\tilde{t}, \tilde{u}\right) = 0$. Then at the point $\tilde{t}, \tilde{u}$ we have

$$u^{k+1} - u^k = (t_{i+1} - t_i) \cdot \partial f(\tilde{t}, \tilde{u})/\partial t \cdot \overline{\nabla} f u^k \tag{11}$$

Let's enter two parameters hk and αk, that satisfy the condition $\partial fk/\partial t = hk(\|\overline{\nabla} ufk\| + \alpha k)$. There is a value $\overline{h}_k > 0$, that at the coefficient $|h_k| \geq \overline{h}_k$ calculation method

$$uk + 1 = uk + hk\Delta t\overline{\nabla} fuk(||\overline{\nabla} ufk|| + \alpha k) \tag{12}$$

implements a sequence of values ui, deviating from the point of stationarity $\tilde{u}$ so that component characters $(f(t_{k+1}, u^{k+1}) - f(t_k, u^k))/(u_i^{k+1} - u_i^k)$, $i = 1, ..n$ of gradient ratings $\overline{\nabla} ufk$ are opposite. Thus, the result uk oscillates around stationarity points, where $\nabla uf(\tilde{t}, \tilde{u}) = 0$.

The value can be rated as follows

$$\overline{h}k = |\partial fk/\partial t|/(||\overline{\nabla} ufk|| + \alpha k), \tag{13}$$

from where $\alpha k = |\partial fk/\partial t/\overline{h}k| - ||\overline{\nabla} ufk||$.

Allowable parameter values of αk and hk can be selected in a fairly wide range, without having accurate estimates of values $\partial fk/\partial t$. The closer the parameter value αi to a value of 0, the spread Δuk there will be more. Reducing the parameter hi reduces the spread Δuk.

To determine the value of the multiplier hk in the process of calculations you can use the above statement, increasing it by a fixed amount $\beta < \overline{h}k$, when $(\overline{\nabla} ufk - 1, \overline{\nabla} ufk) > 0$ and reducing, in the opposite case:

$$hk + 1 = h0 + \beta si(\nabla ufk - 1, \nabla ufk). \tag{14}$$

An estimate of the error $\|\Delta u^k\|$ of the solution can be obtained based on the value of the increment β. If you put $\beta \cong \overline{h}k/2$, from (12) follows an approximate estimate of the error $\|\Delta uk\|$ of the solution

$$\|\Delta u^k\| \cong \Delta t|\partial f(t, u)/\partial t| \cdot \|\overline{\nabla} fu^k\|/2. \tag{15}$$

To estimate the upper limit of the error [9], suppose that at each point tk, u^k the objective function is strictly convex: $||\nabla uuf^k|| > 0$. When $|hk| \geq |\partial fk/\partial t\|\overline{\nabla} uf\|\Delta t|$ and under $0 \leq \alpha k \leq h/(\partial fk/\partial t \cdot \Delta t) - \|\overline{\nabla} ufk\|$, the calculation method implements a sequence of values uk, deviating from the points of stationarity alternately for each coordinate by a value Δuk, whose lower limit of the norm

$$\inf\|\Delta u^k\| = |h_k\Delta t\|\overline{\nabla} fu^k\|/\|\nabla_{\mathrm{uu}} f_k\||^{\frac{1}{2}} \tag{16}$$

is achieved when

$$|h_k| = |\partial f_k/\partial t\|\overline{\nabla}_u f_k\|\Delta t| \tag{17}$$

4 Calculation Examples

As an illustration of the work of an extreme regulator, consider the numerical implementation of models in some application regions.

4.1 Speed Control Model of an Object Moving in an Environment with a Time-Variable Load Force

The regulator maximizes the useful power of the motor using only the values of this power, measured at discrete moments in time. At each moment following the measurement, a velocity value close to the optimal value is calculated. The load force is an external parameter that, like the model of the control object, is not used by the regulator.

The simplified object model is represented by the following ratios in discrete time. The control (and at the same time the status parameter) is the speed of movement $u(t)$. Target function—useful power

$$r(t) = f(t, u(t)) = p(t)u(t) - C_0 - cu^2(t) \rightarrow \max_u, \tag{18}$$

where $t = 0, 1, 2...T$;$p(t)$—variable load force; $p(t)u(t)$—full power; cu—resistance force proportional to the velocity u with the coefficient c; power loss from environmental resistance $cu2$, C_0—constant power loss. It is not difficult to see that the exact value of the optimal speed is of the form

$$u(t) = p(t)/2c. \tag{19}$$

The graph of this dependence is used to estimate the deviations of the calculated values from the optimal velocity values.

Parameter of the external environment (unevenly varying load force) $p(t)$ is calculated by the formula $p(t) = dp(t-1)cos(t/3)$, $p(0) = p_0$. Here d is the rate of regular growth, the initial load force is p_0.

The model of the regulator has the following form. Estimating the gradient of the target control function

$$e(t) = (r(t) - r(t-1)) / (u(t) - u(t-1)). \tag{20}$$

The control at the step $t + 1$ is calculated according to the method of approaching the extremum

$$u(t+1) = u(t) + h(|e(t)| + \alpha)/e(t) \tag{21}$$

where α is the stabilization parameter, h is the multiplier.

The multiplier h is calculated in accordance with the formula (14):

$$h(t+1) = h_0 + \beta sign(e(t) \cdot e(t-1)), \tag{22}$$

where β is the variation parameter of the multiplier h.

In Figs. 1, 2 are the results of calculations for the model of object motion with the parameters $C_0 = 0$; $c = 1$ and extreme regulator with parameters $\alpha = 0{,}3$; $h_0 = 2$, $\beta = 1$. The resistance force with an initial value of $p_0 = 10$ varies with a constant rate of $d = 1.03$. The optimizing speed controller quickly brings the usable power to the optimum power and monitors its optimal level over time within the margin of error of the method.

The range of oscillations around the optimal speed level is due to the discrete nature of the model, the non-stationary behavior of the resistance of the medium and the inaccuracy of the selection of regulator parameters. With an increase in the

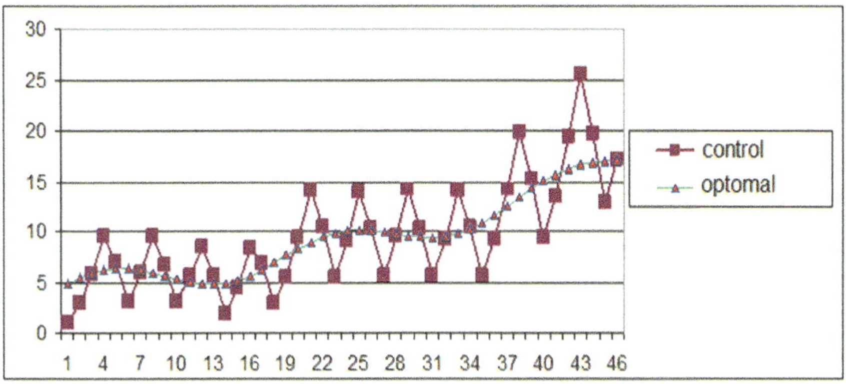

Fig. 1 Results of control speed calculations for extreme regulator

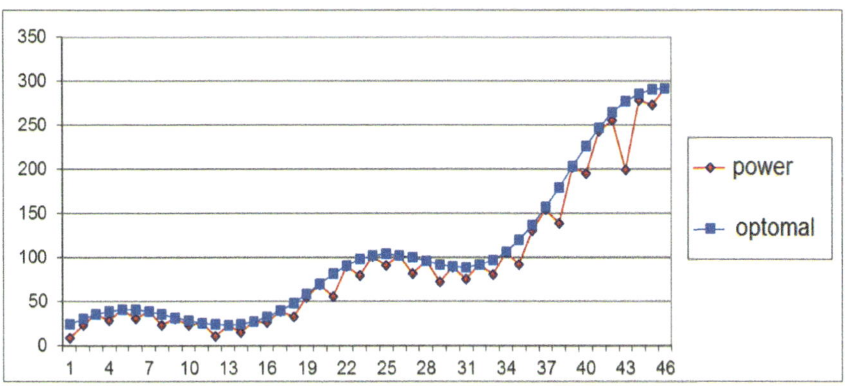

Fig. 2 Results of net power calculations for extreme regulator

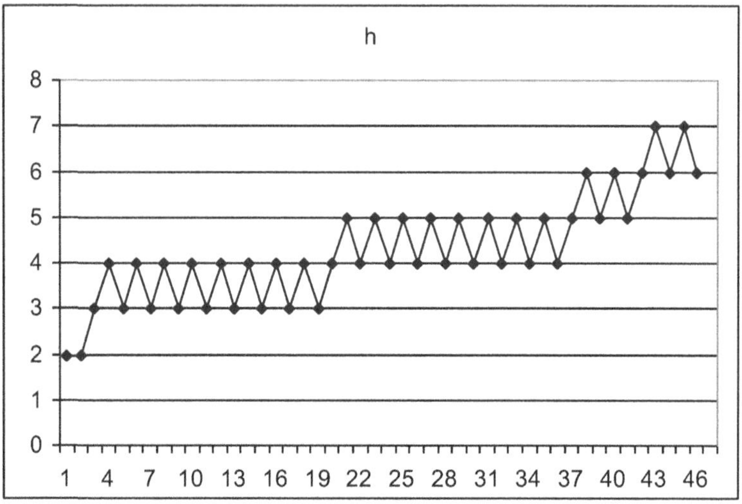

Fig. 3 Dynamics of change in pitch h during operation of the extreme regulator

derivative of the resistance force of the medium over time, the error of the solution increases, which corresponds to the given approximation estimate.

The coefficient αk score was chosen for the sustainability of the process. In this example, the coefficient αk values throughout the trajectory were constant.

On the trajectory, the multiplier h varies so that the level reached by the objective function $f(t, u(t))$ within a few steps of the discrete algorithm is maximized. The value of h in the initial steps has a value of 2, varies from step to step and at the end of the trajectory it is 6 (Fig. 3).

This multiplier growth correlates with the increase in resistance force. Variation of h_0 parameters, within small limits, does not give a qualitative change in the dynamics of the regulator. If the parameter values are reduced to $h_0 = 1$, $\beta = 0, 1$, the rate of approximation to the optimal trajectory will decrease, but the maximum deviation of the objective function from the maximum, which is observed at the initial section of the trajectory and at the points of the greatest curvature of the trajectory, will also decrease (Fig. 4, 5).

In this case, the condition of fluctuation of control at each step around the optimal trajectory is not valid, although the deviation of the objective function from the optimum (Fig. 5) decreases compared to the previous mode.

The value of h varies from step to step with large intervals of monotony and at the end of the trajectory it is 2.5, while the ratio h/β is about the same (Fig. 6).

In the above examples, the value of the parameter h corresponds to the rate of change in the resistance force of the medium. In addition, finite difference approximations of derivative of objective functions $f(t, u)$ by the formula (17) can be applied to the estimation h. The parameter α can then be selected according to the estimate

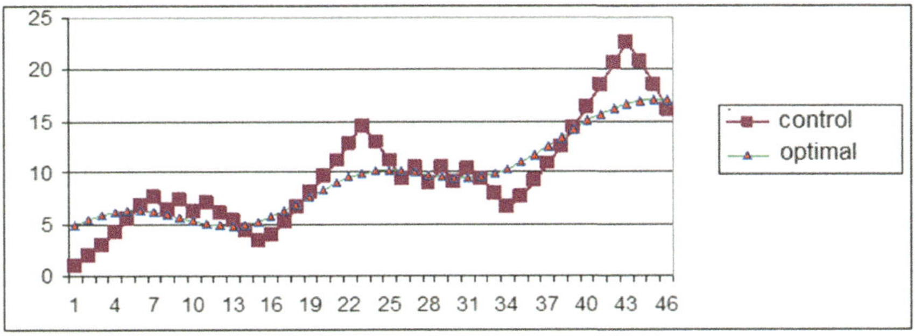

Fig. 4 Results of control speed calculations for extreme regulator

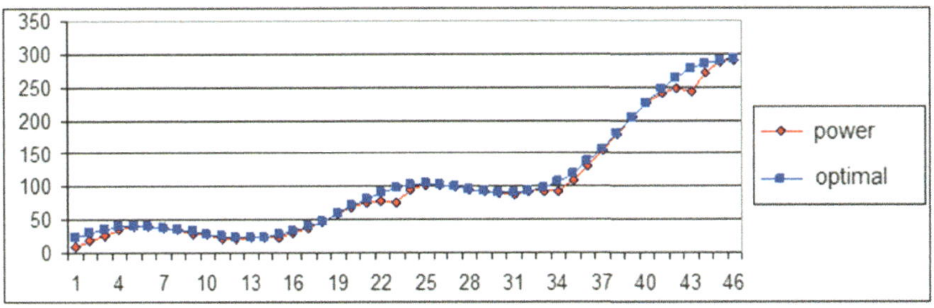

Fig. 5 Results of net power calculations for the extreme regulator

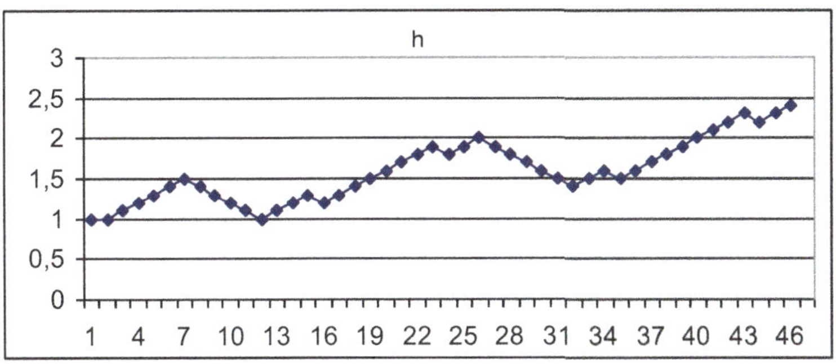

Fig. 6 Dynamics of change in pitch h during operation of the extreme regulator

$\alpha \cong |\partial f/\partial t|0/h0$. Numerical experiments show that the variation of the parameters of the regulator in a sufficiently wide range does not have a significant impact on the qualitative behavior of the calculated trajectory.

4.2 Optimal Controller Design for a Simplified Model of Production

Another approximate method for optimizing a discrete time-varying system is illustrated below by a numerical example of optimal controller design for a simplified model of production. This model is described by the following discrete-time finite difference relations. However, according to the problem statement, only the values of the objective function and control at the previous and current time instants are used for solving.

The control (and simultaneously the state parameter) is the production output $u(t)$. The objective function—profit—has the form

$$r(t) = f(t, u(t)) = p(t)u(t) - C_0 - cu^2(t) \to \max_u, \tag{23}$$

where $t = 0, 1, 2, ...; p(t)u(t)$ gives the income; C_0 are fixed costs; $C_0 + cu^2$ is an estimate of the total costs, including production assets, remuneration of labor and direct expenditures; finally, $p(t)$ specifies the unit price of the products (the environment's parameter)

$$p(t) = dp(t-1), \; p(0) = p_0. \tag{24}$$

where d is the growth coefficient, and p_0 is an initial price.

The marginal profit (the gradient of the objective function with respect to the control variable) is estimated as

$$e(t) = (r(t) - r(t-1)) / (u(t) - u(t-1)). \tag{25}$$

In the next step, the control is calculated using the proposed approximate optimization method:

$$u(t+1) = u(t) + h/((e(t)(|e(t)| + \alpha)). \tag{26}$$

Here α is the stabilization parameter, and h is the coupling coefficient.

Figure 7 shows the simulation results for the production output model with the parameters $C_0 = 1$ and $c = 1$ and the optimal controller with the parameters $\alpha = 0.1$ and $\mathrm{h} = 1$. The product price with the initial value $p_0 = 3$ varies with the constant rate $d = 1.03$. The exact optimal solution has the form (18).

The optimal controller quickly (in one step) approximates the production output the optimal one and, over time, tracks the optimal output within the method's error.

The range of fluctuations around the optimal output is conditioned by the discrete nature of the model, the nonstationary behavior of the product price, and an inaccurate choice of the controller's parameters. With an increase in the time derivative of the price, the solution error grows, which agrees with the approximation estimate presented above. At the initial steps, the value hi has the estimate

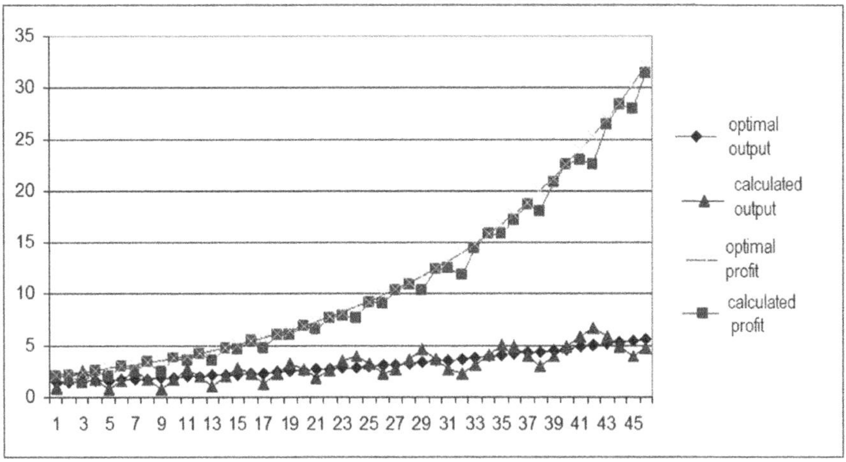

Fig. 7 Simulation results for the optimal controller

$|h_i| = |\partial f / \partial t||\overline{\nabla}_u f||\Delta t|_i \cong 1$ and the sign + 1. An appropriate estimate of the coefficient αi was selected from the stability considerations to satisfy $\|\overline{\nabla}_u f_i\| \gg \alpha_i > 0$. They can be experimentally refined along the trajectory by maximizing the value achieved by the objective function $f(t, u(t))$ during several steps of the discrete algorithm. In the example, the values of αi and hi were constant along the entire trajectory.

The coincidence of the expressions for the objective function (18) and (23) is not essential to illustrate the breadth of the applicability of the extreme regulator method, since they are not explicitly used in the calculation of the control function. It is essential for both cases of target functions to have stationary control points.

5 Conclusion

The approximate optimization method proposed in this chapter is not very critical to the choice of the parameters α and h. They can be determined in a particular way depending on the applied problem under consideration. In the numerical example, the values of the parameters correspond to the rate of increase in the unit price of products. In addition, the parameter h can be estimated using finite-difference approximations for the derivatives of the function $f(t, u)$. In this case, the parameter α can be estimated as $\alpha \cong \|\overline{\nabla}_u f_0\|^2 \times h / \max_t(\partial f / \partial t)$. According to numerical simulations, varying the controller's parameters in a rather wide range has an insignificant effect on the qualitative behavior of the calculated trajectory.

Since the values of the objective function are calculated (or measured) only in the nodes of the discrete time grid and for the corresponding values of the control vectors, it is possible to build a spline approximation of this function of the desired smoothness and apply accurate gradient descent methods for it [8, 9]. However, the accuracy of the result will also remain finite, since such a spline approximation is not the only one. In addition, the computational complexity of this method will significantly exceed the complexity of the proposed approach.

The proposed method can be used in a variety of areas where an accurate model of the object is absent, and measurements or assessments of the state are possible only at discrete moments of time. An example is the management of technological processes [1–5], operational management in the field of economics [10].

References

1. Krasovskij, A.A.: Dynamics of continuous self-tuning systems [Dinamika nepreryvnyh samonastraivayushchihsya system], M (1963)
2. Morosanov, I.S.: Relay extreme systems [Relejnye ekstremal'nye sistemy], M (1964)
3. Kuncevich, V.M.: Impulse self-adjusting and extreme automatic control systems [Impul'snye samonastraivayushchiesya i ekstremal'nye sistemy avtomaticheskogo upravleniya], K (1966)
4. Rastrigin, L.A.: Extreme control systems [Sistemy ekstremal'nogo upravleniya], M (1974)
5. Selezneva, S.V., Konovalova, I.I.: Simulation of the algorithm of the extreme regulator of air supply to the furnace. APRIORI. Series: Nat. Tech. Sci. **3**, 12 (2014). (in Russian)
6. Xi, Y., Wang, F.: Extreme pathway analysis reveals the organizing rules of metabolic regulation. PLOS ONE (2019). https://doi.org/10.1371/journal.pone.0210539
7. Egorov A.I.: Fundamentals of control theory [Osnovy teorii upravleniya]. M.: FIZMATLIT, p. 504 (2004)
8. French, M.: Fundamentals of optimization. Methods, minimum principles, and applications for making things better, 249 (2018)
9. Popkov, A.Y.: Gradient methods for nonstationary unconstrained optimization problems. Autom. Remote Control. **66**(6), 883–891 (2005). https://doi.org/10.1007/s10513-005-0132-z
10. Gusev, V.B.: Unconstrained optimization of a time-varying objective function on a discrete time scale. Control Sci. **1**, 31–36 (2021). https://doi.org/10.25728/cs.2021.1.4

Structural Changes During Electrical Aging of Insulation Materials of Cable Networks

N. K. Poluyanovich and M. N. Dubyago

Abstract The chapter is devoted to the study of the structure of electrical insulating materials (EM) of power cable lines (PLC). As part of the study, a neural network (NN) model was constructed that describes the architecture of the cyber-physical system (CPS) for predicting the EM resource of cable electrical networks. The NN emulator uses the output of the NN instead of the output of the object, which makes it possible to predict the dynamics of the object many steps ahead. An algorithmic solution has been synthesized for automatic selection of parameters and training of the NN with subsequent prediction, which makes it possible to increase the reliability of the CPS by reducing the time for creating the optimal configuration of the NN. The representation of a dynamic object based on NN is fundamentally different from the description of a dynamic object using differential equations. The resulting NN model can be effectively used to analyze the thermal fluctuation processes occurring in the control object—PCL, and to predict the behavior of the object.

Keywords Cyber-physical system · Neural network · Reliability of cable systems

1 Introduction

During operation, the insulation of the cable line (CL) is subjected to thermal, electrical, chemical, mechanical, atmospheric and other types of influences, as a result of which its electrical properties change and, as a result, the technical characteristics of the EM also undergo changes [1–3]. In solving the problems of increasing the volume of energy consumed associated with the development and emergence of new energy-intensive industries, cyber-physical systems (CPS) can play a significant role, capable of monitoring control systems for processes of uninterrupted and efficient supply of consumers with electricity [4]. The causes of aging in polyethylene insulation are inevitable, both technological and operational. The work [5] is devoted

N. K. Poluyanovich (✉) · M. N. Dubyago
Southern Federal University, Russian Federation, Taganrog, Russia
e-mail: nik1-58@mail.ru

A. G. Kravets et al. (eds.), *Cyber-Physical Systems Engineering and Control*,
Studies in Systems, Decision and Control 477,
https://doi.org/10.1007/978-3-031-33159-6_19

to structural changes during electrical aging of cross-linked polyethylene used as insulation for power cables. In [6], the effect of simultaneous combined thermal and mechanical stresses on the overall insulation of unshielded low-voltage power cables is considered. The results showed that combined aging affects both the electrical and mechanical properties of the overall insulation. In [7], the chemical and physical composition and mechanisms of aging of the insulating materials of XLPE cables were studied. The mechanisms of aging of XLPE cable insulation materials have been studied, including a discussion of changes in the polymer chains of cross-linked polyethylene as samples age. In [8], the influence of long-term electrical and thermal stresses on the endurance of the insulating material of XLPE cables, which were subjected to electrical (up to 28 kV/mm), thermal T = 363 K (90 °C) and electrothermal stress, was considered.

2 Physics of the Aging Process of Material Insulation

Common defects are micro-sized air cavities, foreign inclusions, cracks, etc. They appear in the insulation during its manufacture or laying and operation of cables due, for example, to periodic compression and expansion of the material under load conditions. These microvoids are filled with gas due to the generation of water vapor and the presence of volatile products in the crosslinking and insulation extrusion manufacturing processes. In recent years, cables with impregnated paper insulation have been replaced with heat-resistant extruded cables with cross-linked polyethylene (XLPE) insulation. The term “crosslinking” or “vulcanization” refers to the processing of polyethylene at the molecular level. With various crosslinking technologies, a substance is obtained with a greater or lesser number of these bonds, and therefore more or less strong.

The main reason for the aging of polymer dielectrics in an alternating field is partial discharges (PD) [9], which occur in air inclusions. In these places, an inhomogeneity of the electric field arises, the intensity of which inside the defect can be much higher than in the insulating material. The electrical strength of impurities is significantly less than the electrical strength of the insulation; therefore, a local electrical breakdown occurs at the site of the defect. When the cable is operating in the nominal mode, there is a gradual deterioration in the insulating properties. At the level of chemical reactions, the process of influence of external influences on the rate of decay of a molecule can be described using the Arrhenius law:

$$K_i = K \cdot e^{\frac{-w_a}{kT}} \quad (1)$$

where K_i—the average probability of the decay of one molecule per unit time, the variable is responsible for the rate of chemical reactions; K—a coefficient depending on the material of the substances involved in the reaction; w_a—the activation energy; k—the Boltzmann constant.

3 Aging of Insulation in an Electrostatic Field with High Intensity

Breakdown breaks the bonds of a carbon atom with neighboring carbon atoms and forms free hydrogen atoms, which can cause bubbles to form.

$$(-CH_2 - CH_2-)_n \rightarrow C + 2H\cdot, 2H\cdot \rightarrow H_2(\text{bubbles})$$

The resulting unbound hydrogen is knocked out as a positively charged H+ ion, which, in turn, promotes ionization and further destruction of the molecular structure of polyethylene as a whole and subsequently, capturing a free electron, becomes neutral, and carbon becomes a conducting microparticle. In this case, neighboring carbon atoms form bonds with each other.

A variant of the transformation of ionized radicals (H· and H+) is possible, then free hydrogen forms bonds with neighboring carbon atoms in intact molecules and forms end groups. When such a bond is broken under the influence of an electric field, free hydrogen and an end group with a neighboring carbon atom are formed.

$$-CH_2 - CH_2 - \ldots - CH_2 + \cdot H \rightarrow -CH_2 - CH_2 - \ldots - CH_3$$

or

$$-CH_2 - CH_2 - \ldots - CH_3 \rightarrow -CH_2 - CH_2 - \ldots - C_{\overline{H}2} + H^+$$

$$-CH_2 - CH_2 - \ldots - CH_3 + H \rightarrow -CH_2 - CH_2 - \ldots - C_{\overline{H}2} + H_2$$

A simplified model of XLPE insulation as a cut of CH_2 molecules, in which a breakdown can occur under the influence of an electric field, is shown in Fig. 1.

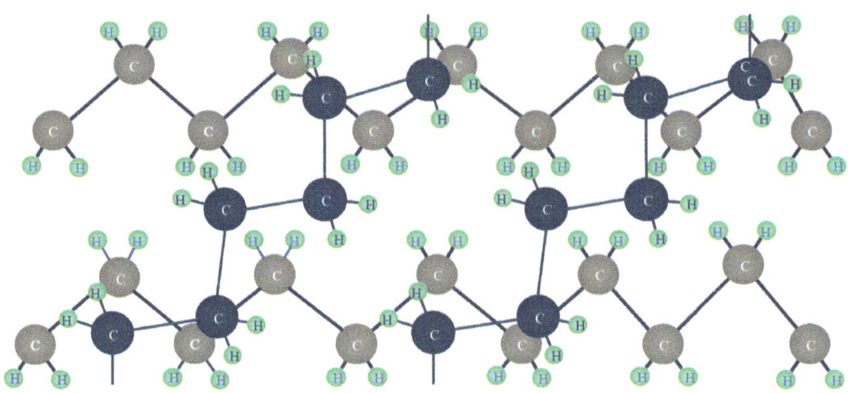

Fig. 1 Structural model of CPE insulation in the form of a cut of molecules

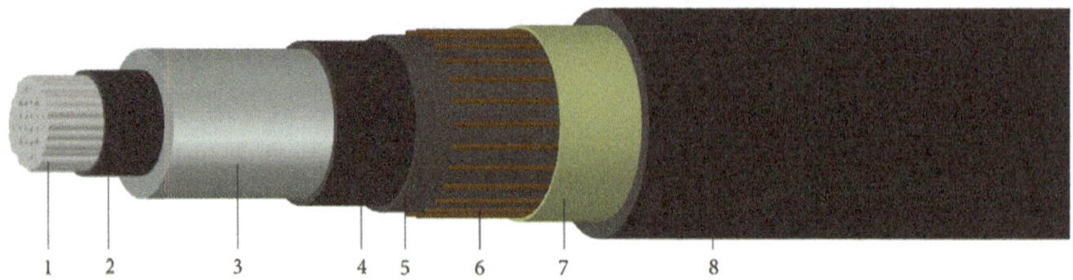

Fig. 2 Cross section of the investigated cable

4 Forecasting of Thermal Fluctuation Processes

The object of research is the electrical insulating material of the power cable APvPg 1x240/35-10, Fig. 2. The subject of the study is the prediction of the service life of insulating materials and the effect of temperature on them.

Physical properties of materials and geometric dimensions of cable elements are considered in [10–12]. The low thermal conductivity of the cable insulation (Fig. 1) leads to a high temperature gradient, so the temperature of the most heated section in the cable section (near the core) differs significantly from the measured temperature, and in transient conditions this difference can increase several times. Therefore, the task of creating neural networks for assessing the throughput, calculating and predicting the temperature of the cores of PCL in real time based on the data of the temperature monitoring system, taking into account changes in the current load of the line and external conditions of heat removal, is relevant.

5 Development of a Neurocomputer System

The cyber-physical system is a multi-level computing platform with many interconnected elements. This is due to the fact that all elements of the CPS are strongly interconnected [13]. The neurocomputer is considered as a hardware-software computer system based on the INN. In the neurocomputer, the arithmetic logic unit (ALU) is implemented on the basis of the INN, with which the training unit is connected (Fig. 3) and calculations are performed in the neural network [6, 14, 15].

6 Synthesis of a Neural Network

The construction of a neural network model (Fig. 3) is solved in two stages: the choice of the type (architecture) of the neural network and the selection of weights (training) of the neural network. When constructing the internal structure of the predictive model, which determines the dependence of the output parameters on the

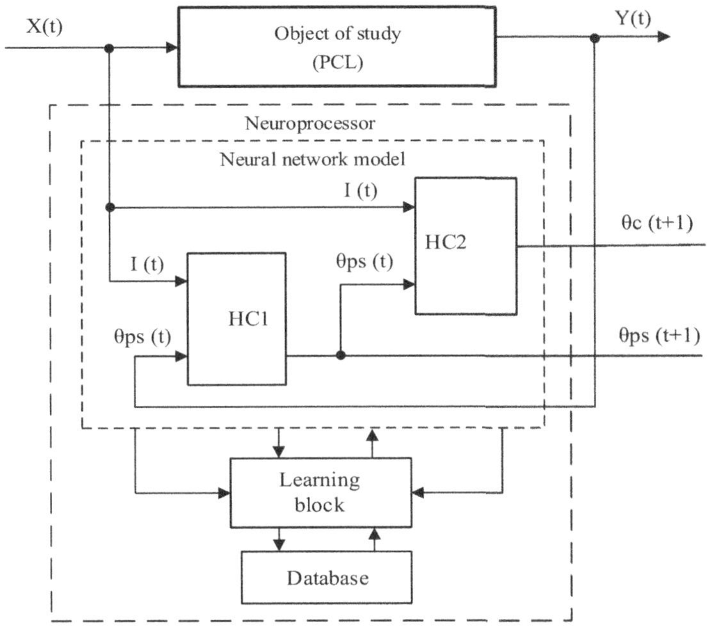

Fig. 3 Generalized structure of the neurocomputer

input parameters, an INN of the "multilayer perceptron" type was chosen; it contained 5 neurons in the input layer, and 2 neurons in the output layer. Mathematically, the functioning of a neuron can be represented by the following system of equations:

$$\begin{cases} u_j = \sum_{i=1}^{n} w_{ij} x_i \\ v_j = u_j + b_j \\ O_j = \psi(v_j) \end{cases} \tag{2}$$

where u_j—linear combination of input actions on a neuron j; w_{ij}—weight of synaptic connection connecting input i and neuron j; xi—neuron input signal; v_j—induced local field of the neuron; b_j—neuron activation threshold; O_j—neuron output; $\varphi(v_j)$—neuron activation function.

The structure of the INN is determined adaptively, according to the developed algorithm 5. The number of intermediate layers was one. The number of neurons in the intermediate layer is also determined by the adaptive algorithm in Fig. 5. For the convenience of calculations, we express the activation threshold in terms of an additional synaptic connection with a weight w0 and a fixed input signal x0 = 1:

$$O_j = \varphi(v_j) = \sum_{i=0}^{n} w_{ij} x_i \tag{3}$$

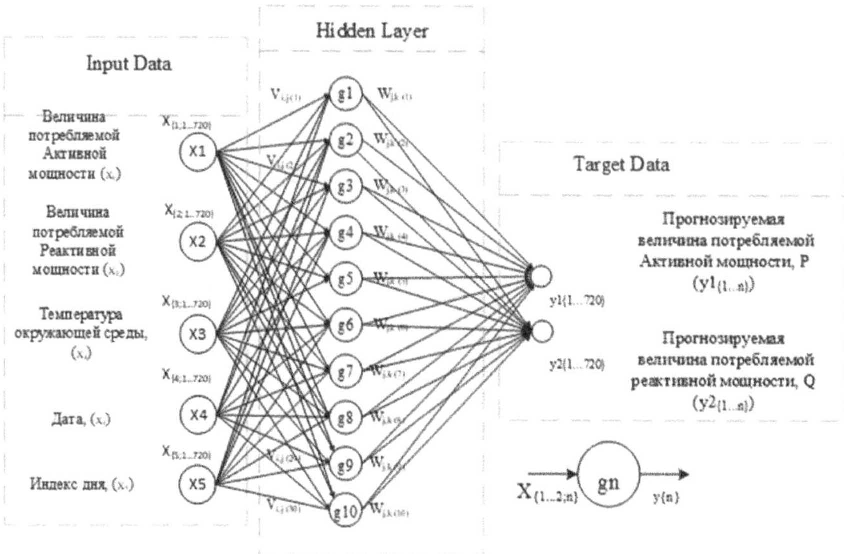

Fig. 4 Structural diagram of the developed NS for predicting the amount of consumed active and reactive power

The neural network presented in Fig. 4, consists of:

- the input layer of sensory elements that receive information directly from the "external environment";
- hidden layers where the main INN calculations take place;
- output layer that forms the final response of the neural network.

All operations in the INN are performed using matrix operations of addition, dot product, etc.

At its core, expression (3) is the scalar product of the vectors of weight coefficients w and the vector of input signals x. If a vector of daily values of factors affecting power consumption is fed to the NN input, then expression (3) can be written as:

$$O_j = \varphi_j \left(\begin{bmatrix} w_1 \\ w_2 \\ \cdots \\ w_n \end{bmatrix}^T \bullet \begin{bmatrix} x_1 \\ x_2 \\ \cdots \\ x_n \end{bmatrix} \right), \tag{4}$$

Then, the overall response of the hidden layer of the INN, consisting of k neurons, can be represented as:

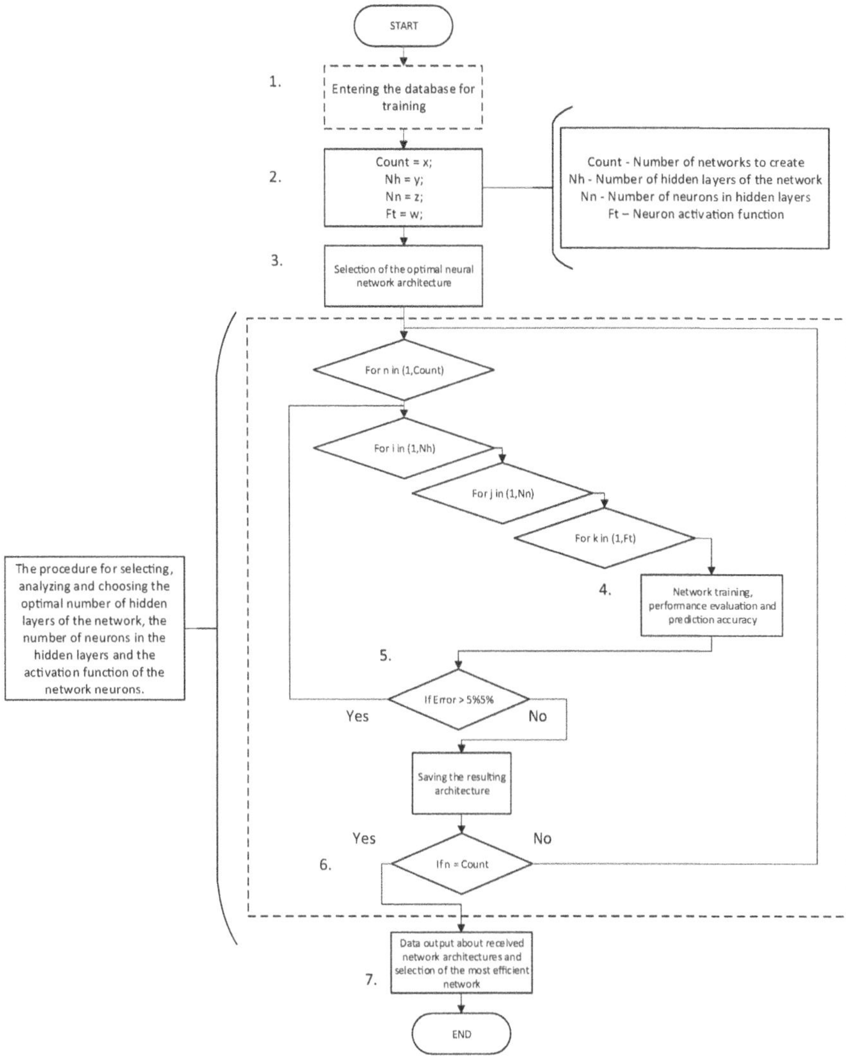

Fig. 5 Algorithm of operation of the adaptive system for predicting the temperature of the PCL

$$O_1 = f(W_{01}, X) = \begin{bmatrix} O_1 \\ O_2 \\ \dots \\ O_k \end{bmatrix} = \varphi_k \left(\begin{bmatrix} w_{01} & w_{02} & & w_{0n} \\ w_{11} & w_{12} & & w_{1n} \\ \dots & \dots & & \dots \\ w_{k1} & w_{k2} & \dots & w_{kn} \end{bmatrix} \bullet \begin{bmatrix} x_1 \\ x_2 \\ \dots \\ x_n \end{bmatrix} \right) \tag{5}$$

where O_1—the layer response vector of the first hidden layer of the INN; W_{01}—the matrix of weight coefficients between the input and the first hidden layer of the neural network of size n * k;

Further, the response O_1 becomes the input signal for the next layer of the hidden layer of the neural network, and, based on expression (5), the response of the next layer is calculated, and so on. The final response of the neural network, which is an hourly forecast of power consumption, is formed based on the expression:

$$Y = \varphi(W_{mh} O_m),$$

where W_{mh} is the weight matrix between the last hidden layer m and the output layer h; Om is the response of the last hidden layer m. The assessment of the accuracy of the INN forecast on a test set with the dimension P is estimated by the average absolute error in percent (MAPE):

$$MAPE = \frac{1}{p}\sum_{i=1}^{p}\frac{|y_i - \widehat{y_i}|}{y_i} \times 100$$

where y_i is the actual value of the load, y_i is the result of the forecast, p is the amount of data. From the analysis of publications, a forecast error of no more than 5% is acceptable.

7 Work Algorithm

To solve the problem of automation of PCL temperature prediction by using a neuro-computer prediction system, an algorithm for the operation of an adaptive system for PCL temperature prediction was developed. The operation algorithm of the adaptive system for predicting the PCL temperature is shown in Fig. 5.

The adaptive algorithm of the PCL temperature prediction system works step by step as follows:

Block 1—Entering the database for training. Loading input parameters, such as the temperature of the protective sheath (θp.s (n)), the effective current of the core (Ic), the number of the cable sample (Nk) from the database of "effective currents and temperatures of the PCL" for subsequent training of the created neural networks.

Block 2—Setting the basic parameters for creating the National Assembly. Determining the range of values of the main network parameters for their subsequent creation, such as the number of neural networks that will be created as part of the learning process (Count), the minimum and maximum number of hidden layers in the created networks (Nh), the minimum and maximum number of neurons in the hidden layers created by the neural networks (Nn), activation functions of neurons in the created neural networks (Ft).

Block 3—Selection of the optimal structure of the National Assembly. Based on the previously set parameters, the construction and enumeration of all possible combinations of neural network architectures is carried out.

Block 4—Training and evaluation of the performance of the NN. For each of the obtained combinations of the number of hidden layers, neurons, and activation functions, the network is trained, prediction quality is tested, and performance is evaluated.
Block 5—Calculation of the forecasting error of the obtained NN. If the prediction accuracy of the received architecture is below the specified criteria, the received architecture is eliminated, the process proceeds to checking the next architecture. If the prediction accuracy meets the specified criteria, the resulting neural network configuration is saved for subsequent work.
Block 6—Checking the number of received NN configurations. The process of creating and training will continue until the specified number of created neural network configurations (Count) is reached.
Block 7—Evaluation of the obtained results and selection of the most efficient NN configuration. If the required number of neural network architectures have been created in the course of work, the process of creation and training stops, one of the most productive neural networks is selected from the obtained combinations and further work is already underway with it.

8 Experimental Studies

For the experimental study of the developed methodology for predicting the resource of insulating materials of cable networks, we have chosen a part of the network infrastructure of the cyber-physical system [4, 16, 17]. The experimental study was in two stages. At the first stage, a study of the parameters affecting the cable network was carried out in order to obtain initial data—a training sample. And a study of "serviceable" samples was carried out. At the second stage, "aged" cable samples were investigated, which are equated to different lifetimes of PCL. Analysis of the results of the study showed that the longer the aging time, the greater the temperature difference between the original and aged sample. As the aging time increases, the temperature difference increases. In this case, the prediction error of the neural network decreases, that is, the accuracy of predicting the temperature of the cable core increases. Comparison of experimental and predictive data for new (training set) and artificially aged (control set) cable samples is given in Table 1.

Table 1 Parameter comparison

Parameter		Cable sample no.		
		1 cable	3 cable	4 cable
Temperature difference Δt, V	Not aged	1.73	2.8	2.3
	Aged	1.2	1.46	1.6
Error, %	Not aged	4.2	5.2	3.8
	Aged	3.8	2.3	3.1

We believe that this is due to the loss of the electrical insulating properties of the material due to the accumulation of fragments of the destroyed structure, which contains, in increasing amounts, inclusions of pure carbon and other conductive inclusions. The reaction of incomplete oxidation of polyethylene with the formation of carbon:

$$\left[\begin{array}{c} H \quad\; H \\ | \quad\;\; | \\ -\,C - C\,- \\ | \quad\;\; | \\ H \quad\; H \end{array}\right]_n + nO_2 \rightarrow 2_n H_2O + 2_n C$$

9 Conclusion

As part of the study, a neural network model was constructed that describes the architecture of a cyber-physical system for predicting the resource of electrical insulating materials of power cable electrical networks. An algorithmic solution has been synthesized for automatic selection of parameters and training of a neural network with subsequent prediction, which makes it possible to increase the reliability of a cyber-physical system by reducing the time for creating an optimal neural network configuration by several times: it is necessary to manually train and check the effectiveness of the resulting configuration;—using the algorithm, this procedure takes no more than 20–30 min.

The considered processes of aging of an electrically insulating material in an electrostatic field with a high intensity and the results of the calculation showed that under the action of an electrostatic field, the aging of an electrically insulating material occurs due to the development of a PD in it.

If there is a general change in the structure of the MI, for example, a change in the chemical structure associated with the development of thermal breakdown, then it is necessary to bear in mind, first of all, the power of the PD.

Acknowledgement The work was supported by a grant from the Russian Science Foundation №22-29-00370 «Development and research of methods for optimization and control of energy conversion processes in power plants of complex type, including those transforming the flow of continuous medium».

References

1. Dubyago, M.N.: Development of an aging model and determination of the residual resource of insulation of power cables. In: Izvestiya SFU. Technical science. no. 4, pp. 107-114 (2014)
2. Dubyago, M.N., Poluyanovich, N.K.: Prediction of residual life of isolating materials in the process of thermal power equipment deterioration. In: International Conference on Advances in Energy, Environment and Chemical Engineering, Changsha, China, pp. 49-54 (2016)
3. Riba, J.R., Alabern, X.: Effects of the circulating sheath currents in the magnetic field generated by an underground power line. In: International conference on renewable energies and power quality, Palma de Mallorca, 5-7 April 2006
4. Poluyanovich, N.K. Dubyago, M.N.: Study of characteristics of partial discharge for assessment of condition of electrical insulating materials of power supply system. In: 2013 4th International Conference on Applied Mechanics and Mechanical Engineering, Singapore, Applied Mechanics and Materials Switzerland, vol. 459, pp. 70–75 (2014)
5. Hedir, A., Moudoud, M., Rondot, S., Slimani, F., Mekious, M., Loucif, A.: Observations on Structural Changes under Electrical Aging of XLPE Used as High Voltage Power Cables Insulation. In: 11 ème Conférence Nationale sur la Haute Tension (2019)
6. Rami S.-A., Ehtasham, M., Tamus, Z.-A.: Assessment of the electrical and mechanical state of low-voltage unshielded nuclear power cables under simultaneous thermal and mechanical stresses: application of non-destructive testing methods (2021)
7. Shuaishuai, L.: Identification of composition, aging mechanisms and non-destructive aging indicator of insulating materials of cables made of commercial filled cross-linked polyethylene (2017)
8. Tzimas, A., Rowland, S., Dissado, L., Fu, M., Nilsson, U.: Influence of long-term electrical and thermal stresses on the endurance of the insulating material of the cable. In: IEEE Transactions on Dielectrics and Electrical Insulation (2009)
9. Dubyago, M., Poluyanovich, N.: Partial Discharge Signal Selection Method for Interference Diagnostics of Insulating Materials. In: 2019 Radiation and Scattering of Electromagnetic Waves, pp. 124–127 (2019)
10. Poluyanovich, N., Medvedev, M., Dubyago, M., Azarov, N., Ogrenichev, A.: Neural network model for evaluating thermofluctuation processes in cable systems using a multi-stage forecasting method. In: Studies in Systems, Decision and Control, vol. 350, pp. 349-359.
11. Poluyanovich, N., Dubyago, M., Azarov, N., Ogrenichev, A.: Neural network analysis of electromagnetic field effect on the thermofluctuation characteristics measurement in the power cable insulating materials. In: Studies in Systems, Decision and Control. Cyber-Physical Systems: Intelligent Models and Algorithms. Springer Nature Switzerland AG 2022, vol. 417, pp. 67–80 (2022)
12. Poluyanovich, N., Medvedev, M., Dubyago, M., Azarov, N., Ogrenichev A.: Neural network model for evaluating thermofluctuation processes in cable systems using a multistage forecasting method. In: Kravets A.G., Bolshakov A.A., Shcherbakov M.V. (eds) Cyber-Physical Systems. Studies in Systems, Decision and Control, vol 350. Springer, Cham (2021)
13. Khaitan et al.: Design techniques and applications of cyber-physical systems: a survey. IEEE Syst. J. (2014)
14. Poluyanovich, N., Medvedev, M., Dubyago, M., Azarov, N., Ogrenichev, A.: Estimation of cable lines insulating materials resource using multistage neural network forecasting method. In: Lecture Notes in Computer Science (including subseries Lecture Notes in Artificial Intelligence and Lecture Notes in Bioinformatics), pp. 288–296 (2020)

15. Poluyanovich, N., Azarov, N., Dubyago, M.: Neural network method for monitoring thermofluctuation processes in cable lines taking into account the interference influence. In: Conference Proceedings - 2021 Radiation and Scattering of Electromagnetic Waves, RSEMW 2021 this link is disabled, pp. 455–459 (2021)
16. Kravets, A.G., Vitaly E.: The software cache optimization-based method for decreasing energy consumption of computational clusters. Energies **15**(20), 7509 (2022). https://doi.org/10.3390/en15207509
17. Kravets, A.G., Kolesnikov, S., Salnikova, N., Lempert, M., Poplavskaya, O.: The study of neural networks effective architectures for patents images processing. Commun. Comput. Inf. Sci. **1084**, 27–41 (2019)

Improving the Architecture of Fuzzy Automated Systems Based on the State Observer Algorithm

Artur Sagdatullin

Abstract The purpose of the chapter is to optimize the architecture of fuzzy automated systems based on the state monitoring algorithm. To achieve this, the following problems are solved: the study of a simulation description based on the discount state of authorized systems; development of an architectural solution based on the state monitoring algorithm. An algorithmic implementation of the state observer is proposed. The results showed that the proposed system allows expanding the functionality of the systems and increasing their performance.

Keywords Fuzzy logic · Information system · Software and algorithmic implementation · Software · State observer · Discrete representation

1 Introduction

The analysis of automated control systems, as a rule, begins with the study of input information and a verbal representation of the control object [1–4]. Such strategies include receiving, processing signals, and other information based on feedback control. This architecture allows you to configure control systems and process the results practically in real-time.

The significant coupling of objects and components of the system can be a considerable difficulty, both in software development and in the implementation of automated systems. Therefore, this research work aims to optimize the architecture of fuzzy automated systems [5] based on the state observer algorithm [6, 7].

In the scientific work, the following tasks are solved [8–12]:

1. Simulation study of automated systems based on a discrete description.
2. Development of an architectural solution based on the state observer algorithm.

A. Sagdatullin (✉)
Intelligent Oil and Gas Technologies, K. Marx St., 10, Kazan 420111, Russia
e-mail: saturn-s5@mail.ru

A. G. Kravets et al. (eds.), *Cyber-Physical Systems Engineering and Control*,
Studies in Systems, Decision and Control 477,
https://doi.org/10.1007/978-3-031-33159-6_20

2 Mathematical Description of Dynamic Systems Under Disturbing Influences

Theoretically, the system under consideration can be described by the following equations:

$$\begin{gathered}(\dot{x_{11}}, \dot{x_{12}}, \ldots, \dot{x_{1n}})(u_{11}, u_{12}, \ldots, u_{1n}),\\ (\dot{x_{21}}, \dot{x_{22}}, \ldots, \dot{x_{2n}})(y_{21}, y_{22}, \ldots, y_{2n}),\\ \ldots \quad \ldots \quad \ldots \quad \ldots \quad \ldots \quad \ldots \quad \ldots \quad \ldots,\\ (\dot{x_{k1}}, \dot{x_{k2}}, \ldots, \dot{x_{kn}})(y_{k1}, y_{k2}, \ldots, y_{kn}),\end{gathered} \tag{1}$$

where x(t) is the external state vector, u(t) is the control signals of the control system, and y(t) is the output signals of the control system.

Equation (1) can describe the states of dynamic systems under disturbing influences. The simulation description can be presented based on the canonical form and the discrete state of the system (2) [13, 14]:

$$\vec{X}[n+1] = A\vec{X}[n] + B\vec{U}[n] + f\vec{X}[n] + g\vec{X}\vec{U}[n], \vec{Y}[n] = h\vec{X}[n], \tag{2}$$

where U is the vector of control signals, X is the state vector of the system, dimension (n, 1), f is the function of unobservable disturbances in the control process, g is the function of disturbing influences on the control object, A is the matrix of system coefficients, B is a matrix of input coefficients, h is a matrix of coefficients of output signals belonging to R^n.

Part of the expression $f\overrightarrow{X}[n] + g\overrightarrow{X}\overrightarrow{U}[n]$ can be combined under the joint coefficient ED, where E is the distribution matrix of input parameters and D is the input observed vector. Initial state: $X(t) = x_0$.

$$\vec{X}[n+1] = A\vec{X}[n] + B\vec{U}[n] + ED[n], \ \vec{Y}[n] = C\vec{X}[n]. \tag{3}$$

where:

$$A = \begin{bmatrix} 0 & 1 & 0 \\ 0 & 0 & 1 \\ 0 & 0 & 0 \end{bmatrix}, B = \begin{bmatrix} 0 \\ 1 \\ 0 \end{bmatrix}, E = \begin{bmatrix} 0 \\ 0 \\ 1 \end{bmatrix}, C = [1 \ 0 \ 0]$$

The representation of state vectors describes the characteristics of the control object for closed-loop systems with feedback, the tasks of which are to find the optimal location of the roots of the system equation by adjusting its parameters [15]. Digital representation is the direct form of implementing state observers as software modules [16, 17]. Thus, a discrete description will be implemented for digital representation. Therefore, to describe such systems, it is required to represent the discrete state of the system (Fig. 1).

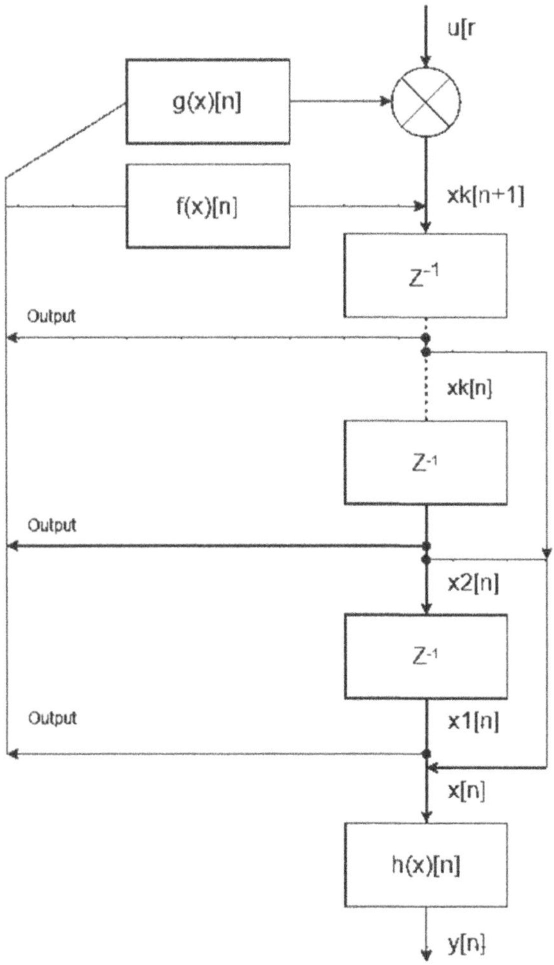

Fig. 1 Functional diagram for representing an object based on a discrete description of the state of the system

3 Algorithm of Control Object with the State Observer

The main problem of monitoring parameters can be formulated as follows: at any given time, the system's state (A, B, C, D) allows you to obtain an estimate of the system's state $\hat{\vec{X}}(t)$ based on the measurement of input and output $\vec{X}[n] - \hat{\vec{X}}(t) \approx 0$. This expression shows that under known initial conditions, the state estimate is approximately equal to the actual value of the output signal. Figure 2 shows the functional representation in the form of a block diagram.

The applied description allows the modeling of systems and the numerical calculation of the schemes under consideration based on the Euler and Runge–Kutta algorithms. The analysis of processes and control systems is also based on discrete

Fig. 2 Block diagram of a control object with the state observer

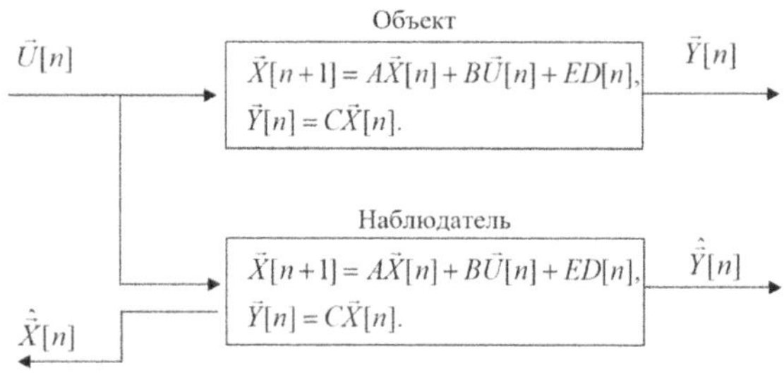

representation. This model shows the connectivity of objects, the internal structure of which can consist of tens and hundreds of components.

Modules for receiving and processing information, and control components, becoming highly interconnected, lose flexibility with changes in algorithms. This behavior can be avoided by optimizing the architecture of fuzzy automated systems [18] based on the state observer algorithm [19–21]. A state observer is a helper component that allows interaction through its methods and properties.

The main advantages of this approach include:

- Reducing the multi-connectivity of the components of the automated system;
- Ability to reuse components;
- Each component is responsible for a specific or abstract task;
- Adding new components to the control system does not increase the connectivity of objects.

To implement the state observer, consider events about the system's state are formed: pressure, velocity, and temperature.

During the system's operation, the automation project can be supplemented with new modules, and several subsystems may need information about the parameters of the equipment. To implement the transfer of information about the current state, it is necessary to implement multiple calls for the essential events. In this case, events are considered that represent real-time data on the state of the pumping unit.

An additional link L is introduced to describe the observer, abstract from the implementation of both the control controller and the object, and the parameters intended for observation.

$$\hat{\vec{X}}[n+1] = A\hat{\vec{X}}[n] + B\vec{U}[n] + ED[n] + L(\vec{Y} - \hat{\vec{Y}}),\ \vec{Y}[n] = C\hat{\vec{X}}[n]. \quad (4)$$

Expression (4) can be represented as follows:

$$\hat{\vec{X}}[n+1] = (A - LC)\hat{\vec{X}}[n] + B\vec{U}[n] + ED[n] + L\vec{Y},\ \vec{Y}[n] = C\hat{\vec{X}}[n]. \quad (5)$$

The obtained equations of the nth order represent a display of a dynamic system, the initial state of which can be described as: $\hat{X}(0)$. The system's initial state is equivalent to the initial estimate of this and is equal to it when the observer is tuned correctly. Let us describe the structure of the control object, representing the pumping unit, in the following form:

```
class PsUnit
    {
        public string Type {get; set;}
        public float Speed {get; set;}
        public float Temperature {get; set;}
        public float Pressure {get; set;}
        private Process process{get; set;}
        public PsUnit(string type, Process process, float
speed, float pressure, float temperature){
            Type = type;
            Speed = speed;
            Temperature = temperature;
            Pressure = pressure;
            this.process = process;
        }
        public void Event() => process.Dispatch(
            new PsUnitEventParams(this.Type, ++Speed,
++Temperature, ++Pressure));
    }
```

The proposed algorithm consists of several entities: a process, abstract parameters containing a method for their display, implementation of parameters such as the type of pumping equipment, rotation speed, pressure, and temperature of bearings. The process reflects the event, and the dispatcher is called when this event fires new data. The dispatcher is represented as an auxiliary link and implements a method for calling all events that have occurred recently.

In this example, a pumping unit represents a separate class, the execution of specific equipment, “Pump_1”. The implementation of this class includes setting the pump parameters and a method that implements sending events when they change.

The following class can represent the algorithm of the recipient of information:

```
class Receiver
    {
        public Process process {get; set;}
        public Receiver(Process process)
        {
            this.process = process;
            process.PsUnitEvens += PsEvent;
        }
        private void PsEvent(object sender, UnitEvent-
Params e )
        {
            if (e is PsUnitEventParams psUnitArgs) {
                Console.WriteLine($"Data of {psUni-
tArgs.PsUnitType} changed");
            }
        }
    }
```

4 Results

The Receiver subsystem is implemented to acquire events, which is responsible for obtaining new data from the equipment and issuing information about their change. Algorithm results:

```
Data of Pump_1 changed {50, 150, 20}
Data of Pump_1 changed {49, 150, 20}
Data of Pump_1 changed {48, 150, 20}
Data of Pump_1 changed {47, 150, 20}
Data of Pump_1 changed {46, 150, 20}
Data of Pump_1 changed {45, 150, 20}
Data of Pump_1 changed {44, 150, 20}
Data of Pump_1 changed {43, 150, 20}
Data of Pump_1 changed {42, 150, 20}
```

The data obtained above show that the observer functions as intended: the developed software module directly scans the parameters of the object that need to be evaluated, and, with each change in these parameters of the control object, the system generates an event for changing a particular parameter and issues a signal that being has changed, which is what the messages above show. According to the data obtained, it can be seen that the implemented algorithm works correctly and allows you to separate the entities of receiving and sending equipment events to different subsystems. This allows to expand the functioning of automated systems and increase their productivity. Thus, a method is proposed for optimizing the architecture of fuzzy automated systems based on the state observer algorithm. The results of the program's work and the algorithm showed that it is possible to build high-performance systems on its basis and divide control classes into various subsystems.

5 Conclusion

In this research work an algorithmic implementation of the state observer is proposed the following problems are solved: the study of a simulation description based on the discount state of authorized systems; development of an architectural solution based on the state monitoring algorithm. The results showed that the proposed system allows expanding the functionality of the systems and in-creasing their performance.

References

1. Astrom, K.J., Hagglund, T.: Advanced PID Control. Research Triangle Park, North Carolina: The Instrumentation, Systems, and Automation Society, p.354. (2006)
2. Zazeh, L.A.: A computational approach to fuzzy quantifiers in natural languages. Comput. Math. **9**, 149–184 (1983)
3. Chen, C.: Linear System Theory and Design. Oxford University Press (2009)
4. Brunton, L.S., Kutz J. N.: Chapter 8: linear control theory. In: Data Driven Science and Engineering: Machine Learning, Dynamical Systems, and Control. Cambridge. https://doi.org/10.1017/9781108380690.009.
5. Sagdatullin, A.: Application of fuzzy logic and neural networks methods for industry automation of technological processes in oil and gas engineering. In: Proceedings 2021 3rd International Conference on Control Systems, Mathematical Modeling, Automation and Energy Efficiency, SUMMA 2021, pp. 715–718. (2021). https://doi.org/10.1109/SUMMA53307.2021.9632251
6. Luenberger, D.: An introduction to observers. IEEE Trans. Autom. Control **16**, 596–602 (1971)
7. Trentelman, H.L., Antsaklis, P.: Observer-based control encyclopedia of systems and control (2014). https://doi.org/10.1007/978-1-4471-5102-9_199-1
8. Sagdatullin, A.: Study of the energy consumption of borehole systems with rod pumps on the basis of simulation of an automated electromechanical system. In: Proceedings—2021 3rd International Conference on Control Systems, Mathematical Modeling, Automation and Energy Efficiency, SUMMA 2021, pp. 1075–1078. (2021). https://doi.org/10.1109/SUMMA53307.2021.9632090
9. Gao, Z., Ho, D.W.: Proportional multiple-integral observer design for descriptor systems with measurement output disturbances. IEE Proc. Control Theory Appl. **151**(3), 279–288 (2004)
10. Chen, W.-H., Yang J., Guo L., Li, S.: Disturbance observer based control and related methods: an overview. IEEE Trans. Ind. Electron. **63**(2), 1083–1095 (2016)
11. Alessandri, A., Coletta, P: Design of Luenberger observers for a class of hybrid linear systems. In: International Workshop on Hybrid Systems: Computation and Control. Berlin (2001)
12. Li, L., Song, G., Ou, J.: Nonlinear structural vibration suppression using dynamic neural network observer and adaptive fuzzy sliding mode control. J. Vib. Control. **16**(10):1503–1526 (2010)
13. Kim, K., Rew, K.: Reduced order disturbance observer for discrete-time linear systems. Automatica **49**(4), 968–975 (2013)
14. Sagdatullin, A.: Analysis of a discrete object control of oil and gas pumping system under uncertainty and unfull information conditions. In: Proceedings of 2020 International Russian

Automation Conference, RusAutoCon, pp. 405–409, 9208112. (2020). https://doi.org/10.1109/RusAutoCon49822.2020.9208112
15. Di Giamberardino, P., Iacoviello, D.: An improvement in a local observer design for optimal state feedback control: the case study of HIV/AIDS diffusion. In Proceedings of the 16th International Conference on Informatics in Control, Automation and Robotics (ICINCO), pp. 100–111. (2019). https://doi.org/10.5220/0007934501000111
16. Sagdatullin, A., Degtyarev, G.: Development of a cyber-physical subsystem for support and decision making of managing oil production and transportation processes under uncertainty conditions. Stud. Syst. Decis. Control **342**, 145–154 (2021). https://doi.org/10.1007/978-3-030-66081-9_11
17. Sagdatullin, A.: Functioning and development of a real-time information system for the oil treatment technological process control. In: Proceedings 2020 2nd International Conference on Control Systems, Mathematical Modeling, Automation and Energy Efficiency, SUMMA 2020, pp. 847–852, 9280577. (2020). DOI https://doi.org/10.1109/SUMMA50634.2020.9280577
18. Sagdatullin, A., Degtyarev, G.: Development of a cyber-physical system for neurofuzzy prediction of the concentration of the contained prime during transportation of oil wells emulsion. Stud. Syst. Decis. Control. **417**, 169–180 (2022). https://doi.org/10.1007/978-3-030-95116-0_14
19. Kumar, V.E. Jerome, J., Ayyappan, S.: Comparison of four state observer design algorithms for MIMO system. Arch. Control. Sci. **23**(LIX)(2), 243–256 (2013). 0.2478/acsc-2013-0015
20. Peet, M.M., Gu, K.: Synthesis of full-state observers for time-delay systems using SOS. In: 23rd International Symposium on Mathematical Theory of Networks and Systems Hong Kong University of Science and Technology, Hong Kong, (2018)
21. Chou, F.-I., Cheng M.-Y.: Optimal observer design of state delay systems. Math. Probl. Eng. **2019**(8493298), 8 (2019). https://doi.org/10.1155/2019/849329

Development of an Algorithm for Interpretation of Input Parameters of Fuzzy Logic Controller for Cyber-Physical Real-Time Systems

Artur Sagdatullin and Gennady Degtyarev

Abstract The purpose of the chapter is to develop an algorithm for interpreting input parameters by a discrete membership function of a fuzzy set for embedded and real-time systems. To achieve the goal set in the scientific work, the following tasks are solved: the study of fuzzy logical inference in automatic control systems: analysis of the construction of a fuzzy set, representation of a discrete platoon value; development of an algorithm and software implementation for real-time systems. Methods for setting input variables for fuzzy logic systems are considered, an algorithmic implementation of the representation of the input value as an element of a fuzzy set and automatic calculation of the membership function are proposed.

Keywords Fuzzy logic · Information system · Software and algorithmic implementation · Software · Discrete set

1 Introduction

Automated real-time control systems are programmable multimodal information systems for solving problems of operational control of objects and technological processes for a time not exceeding Δt. Automated real-time systems play a crucial role in implementing the control action on the technological process. Such important process parameters as costs, reliability, and efficiency depend on control quality. In studying methods and algorithms for information input, the interaction of information with external factors is very relevant since the complexity of control and automation systems grows over time, and information uncertainty only increases [1, 2].

Uncertainties of various types introduce an element of fuzziness into the information sections of automated systems [3, 4]. The reliability of the operation of real-time systems, coupled with the presence of many sources of its receipt, becomes more and more unpredictable [5].

A. Sagdatullin (✉) · G. Degtyarev
Intelligent Oil and Gas Technologies, K. Marx st., 10, Kazan 420111, Russia
e-mail: saturn-s5@mail.ru

A. G. Kravets et al. (eds.), *Cyber-Physical Systems Engineering and Control*,
Studies in Systems, Decision and Control 477,
https://doi.org/10.1007/978-3-031-33159-6_21

The research work aims to study and program the implementation of multidimensional fuzzy inference for real-time systems [6].

To achieve the goal, the following tasks are solved:

1. The study of fuzzy logical inference in automatic control systems: analysis of the construction of a fuzzy set, representation of a discrete input value.
2. Development of an algorithm and software implementation for real-time systems.

2 Mathematical Description of Fuzzy Logic Input Variables and Its Interpretation Equations

Algorithmic and software tools currently achieve the improvement of automated real-time systems. One such tool is fuzzy logic and its applications [7, 8]. While in Boolean algebra, values are strictly limited to zero and one, fuzzy logic, in contrast to the hard binding to *[0, 1]*, defines intermediate values. This means that the representation of physical quantities is reduced to a logical and linguistic description of variables that occurs in natural language, for example, control error *["high," "medium," "zero"]*. Fuzzy logic is based on representing fuzzy sets *A, B, C, ..., N* and the membership of a crisp physical quantity in this set: $x_1, x_2, \dots x_n$ in the range *[0, 1]*. The membership function $\mu(x)$ of a clear physical value to a fuzzy quantity can be described as [9, 10]:

$$\mu_A(x) = \begin{cases} 0,\ if\ x < a_1, \\ L\_S(x - a_1),\ if\ a_1 < x < a_2, \\ L\_S = \dfrac{1}{a_2 - a_1} \\ 1,\ if\ a_2 < x < a_3, \\ R\_S(x - a_3),\ if\ a_3 < x < a_4, \\ R\_S = \dfrac{1}{a_4 - a_3}, \\ 0,\ if\ x > a_4. \end{cases} \quad (1)$$

where a_1, a_2, a_3, a_4 are the coefficients of the generated set.

Figure 1 shows a graphical representation of the trapezoidal membership function $\mu(x)$ for the input parameter "pump rotation speed" of the real-time system. However, mathematical models describing the dynamics of systems can be very complex, and managing them in real-time can take significant resource intensity, exceeding the capabilities of standard controllers. Also, it can be pretty challenging to choose a working mathematical model, which, in conditions of multiply connected variables and multidimensionality, can significantly affect the quality of the output parameters. At the same time, the processing of input parameters also plays an important role.

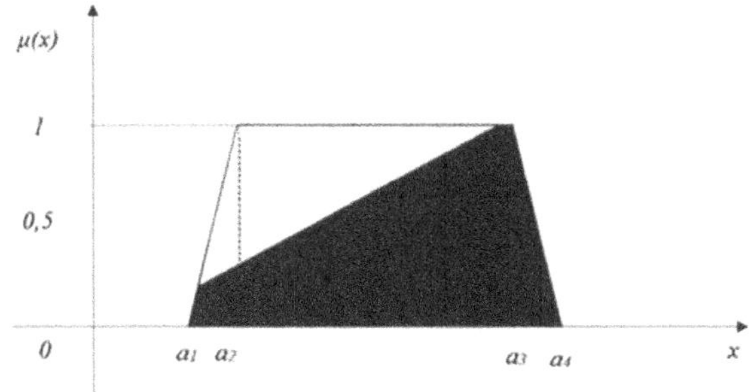

Fig. 1 Trapezoidal membership function $\mu(x)$

Processing input parameters in a fuzzy logic system usually includes several successive stages: fuzzification, rule base, inference, and defuzzification [11, 12]. This sequence allows you to transform crisp input values into fuzzy variables according to the established rules.

Mathematically, the fuzzy set A can be expressed as a set of objects of type x by the following expressions:

$$A = \{\dot{x}_n,\ \mu_A(\dot{x}_n) \mid \dot{x}_n \in X\},$$

$$A = \frac{\mu_A(x_1)}{x_1} + \frac{\mu_A(x_2)}{x_2} + \ldots + \frac{\mu_A(x_3)}{x_3} \tag{2}$$

$$\mu_A(x_n) = 1 - \mu_A^c(x_n),$$

where $\mu_A(x)$—fuzzy set membership function A, which conducts a one-to-one correspondence of an element of the universal set of values X and its membership level in the range from zero to one, $\mu_A^c(x_n)$—membership functions of a complementary fuzzy set.

The universe set X can contain both classical objects of continuous space and discrete ones. For example, imagine that there is $X =$ *{"Velocity", "Pressure", "Head (or flow rate)"}* and there is a fuzzy set describing these quantities:

$$A = \{(\text{"}Velocity\text{"},\ 0,5,\ (\text{"}Pr\,essure\text{"},\ 0,8),\ (\text{"}Head\text{"},\ 0,1)\},$$

$$A^c = \{(\text{"}Velocity\text{"},\ 0,5,\ (\text{"}Pressure\text{"},\ 0,2),\ (\text{"}Head\text{"},\ 0,9)\}.$$

The universe was interpreted as discrete with unordered objects in the case presented above. In this case, to represent a discrete set, consider the following example:

$X =$ *{0, 1, 2}*, where the universal set displays the number of pumps in operation. Then, the fuzzy set $A =$ "*number of pumps in operation*" can be described:

$$A = \{(\text{“0”},\ 0,1,\ (\text{“1”},\ 0,8),\ (\text{“2”},\ 0,3)\},$$

$$A^c = \{(\text{“0”},\ 0,9),\ (\text{“1”},\ 0,2),\ (\text{“2”},\ 0,7)\}.$$

A discrete ordered set A with a membership function to the universal set X is presented. Complementary sets, in this case, are given to specify a fuzzy set as a pair—membership function and value. The study showed that fuzzy sets that describe input values could be represented as discrete and continuous sets described by various membership functions [13]. The description for the pump rotation speed parameter is considered.

This implementation allows the following transformations:

- a value vector representing clear input parameters:

$\overrightarrow{X} = \begin{bmatrix} x_1 \\ x_2 \\ x_3 \\ \dots \\ x_n \end{bmatrix}$, parameters reflecting the new state of the system for the period of Δt.

- a value vector representing crisp output parameters:

$\overrightarrow{Y} = \begin{bmatrix} y_1 \\ y_2 \\ y_3 \\ \dots \\ y_n \end{bmatrix}$, parameters reflecting the change in the state of the system over the period of Δt.

3 Algorithm for Representing the Input Value as an Element of a Fuzzy Set and Automatic Calculation of the Membership Function

An important feature of fuzzy logic is the use of logical operations that must be performed even in the following cases:

$$\{\dot{x}_n,\ \mu_A(\dot{x}_n)\} \to 1, \tag{3}$$

$$\{\dot{x}_n,\ \mu_A(\dot{x}_n)\} \to 0.$$

Consider two sets A_1, A_2, which are subsets of the same general set. The operation of combining two sets can be represented as:

$$A_1 \cup A_2 \to \max[\mu_{A_1}(x), \mu_{A_2}(x)]. \tag{4}$$

Logical mapping:

$$A_1 \cup A_2 \to \{x|(x \in A_1) \vee (x \in A_2)\}. \tag{5}$$

The merged set represented by the expression above contains all the elements of both the first and second sets, for example:

$$A_1 \to \left\{\frac{x_{11}}{\exists x_{11}(\mu_{11})}, \frac{x_{12}}{\exists x_{12}(\mu_{12})}, \frac{x_{13}}{\exists x_{13}(\mu_{13})}, \ldots, \frac{x_{1n}}{\exists x x_{1n}(\mu_{1n})}\right\}. \tag{6}$$

$$A_2 \to \left\{\frac{x_{21}}{\exists x_{21}(\mu_{21})}, \frac{x_{22}}{\exists x_{22}(\mu_{22})}, \frac{x_{23}}{\exists x_{23}(\mu_{23})}, \ldots, \frac{x_{2n}}{\exists x_{2n}(\mu_{2n})}\right\}. \tag{7}$$

where

$$\begin{aligned}
&\exists x_{11}(\mu_{11}) = \exists x_{21}(\mu_{21}),\\
&\exists x_{12}(\mu_{12}) = \exists x_{22}(\mu_{22}),\\
&\exists x_{11}(\mu_{13}) = \exists x_{23}(\mu_{23}),\\
&\quad \ldots\ \ldots\ \ldots\ \ldots\\
&\exists x_{1(n-1)}(\mu_{1(n-1)}) = \exists x_{2(n-1)}(\mu_{2(n-1)}),\\
&\exists x_{1n}(\mu_{1n}) = \exists x_{2n}(\mu_{2n}),
\end{aligned}$$

From where follows:

$$\mu(A_1 \cup A_2)(x) \to 1, \quad if \quad \begin{cases} x \in A_1, \\ x \in A_2. \end{cases} \tag{8}$$

$$\mu(A_1 \cup A_2)(x) \to \left\{\begin{array}{l} \dfrac{\max(x_{11}, x_{21})}{\exists x_{11}(\mu_{21})}, \dfrac{\max(x_{12}, x_{22})}{\exists x_{12}(\mu_{21})}, \dfrac{\max(x_{13}, x_{23})}{\exists x_{13}(\mu_{31})}, \\ \ldots, \dfrac{\max(x_{1(n-1)}, x_{2(n-1)})}{\exists x_{1(n-1)}(\mu_{2(n-1)})}, \dfrac{\max(x_{1n}, x_{2n})}{\exists x_{1n}(\mu_{2n})} \end{array}\right. \tag{9}$$

The intersection of sets can be expressed by the following expression, including elements contained in both the first and second sets:

$$A_1 \cap A_2 \to \min[\mu_{A_1}(x), \mu_{A_2}(x)]. \tag{10}$$

Logical mapping:

$$A_1 \cap A_2 \rightarrow \{x|(x \in A_1) \wedge (x \in A_2)\}. \tag{11}$$

The set obtained from the expression above can be represented in the following form:

$$\mu(A_1 \cap A_2)(x) \rightarrow \left\{ \begin{array}{l} \frac{\min(x_{11}, x_{21})}{\exists x_{11}(\mu_{21})}, \frac{\min(x_{12}, x_{22})}{\exists x_{12}(\mu_{21})}, \frac{\min(x_{13}, x_{23})}{\exists x_{13}(\mu_{31})}, \\ \ldots, \quad \frac{\min(x_{1(n-1)}, x_{2(n-1)})}{\exists x_{1(n-1)}(\mu_{2(n-1)})}, \frac{\min(x_{1n}, x_{2n})}{\exists x_{1n}(\mu_{2n})} \end{array} \right. \tag{12}$$

The complementary set includes all elements not included in the set *A*:

$$\mu(\tilde{A})(x) \rightarrow 1 - \mu(A)(x). \tag{13}$$

$$if \quad \begin{cases} \mu(\tilde{A})(x) = 1, \ x \notin A, \\ \mu(\tilde{A})(x) = 0, \ x \in A. \end{cases}$$

As a result, a fuzzy set can be represented as an addition to the classical concept of a set, in which characteristic functions can take values in a given range belonging to [0, 1]. Algorithmically, such a representation can be implemented in the form shown in Fig. 2.

Based on the above expressions (1)–(5), it can be noted that the mathematical properties of fuzzy sets, such as intersections, complementarity and union, can also be described by the following dependencies:

$$\mu(A_1 \cap A_2) \rightarrow \mu_{A_1}(x)\mu_{A_2}(x) \tag{14}$$

$$\mu(A_1 \cup A_2) \rightarrow \mu_{A_1}(x) + \mu_{A_2}(x) - \mu_{A_1}(x)\mu_{A_2}(x) \tag{15}$$

The algorithm for representing the input value as an element of a fuzzy set and automatically calculating the membership function can be expressed programmatically in the following form:

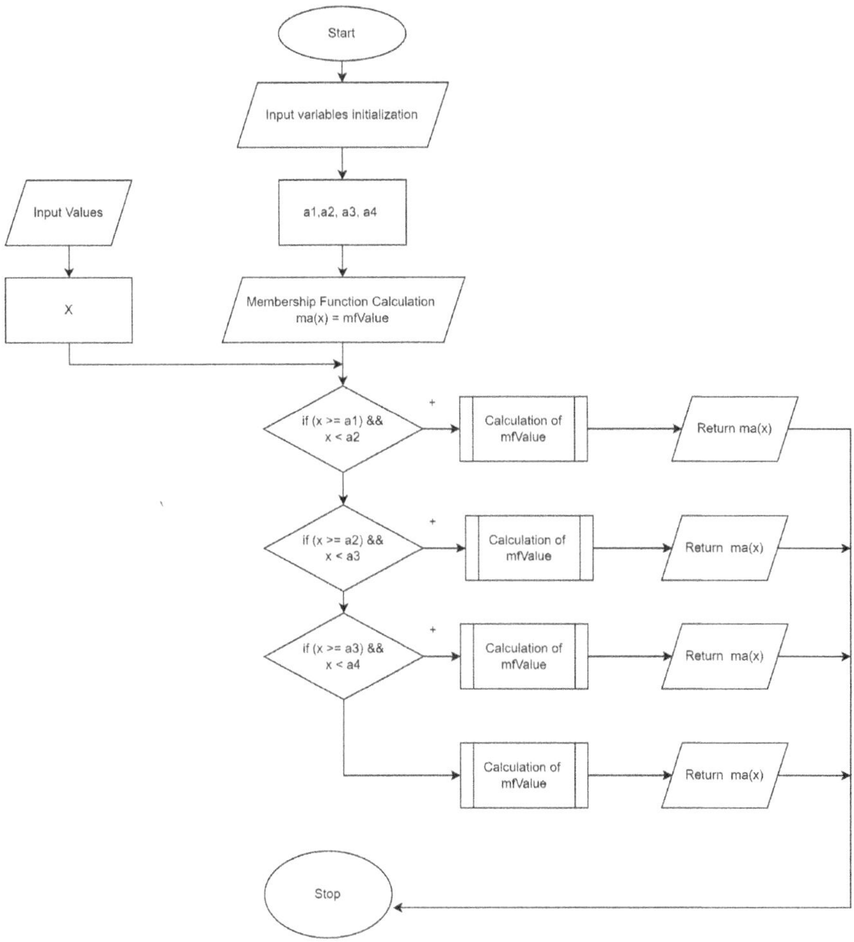

Fig. 2 Algorithm for representing the input value as an element of a fuzzy set and automatic calculation of the membership function

```
Algorithm
Data: a1, a2, a3, a4, x
Result: mfValue
L_S ← 0;
R_S ← 0;
m ← 1
while (m != 0) do
        L_S = 1 / (a2 – a1)
        R_S = 1 / (a4 – a3)
        case 1: x <= a1 || x >= a4
                mfValue = 0
        case 2: x >= a1 & X <= a2
                mfValue = L_S · (x – a1)
        case 3: x >= a2 & x <= a3
                mfValue = 1
        case 4: (x >= a3) & (x <= a4)
                mfValue = R_S · (x - a4)
        return mfValue
```

Embedded systems and real-time systems are subject to increased requirements for the execution time of calculations [14]. These requirements allow you to establish a connection between the event and the subsequent calculation that occurred in the system. *A* distinction is made between soft real-time systems, in which the mismatch Δt will slightly affect the system's state, and hard real-time systems, in which the execution of the calculation after a certain delay is not taken into account. The basis of the real-time systems considered in work are the processes described in time:

$$\dot{P}(t) = \begin{bmatrix} p_1(t) \\ p_2(t) \\ p_3(t) \\ \dots \\ p_n(t) \end{bmatrix} \rightarrow \begin{bmatrix} q_1 \rightarrow q_{11} \gg q_{12} \gg q_{13} \dots \gg q_{1n} \\ q_2 \rightarrow q_{21} \gg q_{22} \gg q_{23} \dots \gg q_{2n} \\ q_3 \rightarrow q_{31} \gg q_{32} \gg q_{33} \dots \gg q_{3n} \\ \dots \\ q_n \rightarrow q_{n1} \gg q_{n2} \gg q_{n3} \dots \gg q_{nn} \end{bmatrix} \tag{16}$$

where P is the vector of processes in the system $p_1(t)$, $p_2(t)$, $p3(t)$... $p_n(t)$, depending on the execution time; $q_1(t)$, $q_2(t)$, $q_3(t)$... $q_n(t)$—events triggered within one process; $q_{11}(t)$, $q_{21}(t)$, $q_{31}(t)$... $q_{nn}(t)$—sub processes launched within one event.

Processes and execution events in real-time systems are interconnected by common computing resources expended on a specific event stream $q_1 \rightarrow q_{11} \gg q_{12} \gg q_{13} \dots \gg q_{1n}$ Unrelated events or processes can also have utterly independent execution, while the system is designed in such a way that there is no separation of resources:

$$
\begin{aligned}
&[p_1(t)] \to [q_1 \to q_{11} \gg q_{12} \gg q_{13} \ldots \gg q_{1n}],\\
&[p_2(t)] \to [q_2 \to q_{21} \gg q_{22} \gg q_{23} \ldots \gg q_{2n}],\\
&[p_3(t)] \to [q_3 \to q_{31} \gg q_{32} \gg q_{33} \ldots \gg q_{3n}],\\
&[\ldots] \to [\ldots],\\
&[p_n(t)] \to [q_n \to q_{n1} \gg q_{n2} \gg q_{n3} \ldots \gg q_{nn}].
\end{aligned} \tag{17}
$$

The representation of this algorithm can be represented programmatically in the C++ programming language. Below is a code block that reflects the structure and methods of the algorithm shown in Fig. 2.

Code Snippet

```
#ifndef FLSET_H
#define FLSET_H
class FlSet {
private:
    float a1;
    float a2;
    float a3;
    float a4;
    float mfValue;
public:
    FlSet();
    FlSet(float a1, float a2, float a3, float a4);
    ~FlSet();
    float getMfValue(float x);
    float setMfValue(float mfValue);
    bool trapMf(float x);
};
#endif
where:
struct flSetList {
  FlSet * flSet;
  flSetList *next;
};
```

Memory allocation in microcontroller systems is an important issue. Restrictions on RAM, internal memory, and computing power can lead to significant delays or exit systems from real-time modes. The functional design of the software interface for the tasks under consideration can be represented in the form of a figure (Fig. 3).

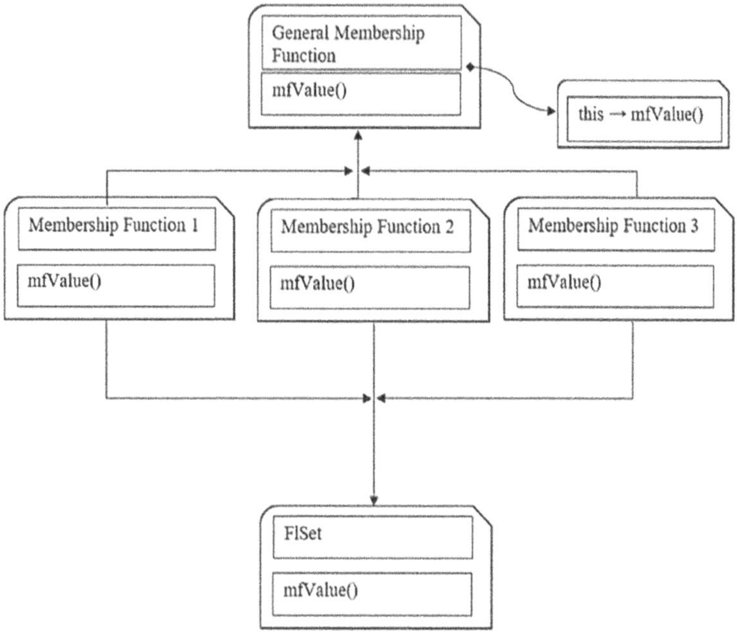

Fig. 3 The functional design of the software interface of the proposed system

4 Results

The result of the proposed system for interpreting the input parameters was to provide startup without significant memory allocation. The reduction in memory consumption was more than 50%, depending on the number of input variables. Organization of data sets using the built-in features of the programming language, such as pointers and references, allows you not to perform repeated data copy operations when passing them to the parameter interpretation system. This approach enables embedding large sets of discrete sets under consideration without significantly increasing the required memory. For real-time systems means an increase in the functions performed for interpretation and inference of correspondence with the most excellent efficiency.

5 Conclusion

A discrete ordered set A with a membership function to the universal set X is presented. Complementary sets, in this case, are given to specify a fuzzy set as a pair—membership function and value. The study showed that fuzzy sets that describe input values could be represented as discrete and continuous sets described by various membership functions. The description for the pump rotation speed parameter is considered.

Thus, methods for setting input variables for fuzzy logic systems are considered, an algorithmic implementation of the representation of an input value as an element of a fuzzy set and automatic calculation of the membership function is proposed, and features of the implementation of multidimensional fuzzy logical input in discrete form for real-time systems are considered.

References

1. Li, D: Uncertainty reasoning based on cloud models in controllers. Comput. Math. Applic. **35**(3), 99–123 (1998)
2. Sagdatullin, A., Degtyarev, G.: Development of a cyber-physical subsystem for support and decision making of managing oil production and transportation processes under uncertainty conditions. Stud. Syst. Decis. Control. **342**, 145–154 (2021)
3. Kim, Y.M., Mendel, J.M.: Fuzzy basis functions: comparison with other basis functions. USCSIPI Report, no. 299 (1995)
4. Sagdatullin, A., Degtyarev, G.: Development of a cyber-physical system for neurofuzzy prediction of the concentration of the contained prime during transportation of oil wells emulsion. Stud. Syst. Decis. Control. **417**, 169–180 (2022)
5. Sagdatullin, A.: Functioning and development of a real-time information system for the oil treatment technological process control. In: Proceedings—2020 2nd International Conference on Control Systems, Mathematical Modeling, Automation and Energy Efficiency, SUMMA 2020, pp. 847–852, 9280577. (2020)
6. Pal, A., Karim, I.U., Mondal, B., Raha, S.: A similarity based fuzzy system as a function approximator. Int. J. Intell. Sci. **8**(4), (2018). https://doi.org/10.4236/ijis.2018.84005
7. Moreno, G., Pascual, V.: Programming with fuzzy-logic and mathematical functions. In: Proceedings of the 6th. International Whorshop on Fuzzy Logic and Applications, WILF'05, University of Milan, Crema (Italy), pp. 89–98. Springer LNCS 3849 (2006)
8. Negoiţă, C.V., Ralescu, D.A.: Fuzzy automata, fuzzy languages, and fuzzy algorithms. applications of fuzzy sets to systems analysis. Pp. 122–151. https://doi.org/10.1007/978-3-0348-5921-9_6
9. Baruah, H.K.: Fuzzy membership with respect to a reference function. J. Assam Sci. Soc. **40**(.3), 65–73 (1999)
10. Gayathri Monicka, J.: Performance evaluation of membership functions on fuzzy logic controlled ac voltage controller for speed control of induction motor drive. Int. J. Comput. Appl. (0975–8887) **13**(5) (2011)
11. Zadeh, L.A.: A computational approach to fuzzy quantifiers in natural languages. Comput. Math. **9**, 149–184 (1983)
12. Rubaai, A.: Fuzzy Logic in Electric Drives, in Power Electronics Handbook, 2nd edn. (2007)
13. Sagdatullin, A.: Study of the energy consumption of borehole systems with rod pumps on the basis of simulation of an automated electromechanical system. In: Proceedings of the 2021 3rd International Conference on Control Systems, Mathematical Modeling, Automation and Energy Efficiency, SUMMA, pp. 1075–1078. (2021)
14. Sagdatullin, A.: Application of fuzzy logic and neural networks methods for industry automation of technological processes in oil and gas engineering. In: Proceedings of 2021 3rd International Conference on Control Systems, Mathematical Modeling, Automation and Energy Efficiency, SUMMA, pp. 715–718 (2021)

Monitoring the State of Vehicles with Dangerous Goods in Cyber-Physical Systems

E. S. Soldatov and A. S. Soldatov

Abstract The issues of providing wireless communication for monitoring the state of vehicles used in transportation of dangerous goods are analyzed. A cyber-physical system is considered that includes heterogeneous networks for monitoring the state of transport equipment carrying dangerous goods and equipped with appropriate sensors. The monitoring system using data networks providing a reliable connection ensures the ability to monitor the state of vehicles in real time. This is made possible by installing an autonomous power supply on each of the tank containers connected to the wireless sensor network and by using internet-of-things telemetry units. An example of using a sensor network for a cryogenic tank container equipped with an autonomous power supply and telemetry system is given. In particular, it is possible, based on the obtained data, to calculate the deviations of the pressure growth rate in the tank container from the nominal value. The widespread use of the proposed technical solutions on various modes of transport will significantly improve the safety of transportation of fire-explosive and toxic cargo.

Keywords Sensor network · Multimodal transportation · Internet of things · Dangerous goods · Monitoring of thermodynamic processes · Non-drainage storage · Tank container

1 Introduction

In industry and other sectors of the economy, there remains an increased interest in the use of tank containers for the transportation of dangerous goods. In particular, the ratio of cryogenic tank containers increases substantially for the long-term storage and transportation of cryogenic products. Such containers are suitable for use in multimodal transport, i.e. with their help it is possible to deliver one shipment

E. S. Soldatov (✉) · A. S. Soldatov
Saint Petersburg Federal Research Center of the Russian Academy of Sciences, 39, 14th Line V.O, Saint Petersburg 199178, Russian Federation
e-mail: volshebnoekoltso@mail.ru

A. G. Kravets et al. (eds.), *Cyber-Physical Systems Engineering and Control*, Studies in Systems, Decision and Control 477,
https://doi.org/10.1007/978-3-031-33159-6_22

of dangerous goods by various modes of transport without refueling and without intermediate losses. It should be noted that the average annual growth rate of the global tank container market is 11.8% [1, 2].

Modern systems for monitoring the state of equipment in transport allow you to track the parameters of the state of a certain number of transport units in operation. For example, navigation information is received and analyzed in real time, warnings are received about a product leak through valves based on readings from a temperature sensor, and warnings are generated for the driver of a vehicle about an emergency [3–6].

The disadvantage of existing technical solutions for transport monitoring systems is the limited functionality in terms of organizing communication channels for transmitting data from a vehicle in poor network coverage areas. Of special note is task of organizing data transfer between the airplane and monitoring server during air transportation [7–10].

To solve the problem of the lack of a stable Internet connection during road, sea or air transportation, both additional communication devices for connecting to satellite networks and technologies based on the Internet of things (IoT) can be used. IoT communication units allow to connect transport containers with dangerous goods into a single information network [11–15].

In order to provide a stable connection for all the monitoring objects in cyber-physical system the conception of sensor networks can be used to solve the problem of lack of sustainable Internet coverage. The most common sensor networks is networks based on ZigBee technology. ZigBee is the name for a set of top layer network protocols using low power and small radio transmitters based on the IEEE 802.15.4 standard. ZigBee is targeted at applications that require long battery life and high data security at low data rates [16, 17].

The main advantage of ZigBee technology is the relatively low power consumption which is especially important for multimodal transport units have to be additionally equipped with the autonomous power supply system's [18, 19].

2 About Wireless Networks for Monitoring the State of Transport with Dangerous Goods in Poor Network Coverage Areas

The scheme of data transfer in a heterogeneous network for monitoring the state of dangerous goods transported by various modes of transport is shown on Fig. 1. The solution of the problem of data transmission to the monitoring server does not cause difficulties, when the vehicle within the area of stable wireless connection. It can be organized, for example, via cellular network, 3G/4G or other conventional means of getting internet connection [20–23].

When transported by road 1 in areas of reliable mobile network coverage 4, data transmission and analysis can be carried out according to known algorithms

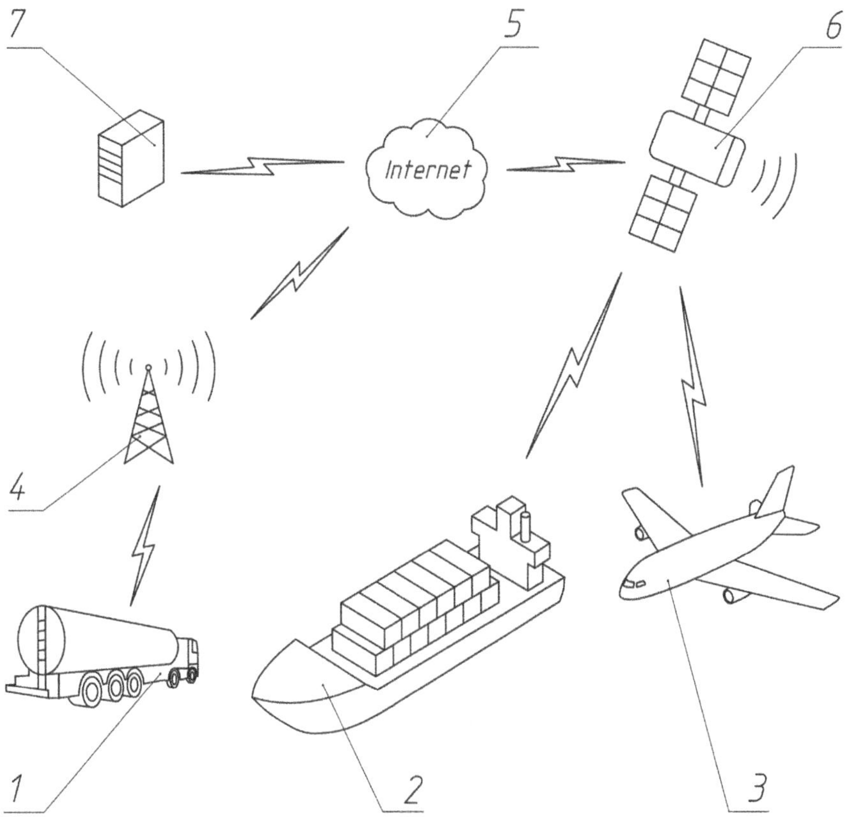

Fig. 1 Data transfer in the monitoring network during transportations of dangerous goods by various modes of transport (1—tank truck; 2—sea transport; 3—airplane; 4—mobile phone network; 5—Internet; 6—satellite network; 7—dispatch server)

using standard Internet 5 protocols. At the same time, in the case of transportation by air transport 3 or sea transport 2, the organization of a communication channel for receiving and transmitting information from the vehicle requires the installation of additional devices on board, since mobile communications become unavailable (without the installation of additional special communication facilities on board). Organization of information exchange between the vehicle and dispatch server 7 is possible by means of satellite data communication networks 6 or special on-board equipment that organizes separate data transmission channels.

3 Technical Tools for Monitoring the State of Multimodal Storage Systems for Cryogenic Products

The period of non-drainage storage of any of the cryogenic products is limited due to the increase in pressure in the tank container due to the inevitable heat inflow from the environment through the insulation. At the same time, the problem of pressure growth in the vessel is exacerbated by the temperature stratification of the product, which occurs as a result of the process of natural convection in a stationary tank container [24, 25].

The widespread monitoring systems for vehicles for the transportation of cryogenic products are not designed to track the condition of a tank container during multimodal transportation, when several modes of transport are used, in addition to road transport. Also commercially available transport tanks and tank containers, as a rule, are not equipped with vacuum pressure sensors in the heat-insulating cavity, which makes it impossible to assess the technical condition of the tank during operation. The vacuum level can only be measured after the tank has been emptied and operation has been temporarily terminated for the duration of maintenance. Thus, without information about the parameters that affect the rate of pressure growth in a tank container, it is not possible to estimate the time of non-drainage storage of a cryogenic product. During long-term transportation and storage, when for various reasons there is no consumption of a liquid product from the tank, this can lead to losses due to automatic discharge through safety devices. When operating a tank container for the transportation of combustible media such as LNG, hydrogen, ethylene, this additionally creates an explosive and fire hazardous situation.

Thus, an important task is to include as many tanks as possible to cyber-physical system for monitoring the condition of tank containers, which, among other things, allow to monitor the deterioration of the condition of the heat-insulating cavity, as well as provide the possibility of a more accurate assessment of the rate of pressure growth in the tank, taking into account the additional heat flow to the product due to the deterioration of thermal insulation.

The operation of the tank container with additional telemetry equipment based on IoT conception is illustrated by the Fig. 2.

In order to be able to monitor the status of each of the tank containers 1, it is necessary, first of all, to equip them with pressure 3 and level 2 sensors, as well as provide them with independent telemetry modules (not directly connected to the vehicle telemetry modules, as is the case with ordinary tank trucks). Navigation information about the tank container, as well as information from pressure and level sensors enters the data transmission unit 4. Also, a data transmission unit (IoT gateway) needs an autonomous power supply system, which will ensure its operation regardless of the mode of storage or transportation, type of transport, number of loading and unloading operations, etc.

Power supply to the IoT gateway is provided from the battery pack 6 through the voltage regulator 5. It is possible to recharge the batteries while the vehicle with the tank container is parked through the charge controller 7 from the external electrical

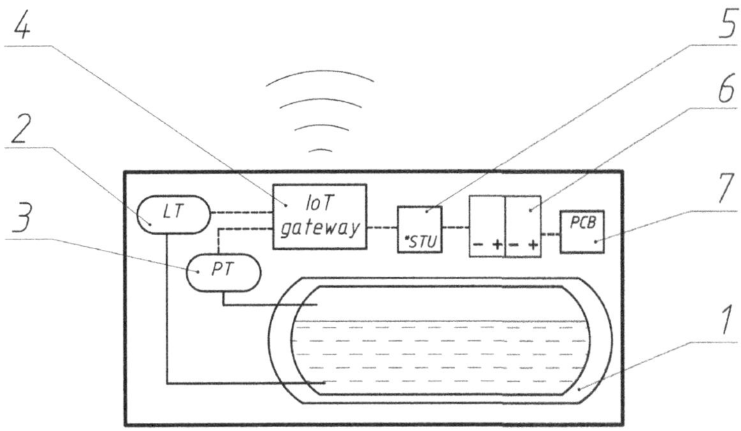

Fig. 2 Tank container with additional equipment for telemetry and autonomous power supply system (1—cryogenic tank container with vacuum thermal insulation; 2—level transmitter (differential pressure sensor between the gas and liquid phases of the tank); 3—pressure transmitter; 4—Internet-of-things transmittion unit; 5—voltage regulator; 6—battery pack; 7—protection circuit board)

network. Thus, the part of the cyber-physical system located within the tank container is autonomous, which makes it possible to ensure the operability of the monitoring during multimodal transportation.

The operation of the sensor network for monitoring multimodal transport is illustrated by the Fig. 3.

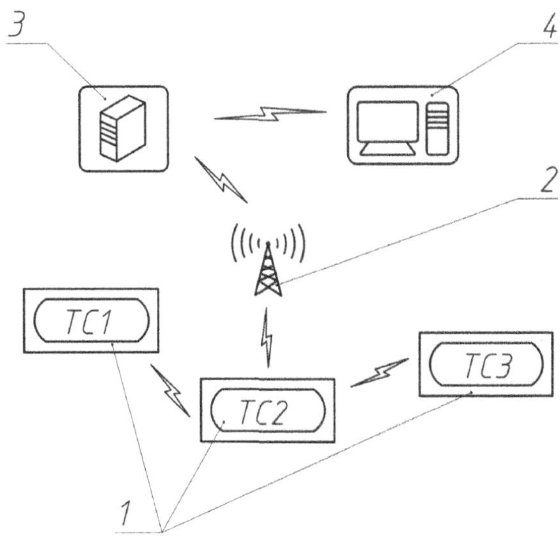

Fig. 3 Data transfer in the sensor network for monitoring the state of multimodal containers with dangerous goods (1—tank container; 2—ground-based network; 3—dispatch server; 4—operator workplace)

The use of a sensor network allows to connect various tank containers to the monitoring system, regardless of their distance from the server or ground-based communication units, thus providing a stable wireless connection within the monitoring network. IoT data transmission units for all tank containers, the status of which needs to be monitored in real time, exchange information between each other in order to provide a connection to monitoring server and to the dispatch center, while calculating the deviation of the pressure growth rate in the tank container from the nominal value is made depending on the current sensor readings and current navigation information.

After receiving data from sensors the current value of the deviation of the pressure growth rate from the nominal value is calculated in the computing module. The operation of the computing module is based on the principle of a lookup table. This technical solution makes it possible to avoid the need to perform time-consuming calculations of the cryogenic product storage parameters each time on demand. The computing module with a given periodicity gives the nearest to the measured values of pressure and liquid level. In this case, each tabular pressure-level pair corresponds to a certain nominal value of the pressure growth rate.

The deviation of the pressure growth rate from the nominal value for the transport storage mode (the container is in motion in accordance with the navigation data, the "Movement" mode) is determined by the formula:

$$\delta_{mov} = \left(\frac{p_i}{p_{i-1}} - G_{rat} \right) \cdot 100\%$$

where p_i is the current measured pressure value, p_{i-1} is the previous measured value, G_{rat} is the nominal value of the pressure rise factor for the current pressure p_i and liquid level L_i. The values of L_i and p_i are determined according to the readings of the pressure sensors and the pressure difference between the gas and liquid phases.

In the case when, in accordance with the navigation information, the vehicle speed is equal to zero ("Parking" mode), the required value of the pressure increase rate deviation is determined by the formula:

$$\delta_{st} = \left(\frac{p_i}{p_{i-1}} - G_{rat} \cdot K_{str} \right) \cdot 100\%$$

where K_{str} is a coefficient that takes into account the degree of temperature stratification in the stationary storage mode of the product, depending on the geometric and operational characteristics of the tank container.

If the pressure growth rate deviation exceeds the critical value, an emergency message is generated and sent to the operator workplace 4 of the dispatch center. Receipt of an alarm message indicates a technical malfunction of the tank container, which led to a significant increase in vacuum pressure in the heat-insulating cavity.

The operator workplace software contains a module for visualizing the results of calculating the deviation in the pressure growth rate and other monitored thermophysical parameters, as well as a module for receiving and processing emergency messages. If necessary, information about a significant change in the technical condition of the tank container is sent from the operator workplace to the driver of the vehicle or the personnel of the operating organization. In this case, it becomes possible to take measures to prevent a further pressure increasing or to safely depressurize the tank container. The operator workplace software can also be used to refine the assessment of losses during storage and multimodal transportation of cryogenic products, as well as to collect statistical data to prepare recommendations for increasing storage time.

Information from all the transport tanks within the cyber-physical system is accumulated on a control panel of the dispatch workplace, for example:

- data on the pressure and level of the liquid product in each tank,
- the technical condition of the thermal insulation,
- the storage mode at a given time (stationary or transport),
- the predicted time until the moment opening of safety valves (non-drainage storage time).

Based on current equipment-state information picture (see Fig. 4), the operator responsible for remote monitoring of the state of cryogenic tanks, can make decisions for:

- sending a message to the logistics service about the need to refuel the tank (when the liquid level drops below 30%;
- informing the responsible person when changing the status from «OK» to «ATTENTION»);
- immediately informing the responsible person in case of an emergency message (situations when the pressure in the tank exceeds 1,15 of maximum, there is no vacuum in the heat-insulating cavity, the liquid level exceeds 98%, etc.).

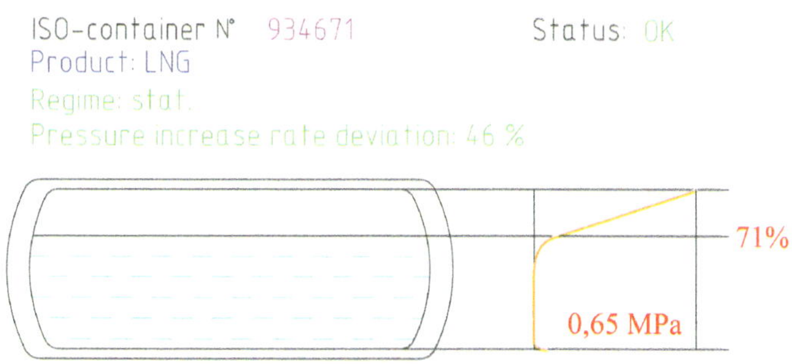

Fig. 4 The typical information picture on the state of tank container on the operator workplace display

4 Conclusion

To solve the problem of lack of Internet connection during multimodal transportation of dangerous goods, it is advisable to use satellite networks, in particular for sea and air transport. Using of sensor networks and autonomous telemetry equipment for monitoring the state of container transport with dangerous goods will significantly increase the level of safety. This is achieved through preventive informing about changes in the technical condition of the vehicle during operation by assessing the deviation of the pressure growth rate from the nominal value. Taking preventive measures by the operating personnel makes it possible to prevent leakage of the product and prevent the formation of explosive and toxic mixtures in the air.

Acknowledgements This research was supported by the grant of the President of the Russian Federation for leading scientific schools of the Russian Federation NSh-122.2022.1.6.

References

1. Weltschev, M., Schwarzer, S., Otremba, F.: Comparison of the operating life of tank containers, tank vehicles and rail tank cars for the carriage of dangerous goods in practice, analysis of causes of damage. Chem. Eng. Trans. **31**, 559–564 (2013)
2. Hoffmann, N., Stahlbock, R., Stefan, V.: A decision model on the repair and maintenance of shipping containers. J. Shipp. Trade Springer **5**(1), 1–22 (2020)
3. Isermann, R.: Vehicle dynamics modeling. In: Automotive Control. ATZ/MTZ-Fachbuch. Springer, Berlin, Heidelberg (2022)
4. Larkin, E., Bogomolov, A., Privalov, A.: Discrete model of mobile robot assemble fault-tolerance. Lecture Notes in Computer Science 11659LNAI:204–215 (2019)
5. Soldatov, E., Bogomolov, A.: Issues of energy-efficient storage of fuel in multimodal transport units. Smart Innov. Syst. Technol. **232**, 393–402 (2022). https://doi.org/10.1007/978-981-16-2814-6_34
6. Jablonski, A.: Principles of condition monitoring systems. In: Condition Monitoring Algorithms in MATLAB. Springer Tracts in Mechanical Engineering. Springer, Cham, pp. 11–29 (2021)
7. Xiong, M., Wang, H.: Digital twin applications in aviation industry: a review. Int. J. Adv. Manuf. Technol **121**(9–10), 1–16 (2022)
8. Soldatov, A., Scherbakov, I., Osipov, A.: Cyber-physical approach to building a flight experiment control system. Lect. Notes Netw. Syst. **597**, 495–507 (2023)
9. Strelets, D., Kornev, A., Kataev, A., Shubin, I., Ivanov, I., Abduhanov, P., Ostanko, D.: A combined artificial neural network for modeling behavior of multiparameter systems. Aviat. Ind. **3**, 4–8 (2017)
10. Soldatov, E., Bogomolov A.: Decision support models and algorithms for remote monitoring of the equipment state. In: CEUR Workshop Proceedings, iss. ITIDMS 2021 - Proceedings of the International Scientific and Practical Conference Information Technologies and Intelligent Decision Making Systems, pp. 1–8. (2021)
11. Balyk, O., Zolotaeva, M., Bogomolov, A., Soldatov, A.: Cyber-physical test facility for certification of robotic unmanned aerial systems. Lect. Notes Netw. Syst. **596**, 385–396 (2023)
12. Larkin, E., Bogomolov, A., Akimenko, T., Privalov, A.: Digital control of continuous production with dry friction at actuators. Smart Innov. Syst. Technol. **232**, 427–436 (2022)

13. Tobin, D., Bogomolov, A., Golosovskiy, M.: Model of organization of software testing for cyber-physical systems. Stud. Syst. Decis. Control. **418**, 51–60 (2022)
14. Costa, F., Genovesi, S., Borgese, M., Michel, A.: A review of RFID sensors, the new frontier of internet of things. Sensors. **21**(9), 3138 (2021)
15. Larkin, E., Akimenko, T., Bogomolov, A., Krestovnikov, K.: Mathematical model for evaluating fault tolerance of on-board equipment of mobile robot. Smart Innov. Syst. Technol. **187**, 383–393 (2021)
16. Pan, M., Tseng, Y.: ZigBee and their applications. In: Mahalik, N.P. (eds.) Sensor Networks and Configuration. Springer, Berlin, Heidelberg (2007)
17. Larkin, E., Akimenko, T., Bogomolov, A.: Modeling the reliability of the onboard equipment of a mobile robot. Izvestiya of Saratov University. New Series. Series: Math. Mech. Inform. **21**(3), 390–399 (2021)
18. Yu, W., Patros, P., Young, B., Klinac, E., Walmsley, T.: Energy digital twin technology for industrial energy management: classification, challenges and future. Renew. Sustain. Energy Rev. **161**, 112407 (2022)
19. Adamec, V., Schullerova, B., Adam, V.: Issues of hazardous materials transport and possibilities of safety measures in the concept of smart cities. In: Leon-Garcia, A., Lenort, R., et al. (eds.) Smart City 360. Lecture Notes of the Institute for Computer Sciences, Social Informatics and Telecommunications Engineering, vol. 166, pp. 790–799. Springer, Cham. (2016)
20. Choudhary, A., Ahmad Khan, D.: Introduction to conditioning monitoring of mechanical systems. In: Malik, H., Iqbal, A., Yadav, A. (eds.) Soft Computing in Condition Monitoring and Diagnostics of Electrical and Mechanical Systems. Advances in Intelligent Systems and Computing, vol. 1096, pp. 205–230. Springer, Singapore. (2020)
21. Elattar, M., Wendt, V., & Jasperneite, J. Communications for cyber-physical systems. In: S. Jeschke, C. Brecher, H. Song, D. B. Rawat (Eds.), Springer Series in Wireless Technology. Industrial Internet of Things: Cybermanufacturing Systems, pp. 347–372 (2017).
22. Oks, S., Fritzsche, A., Moslein, K.: An application map for industrial cyber-physical systems. In: Jeschke, S., Brecher, C., Song, H., Rawat, D.B.: (eds.) Springer Series in Wireless Technology. Industrial Internet of Things: Cybermanufacturing Systems, pp. 21–46 (2017).
23. Barkanyi, A., Chovan, T., Nemeth, S., Abonyi, J.: Modelling for digital twins—potential role of surrogate models. Processes **9**(476), 1–29 (2021)
24. Purwosaputra, A., Artana, K., Dinariyana, A.: Selection over small-scale LNG transportation model in designing an efficient and reliable LNG supply chain using analytical network process. Mar. Syst. Ocean. Technol. (Springer) **17**, 95–112 (2022)
25. Kender, K., Kaufmann, F., et al.: Development of a digital twin for a flexible air separation unit using a pressure-driven simulation approach. Comput. Chem. Eng. **151**(7), 107349 (2021)

Uniform Distribution Law as a Base of Statistical Decision Criteria

S. Efimenko, A. Smetankin, A. Klavdiev, D. Garanin, S. Kolesnichenko, and I. Chernorutsky

Abstract Nowadays, testing the hypothesis of the uniform distribution law, or in other words, identification, is a relevant topic of research. The main goal of the study is to determine the most powerful goodness-of-fit test. The uniform distribution law is a good basis for the statistical decision criteria. At the same time, the uniform distribution law can be easily implemented. The issue of determining the most powerful goodness-of-fit test has recently been attracting much attention. For example, the power of known tests was studied for moderate-sized samples with size $n \geq 10$. Samples of the smaller size are called small samples. Currently, the theory of making statistical decisions using small samples lacks scientific justification and development. The complexity of the problem of determining the best estimates using small amount of statistical data depends on the type of distribution and the sample size. Moreover, such problem cannot be the object of general mathematical theory in the conditions of non-asymptotic formulation.

Keywords The uniform law of distribution · Small samples · Identification · Criteria of the consent · Statistical decisions

1 Theoretical Justification of the Method

The goodness-of-fit test (criterion of uniformity of random variables) can be obtained from the stochastic similarity criterion. Such test is a convolution of particular uniformity criteria, and is used to make a statistical decision [1]. Moreover, the extension

S. Efimenko · A. Smetankin (✉) · D. Garanin · I. Chernorutsky
Peter the Great St. Petersburg Polytechnic University, St. Petersburg, Russia
e-mail: smetanckin.anatoly@yandex.ru

S. Kolesnichenko
The Admiral Makarov State University of Maritime and Inland Shipping, St. Petersburg, Russia

A. Klavdiev
Saint Petersburg Mining University Russian Federation, St. Petersburg, Russia

A. G. Kravets et al. (eds.), *Cyber-Physical Systems Engineering and Control*,
Studies in Systems, Decision and Control 477,
https://doi.org/10.1007/978-3-031-33159-6_23

of the theorem on the ratio of samples of independent random variables is proofed by the following calculations [2, 3].

Theorem *There is a sample of independent random variables* $x_1, x_2, ..., x_n$ *uniformly distributed within a range of* [0; 1]. *The ordered series is* $x'_1 \leq x'_2 \leq ... \leq x'_n$ *compiled from these variables* [4, 5]. *The series* $\upsilon_1 \leq \upsilon_2 \leq ... \leq \upsilon_k$, $k = n - 1$ *is obtained after the division of all of the terms of the ordered series (except* x'_n*)* by x'_n. *This and the following series are treated in the same manner. As a result, the random variable* V_1 *uniformly distributed within a range of* [0; 1] *is obtained* [6–8].

Proof Suppose the sample size is $n = 2$.

The ordered series $x_1 \leq x_2$ (hereinafter the terms of the ordered series are not marked with a dash in order to simplify the notation) is compiled from the sample. Joint probability density of the ordered random variables $x_1 \leq x_2$ takes the following form [9, 10].

$$f_{x_1,x_2}(x_1, x_2) = 2! \prod_{i=1}^{2} f_{x_i}(x_i) = 2!, \tag{1}$$

where $f_{x_i}(x_i) = 1$—density function of i-th sample in the series.

Two statistics (on the number of terms of ordered series) are added

$$\upsilon_1 = \frac{x_1}{x_2}$$

and

$$\upsilon_2 = x_2. \tag{2}$$

The joint density function is the following due to the one-to-one inverse transformations, $x_1 = \upsilon_1\upsilon_2$ and $x_2 = \upsilon_2$, of variables (2)

$$f_{\upsilon_1,\upsilon_2}(\upsilon_1, \upsilon_2) = f_{x_1,x_2}(x_1, x_2)_{x_i = x_i(\upsilon_1,\upsilon_2)} \cdot |J|, \tag{3}$$

where $J = \frac{\partial(x_1,x_2)}{\partial(\upsilon_1,\upsilon_2)} = \begin{vmatrix} \upsilon_2 & 0 \\ \upsilon_1 & 1 \end{vmatrix} = \upsilon_2$—Jacobian.

Then, the joint density function (3) is the following, when considering (1)

$$f_{\upsilon_1,\upsilon_2}(\upsilon_1, \upsilon_2) = 2!\upsilon_2. \tag{4}$$

Now, the density of the variable υ_1 is obtained by eliminating the auxiliary variable υ_2 through the integration of the expression (4) over the domain of υ_2 values [11].

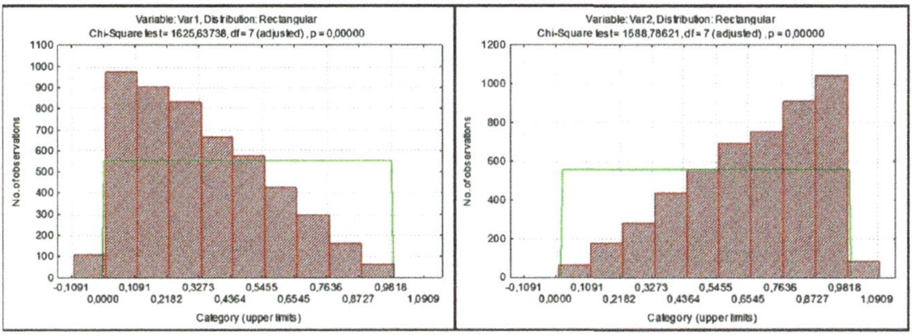

Fig. 1 Histograms of the statistical distributions $x_1 \leq x_2$

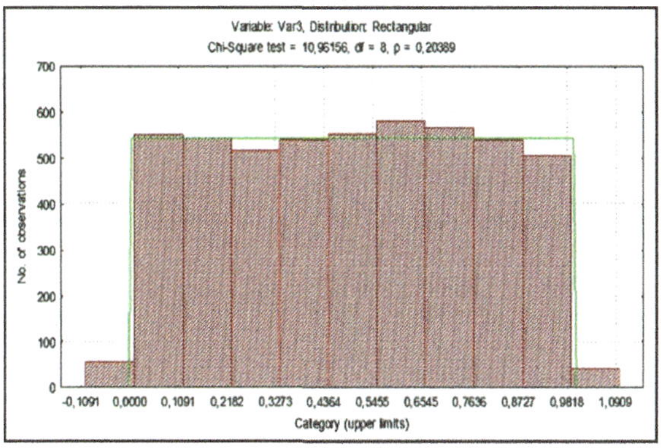

Fig. 2 Histogram of statistical distribution υ_1

$$f_{\upsilon_1}(\upsilon_1) = \int_0^1 f_{\upsilon_1,\upsilon_2}(\upsilon_1, \upsilon_2) d\upsilon_2 = \int_0^1 2\upsilon_2 d\upsilon_2 = 1! \tag{5}$$

The factorial notation is left in order to generalize the results. The result (5) indicates the uniform distribution of the variable υ_1. The histograms of the statistical distributions $x_1 \leq x_2$ are presented in the Fig. 1. It can be assumed that these distributions follow the beta-distribution law [2, 12].

The test showed that the achieved significance values of corresponding hypotheses with the beta-distribution parameters $\alpha = 1$, $\beta = 2$ for the statistics x_1 $(p = 0,66)$ and $\alpha = 2$, $\beta = 1$ for the statistics x_2 $(p = 0,3)$ indicate against their deviation [3, 13].

The result of the testing the uniformity of the statistics υ_1 is presented in the form of the histogram in the Fig. 2 [14, 15]. It can be concluded, that the achieved significance value $(p = 0,2)$ also indicates against the deviation of the hypothesis of its uniform distribution (or beta-distribution with the parameters $\alpha = 1$, $\beta = 1$).

Suppose the sample size is $n = 3$.

In this case, the ordered sample is $x_1 \le x_2 \le x_3$. The following statistics are introduced:

$$\upsilon_1 = \frac{x_1}{x_3},\ \upsilon_2 = \frac{x_2}{x_3} \text{ and } \upsilon_3 = x_3 \tag{6}$$

The inverse transformations of these statistics are the following:

$$x_1 = \upsilon_1\upsilon_3,\ x_2 = \upsilon_2\upsilon_3 \text{ and } x_3 = \upsilon_3.$$

At the same time, [3, 4] the Jacobian of transformation is the following:

$$J = \frac{\partial(x_1, x_2, x_3)}{\partial(\upsilon_1, \upsilon_2, \upsilon_3)} = \begin{vmatrix} \upsilon_3 & 0 & 0 \\ 0 & \upsilon_3 & 0 \\ \upsilon_1 & \upsilon_2 & 1 \end{vmatrix} = \upsilon_3^2.$$

The joint density function of the terms of the ordered series $\upsilon_1 \le \upsilon_2 \le \upsilon_3$, considering the size of the sample for (1) and (3), is the following $f_{\upsilon_1,\upsilon_2,\upsilon_3}(\upsilon_1, \upsilon_2, \upsilon_3) = 3!\upsilon_3^2$.

Therefore, the joint density function of $\upsilon_1 \le \upsilon_2$ is the following:

$$f_{\upsilon_1,\upsilon_2}(\upsilon_1, \upsilon_2) = \int_0^1 f_{\upsilon_1,\upsilon_2,\upsilon_3}(\upsilon_1, \upsilon_2, \upsilon_3)d\upsilon_3 = \int_0^1 3!\upsilon_3^2 d\upsilon_3 = 2!. \tag{7}$$

Two statistics are introduced for the mentioned statistics $V_1 = \frac{\upsilon_1}{\upsilon_2}$ and $V_2 = \upsilon_2$. Their one-to-one inverse transformations are the following: $\upsilon_1 = V_1 V_2$ and $\upsilon_2 = V_2$.

At the same time, the Jacobian of transformation is the following:

$$J = \frac{\partial(\upsilon_1, \upsilon_2)}{\partial(V_1, V_2)} = \begin{vmatrix} V_2 & 0 \\ V_1 & 1 \end{vmatrix} = V_2$$

.

Then, the joint density function of the terms of the ordered series $V_1 \le V_2$ is the following: $f(V_1, V_2) = 2!V_2$. Further, the density function of the statistics V_1 is obtained by the integration over the variable V_2.

$$f(V_1) = \int_0^1 f(V_1, V_2)dV_2 = 2\int_0^1 V_2 dV_2 = 1, \tag{8}$$

It indicates the uniform distribution of the statistics V_1 within a range of [0; 1]. It can be shown that the distribution of the statistics υ_1, υ_2 and their ratio $V_1 = \frac{\upsilon_1}{\upsilon_2}$ also follow the beta-distribution law with the significance values of $p_{\upsilon_1} = 0, 03$, $p_{\upsilon_2} = 0, 33$, and $p_{V_1} = 0, 75$ with the parameters $\alpha = 1, \beta = 2$; $\alpha = 2, \beta = 1$;

and $\alpha = 1$, $\beta = 1$ respectively. Therefore, the statistics V_1 is uniformly distributed within a range of [0; 1].

Now, suppose the sample size is $n - 1$ according to the method of mathematical induction. The ordered series is the following: $x_1 \le x_2 \le \ldots \le x_{n-1}$. The following statistics are introduced:

$$\upsilon_1 = \frac{x_1}{x_{n-1}},\ \upsilon_2 = \frac{x_2}{x_{n-1}}\ \upsilon_{n-1} = x_{n-1}$$

Their one-to-one inverse transformations are the following:

$$x_1 = \upsilon_1\upsilon_{n-1},\ x_2 = \upsilon_2\upsilon_{n-1} \cdots,\ x_{n-1} = \upsilon_{n-1}.$$

In addition, the Jacobian of transformation is expressed as follows [14, 15]:

$$J = \frac{\partial(x_1, x_2, \ldots, x_{n-1})}{\partial(\upsilon_1, \upsilon_2, \ldots \upsilon_{n-1})} = \begin{vmatrix} \upsilon_{n-1} & 0 & \ldots & 0 \\ 0 & \upsilon_{n-1} & \ldots & 0 \\ \ldots & \ldots & \ldots & \ldots \\ \upsilon_1 & \upsilon_2 & \ldots & 1 \end{vmatrix} = \upsilon_{n-1}^{n-2}.$$

The joint density function of the variables of the ordered series $\upsilon_1 \le \upsilon_2 \le \ldots \le \upsilon_{n-1}$, considering the size of the sample for (1) and (3), is the following $f_{\upsilon_1,\upsilon_2,\ldots,\upsilon_{n-1}}(\upsilon_1, \upsilon_2, \ldots, \upsilon_{n-1}) = (n-1)!\upsilon_{n-1}^{n-2}$. Therefore, the joint density function of $\upsilon_1 \le \upsilon_2 \le \ldots \le \upsilon_{n-2}$ is:

$$f_{\upsilon_1,\upsilon_2,\ldots,\upsilon_{n-2}}(\upsilon_1, \upsilon_2, \ldots, \upsilon_{n-2}) = \int_0^1 f_{\upsilon_1,\upsilon_2,\ldots,\upsilon_{n-1}}(\upsilon_1, \upsilon_2, \ldots, \upsilon_{n-1})d\upsilon_{n-1}$$
$$= (n-1)!\int_0^1 \upsilon_{n-1}^{n-2}d\upsilon_{n-1} = (n-2)!. \quad (9)$$

As a result, $n - 2$ variables in the ordered series $\upsilon_1 \le \upsilon_2 \le \ldots \le \upsilon_{n-2}$ are obtained. The statistics are introduced:

$$V_1 = \frac{\upsilon_1}{\upsilon_{n-2}},\ V_2 = \frac{\upsilon_2}{\upsilon_{n-2}}, \cdots, V_{n-2} = \upsilon_{n-2},$$

Their one-to-one inverse transformations are the following:

$$\upsilon_1 = V_1V_{n-2}, \upsilon_2 = V_2V_{n-2}, \cdots, \upsilon_{n-2} = V_{n-2}$$

In addition, the Jacobian of transformation is expressed as follows:

$$J = \frac{\partial(\upsilon_1, \upsilon_2, \ldots, \upsilon_{n-2})}{\partial(V_1, V_2, \ldots V_{n-2})} = \begin{vmatrix} V_{n-2} & 0 & \ldots & 0 \\ 0 & V_{n-2} & \ldots & 0 \\ \ldots & \ldots & \ldots & \ldots \\ V_1 & V_2 & \ldots & 1 \end{vmatrix} = V_{n-2}^{n-3}.$$

The joint density function of the variables of the ordered series $\upsilon_1 \leq \upsilon_2 \leq \ldots \leq \upsilon_{n-2}$, considering the size of the sample for (1) and (3), is:

$$f_{V_1, V_2, \ldots, V_{n-2}}(V_1, V_2, \ldots, V_{n-2}) = (n-2)! V_{n-2}^{n-3}. \tag{10}$$

Therefore, the joint density function of $V_1 \leq V_2 \leq \ldots \leq V_{n-3}$ after the integration of (10) over the V_{n-2} is the following:

$$f_{V_1, V_2, \ldots, V_{n-3}}(V_1, V_2, \ldots, V_{n-3}) = (n-1)!$$

The density of the final statistics is obtained after the transformation of all of the statistics V.

$$f_{V_1}(V_1) = 1, \tag{11}$$

It indicates the uniform distribution of the convolution υ_k—criteria (probability criteria) for the sample of the size $n - 1$, within a range of [0; 1].

2 Demonstration of the Method Capabilities for the Samples of the Size Smaller Than 10

For example, suppose the sample size is $n = 7$. The histograms of the statistical distributions of $\upsilon_1 \leq \upsilon_2$ are presented in the Fig. 3.

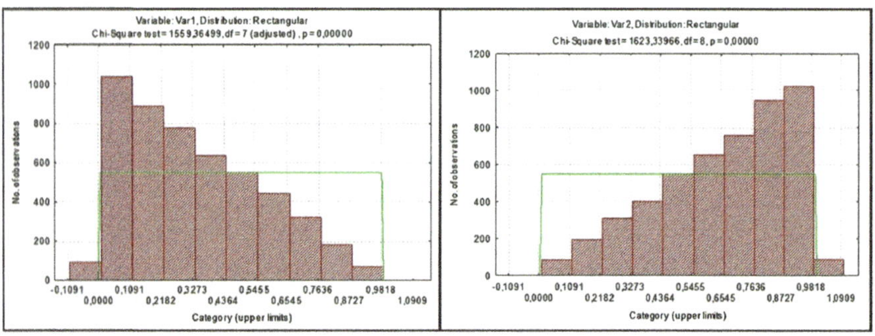

Fig. 3 Histograms of statistical distributions $\upsilon_1 \leq \upsilon_2$

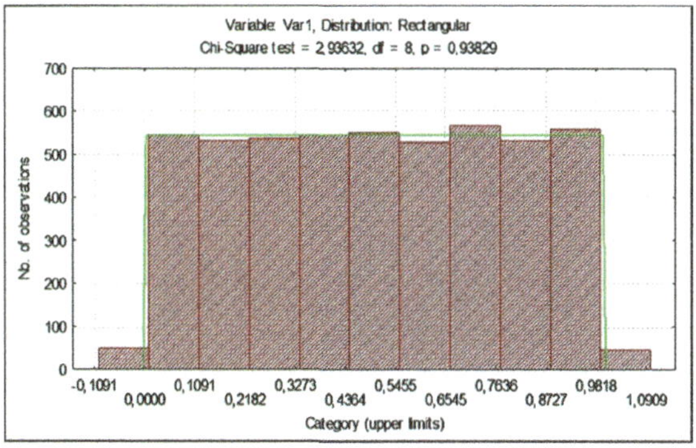

Fig. 4 Histogram of statistical distribution V_1

The complete equivalence of the statistical distributions $x_1 \leq x_2$ (see Fig. 1) can be seen in the Fig. 3. Moreover, the distribution υ_1 on the Fig. 2 looks identical to the distribution V_1 presented in the Fig. 4.

It can be seen in the Fig. 4 that the achieved significance value $p = 0,94$ indicates against the deviation of the hypothesis of its uniform distribution [12].

Therefore, the theorem is proofed theoretically and empirically.

The procedure of decision making becomes more flexible when the method (of achieving the significance value) is implemented. The lesser the value of $p(s)$ is obtained, the more the samples indicate against the tested hypothesis [6]. On the other hand, the lesser the value of s, the higher the probability of the fact that tested hypothesis H is true [6].

The relative p-value is introduced in order to consider them simultaneously:

$$p_H = \left.\frac{p(s)}{s}\right|_H. \tag{12}$$

Then, the efficiency of the H_0 and H_1 differentiation can be decided during the testing of the hypothesis H_0 and when the alternative hypothesis H_1 is considered [13, 16].

$$W = \frac{p_{H_0} - p_{H_1}}{p_{H_0}}. \tag{13}$$

For example, the results of the testing of the hypothesis $H_0{:}F(x) = x,\ \ x \in [0; 1]$ for the uniformity and the alternative hypothesis $H_1{:}F(x) = B_I(1.5, 1.5, 1, 0)$ of type I beta-distribution for the uniformity law are presented in the Fig. 5.

Corresponding results for samples with the size up to $n = 10$ are presented in the Table 1 without the relevant charts.

It can be seen in the table that the efficiency of the differentiation between the hypotheses H_0 and H_1 in the given range n is high when the suggested criterion is implemented. The values of the power of the goodness-of-fit test for its right-hand

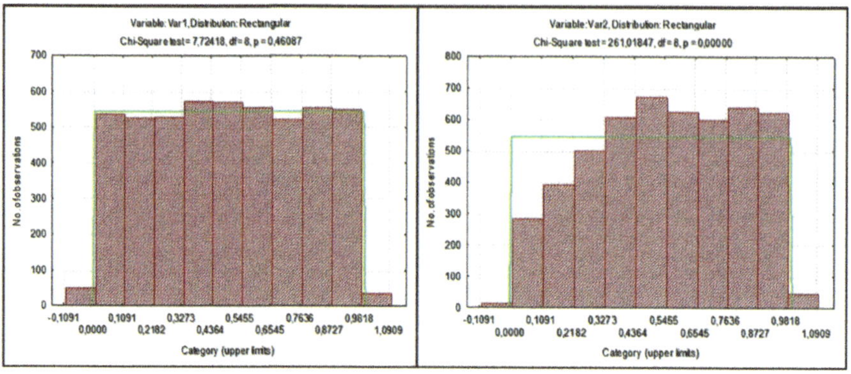

Fig. 5 Testing for the uniformity of the sample with the size $n = 2$

Table 1 Achieved p-values of the hypothesis H_0 in relation to H_1

n	H_0			H_1			W
	$p(s)_{H_0}$	s_{H_0}	p_{H_0}	$p(s)_{H_1}$	s_{H_1}	p_{H_1}	
2	0.46	7.7	0.060	0	261	0	1
3	0.63	6.2	0.102	0	392	0	1
4	0.53	7.1	0.075	0	428	0	1
5	0.37	9.1	0.041	0	479	0	1
6	0.88	3.7	0.238	0	420	0	1
7	0.17	11.5	0.015	0	449	0	1
8	0.77	4.9	0.157	0	546	0	1
9	0.23	10.6	0.022	0	487	0	1
10	0.26	10.1	0.026	0	509	0	1

critical region and for the samples of the sizes $n \leq 10$ using traditional method of estimation are presented in the Table 2.

It can be seen in the Table 2 that the power of the convolution of the probability criterion is bigger than the power of the Zhang criterion Z_A, which is the most preferable amongst the criteria researched in, when $n = 10$. The power of the suggested criterion when the sample size is minimal is bigger than the power of the Zhang criterion Z_A when the sample size is $n = 10$. The suggested criterion gives an advantage over other criteria when the differentiation of such similar hypothesis is required.

Therefore, the simple procedure of testing for the uniformity can be a very useful tool for studying small samples.

Table 2 The power of the convolution of the information criterion for the hypothesis H_1

n	α				
	0.15	0.1	0.05	0.025	0.01
10	0.21	0.14	0.07	0.04	0.01
9	0.21	0.14	0.07	0.04	0.01
8	0.21	0.14	0.07	0.04	0.01
7	0.20	0.14	0.07	0.04	0.01
6	0.20	0.14	0.07	0.04	0.01
5	0.21	0.13	0.07	0.03	0.01
4	0.19	0.13	0.07	0.03	0.01
3	0.18	0.13	0.06	0.03	0.01
2	0.17	0.11	0.06	0.03	0.01

3 The Step-by-Step Algorithm of Method Implementation for the Samples of the Size $n > 10$

1. The sample of the size $n > 10$ should be divided into $k = 5...7$ intervals as for the Kolmogorov criterion or χ^2.
2. Theoretical $F(x)$ and empirical $F_э(x)$ values should be calculated for each of the intervals.
3. Then k partial probability criteria are expressed

$$\vartheta_k = \begin{cases} \frac{F(x)}{F_э(x)}, \; if \; F(x) \le F_э(x); \\ \frac{F_э(x)}{F(x)}, \; if \; F_э(x) > F(x). \end{cases}$$

These criteria are put in order $\upsilon_1 \le \upsilon_2 \le ... \le \upsilon_k$, and the convolution is calculated as shown above.

4 Findings of the Study

The theoretical justification of the application of the uniform distribution law as a base of determining the goodness-of-fit test is given in the article. Such criterion is used for making statistical decisions. The theorem of ratio of small samples of independent random variables of the size $n \le 10$ is presented, justified and proofed. The results of uniformity testing of the hypothesis and the alternative hypothesis for the uniform distribution law are shown. The advantages of the method, its flexibility, simplicity and versatility are displayed. The results of its practical implementation are demonstrated.

5 Conclusions

The issue of determining the most powerful goodness-of-fit test in order to make statistical decisions has recently been attracting much attention [1]. The power of known criteria for moderate-sized samples is usually studied. The complexity of the problem of determining the best estimates using small amount of statistical data depends on the type of distribution and the sample size. Currently, the theory of making statistical decisions using small samples lacks scientific justification and development. Moreover, such problem cannot be the object of general mathematical theory in the conditions of non-asymptotic formulation [6]. In this case it is helpful to use the uniform distribution law for calculating the criteria. At the same time, the uniform distribution law can be easily implemented.

References

1. Abraham, A., Grosan, G.: Swarm intelligence in data mining, p. 267. Springer, Berlin (2006)
2. Cover, T.M., Thomas, J.A.: Elements of Information Theory. Wiley **2**, 748 (2009)
3. Wilcox, R., R.: Inferences about the population mean: empirical likelihood versus bootstrap-t. J. Mod. Appl. Stat. Methods **9**(1) (2010)
4. Smetankin, A., Efimenko, S., Zheleznov, E., Cimai, Y., Chernorutsky, I., Kolesnichenko, S.: On a method for identifying failure models of complex systems. Lecture Notes in Networks and Systems 13–20 (2023)
5. Efimenko, S., Smetankin, A., Liashenko, A., Arutiunian, M., Chernorutsky, I., Kolesnichenko, S.: Method of expansion of mathematical tools of the reliability theory due to the properties of stochastic theory of similarity. Lecture Notes in Networks and System 30–40 (2023)
6. Efimenko, S., Volovik, A., Klavdiev, A.: Nonparametric identification of stochastic models of difficult systems. In: Proceedings of the International 2015 Applied Methods of Statistical Analysis Workshop Novosibirsk State Technical University, pp. 270–275 (2015)
7. Park, S.Y., Bera A.K.: Maximum entropy autoregressive conditional heteroskedasticity model. J. Econ. 219–230 (2009)
8. Qinyuan, W., Yong, D. Neal,X.: Exponential negation of a probability distribution. Soft Comput. **26**, 2147–2156 (2022)
9. Zaiko, A.I.: Using an information criterion to choose the time interval for discretization of signals with a uniform distribution law. Meas. Tech. 146–150 (2002)
10. Smid, S.C., McNeish, D., Miocevic, M., Schoot, R.: Bayesian versus frequentist estimation for structural equation models in small sample contexts: a systematic review. Struct. Equ. Model.: Multidiscip. J. 131–161 (2020)
11. Nechval, N.A., Berzinsh, G., Nechval, K.N.: A new technique of invariant statistical embedding and averaging in terms of pivots for improvement of statistical decisions under parametric uncertainty. Trans. Comput. Sci. Comput. Intell. (2021)
12. Chanwoo, K., Mehul, K., Kwangyoun, K., Dhananjaya, G.: Power-law nonlinearity with maximally uniform distribution criterion for improved neural network training in automatic speech recognition. IEEE Automatic Speech Recognition and Understanding Workshop (ASRU) (2019)
13. Fomby, T.B., Johnson, S.R., Hill, R.C.: Point estimation and tests of hypotheses in small samples. Chapter pp. 26–51(1984)
14. Smetankin, A., Efimenko, S., Garanin, D., Malihina, I., Shilkin, V., Chernorutsky, I.: Modeling the effectiveness of an investment strategy in conditions of insufficient information. Lecture Notes in Networks and Systems, pp. 322–331 (2022)

15. Volcic, A.: Uniform distribution, Benford's law and scale-invariance. Bollettino dell'Unione Matematica Italiana **13**, 539–543 (2020)
16. Nechval, N.A., Berzins, G., Nechval, K.N.: A new technique of invariant statistical embedding and averaging via pivotal quantities for intelligent constructing efficient statistical decisions under parametric uncertainty. Autom. Control. Comput. Sci. **54**, 191–206 (2020)

The manufacturer's authorised representative in the EU is Springer
Nature Customer Service Centre GmbH, Europaplatz 3, 69115 Heidelberg,
Germany. If you have any concerns regarding our products, please
contact ProductSafety@springernature.com

Printed and bound by CPI Group (UK) Ltd, Croydon, CR0 4YY
28/09/2026
02224075-0011